Monographs on Statistics and Applied Probability 107

Statistical Methods for Spatio-Temporal Systems

T0173050

MONOGRAPHS ON STATISTICS AND APPLIED PROBABILITY

General Editors

V. Isham, N. Keiding, T. Louis, S. Murphy, R. L. Smith, and H. Tong

Monographs on Statistics and Applied Probability 107

Statistical Methods for Spatio-Temporal Systems

Edited by

Bärbel Finkenstädt
University of Warwick
Coventry, U.K.

Leonhard Held
Ludwig-Maximilians University
Munich, Germany

Valerie Isham
University College London
U.K.

CRC Press
Taylor & Francis Group
Boca Raton London New York

CRC Press is an imprint of the
Taylor & Francis Group, an **informa** business

A CHAPMAN & HALL BOOK

CRC Press
Taylor & Francis Group
6000 Broken Sound Parkway NW, Suite 300
Boca Raton, FL 33487-2742

First issued in paperback 2019

ISBN-13: 978-1-58488-593-1 (hbk)
ISBN-13: 978-0-367-39011-2 (pbk)

**Visit the Taylor & Francis Web site at
http://www.taylorandfrancis.com**

**and the CRC Press Web site at
http://www.crcpress.com**

Preface

This volume contains a selection of invited papers presented by the contributors at the sixth Séminaire Européen de Statistique (SemStat) held as a summer school of the European Mathematical Society (EMS) at Castle Höhenried in Bernried near Munich in Germany on 12–18 December 2004. The aim of SemStat is to provide scientists who are at early stages of their careers with an opportunity to get quickly to the forefront of knowledge and research in areas of statistical science that are of current major interest. The chosen theme for this SemStat was "Statistical Methods for Spatio-Temporal Systems" and the invited papers were presented in either one or a series of three lectures by leading researchers and scientists in this field. These invited papers correspond to chapters in this book. Around 40 young researchers from various European countries participated in the 2004 SemStat summer school. They gave short presentations of their own research as listed on the following pages.

As with previous SemStat volumes the book concentrates on important statistical methods, theoretical aspects, and topical applications. The structure of the book is not strictly monographic in that each chapter is self-contained, consisting of a long expository article that starts by introducing the subject and progresses swiftly to incorporate recent research trends. The study of other chapters is beneficial but not vital to the understanding of the material in any single chapter. The order of the chapters is not important and readers may directly pick a chapter they are particularly interested in, study a collection of chapters, or read the whole book, ordering the chapters as they prefer. Applied scientists dealing with spatio-temporal data in a variety of research areas should benefit from this book, as should statistical researchers interested in modern statistical methodologies. Lecturers will find a variety of material suitable for graduate lecture courses in a statistical degree programme.

Spatio-temporal systems are systems that evolve over both space and time. The statistical viewpoint is to regard spatio-temporal data as realizations of random variables spread out in space and evolving in time. Statistical or stochastic models, such as point process models, for example, are used to provide the probabilistic backbone facilitating statistical inference from data. Such an approach is described in Chapter 1 where Diggle presents a review of statistical methods for spatio-temporal point process data. These methods are illustrated with specific examples of epidemic data on bovine tuberculosis, gastroenteric disease surveillance, and the foot and mouth outbreak in the U.K. Each of the applications gives rise to different statistical modelling approaches.

The second chapter by Vedel Jensen et al. is concerned with the important issue of modelling randomly growing objects as observed in diverse biological systems such as colonies of bacteria, tumours, and plant populations. The authors describe recent advances in stochastic growth models based on spatio-temporal point processes as well as growth models based on Lévy bases.

Readers who are interested in the use of data transformation tools such as power spectra, wavelets, and empirical orthogonal functions will appreciate the overview presented in Chapter 3 by Guttorp, Fuentes, and Sampson. The authors review methods, illustrating their use in a variety of applications in ecology and air quality. They also develop formal statistical tests verifying important assumptions such as stationarity and separability of a space-time covariance. The latter is also central to the following chapter. Geostatistical approaches to modelling spatio-temporal data rely on parametric covariance models and rather stringent assumptions, such as stationarity, separability, and full symmetry. Chapter 4 by Gneiting, Genton, and Guttorp reviews recent advances in the literature on space-time covariance functions, illustrated using wind data from Ireland.

Readers of the previous SemStat conference volume on *Extreme Values in Finance, Telecommunications, and the Environment* published in this monograph series (Vol. 99, 2004) will be familiar with statistical approaches to rainfall or windspeed data based on extreme value theory. In hydrological applications such as flood risk assessment, the simulation of rainfall data with high spatial-temporal resolution is required. In Chapter 5, Chandler et al. describe some stochastic and statistical models that can be used to provide simulated rainfall sequences for hydrological use. Model construction, inference, and diagnostics are all discussed. Many of the techniques described have applicability in more general space-time settings. This is also the case for the material introduced in Chapter 6 by Higdon, which provides a comprehensive primer on space-time modelling on the basis of Gaussian spatial and space-time models from a Bayesian perspective. Gaussian Markov random field specifications and Bayesian computational inference via Gibbs sampling and Markov chain Monte Carlo are central issues of this chapter. The methods are introduced and illustrated by a variety of examples using data on temperature surfaces, dioxin concentrations, ozone concentrations, and also simulated data from a well-established deterministic dynamical weather model.

It is not possible to cover all aspects of the conference theme in a single volume. Currently there are few direct links between the mathematical approaches to the mechanistic modelling of spatio-temporal systems using, for example, differential equations, pair approximations, or interactive particle systems and the stochastic and statistical modelling approaches as introduced here. A major reason for this is that the complexity and mechanistic realism that can be formulated mathematically is much larger than the model complexity that can be entertained regarding any available data from such systems. Statistical approaches generally start from the viewpoint of the data assuming stochastic or statistical modelling approaches as a vehicle for faciliating inference. It is our hope that the coverage provided by this volume will help readers acquaint themselves speedily with current statistical research issues in modelling spatio-temporal data and that it will enable further understanding and possible advances between the mechanistic and the statistical modelling communities.

The Séminaire Européen de Statistique is an activity of the European Regional Committee of the Bernoulli Society for Mathematical Statistics and Probability. The scientific programme was organised by the SemStat steering group, which, at the time of the conference, comprised Ole Barndorff-Nielsen (Aarhus), Bärbel Finkenstädt (Warwick), Leonhard Held (Munich), Alex Lindner (Munich), Enno Mammen (Mannheim), Gesine Reinert (Oxford), Michael Sørensen (Copenhagen), Ingrid Van Keilegom (Louvain-la-Neuve), and Aad van der Vaart (Amsterdam). The local organization of the meeting was in the hands of Leonhard Held (Munich) and the smooth running was in large part due to the enthusiastic help of Susanne Breitner, Michael Höhle, and Thomas Kneib (Munich). This SemStat was funded as an EMS summer school by the European Union under the sixth European framework (Marie Curie Actions) and some essential additional funding was provided by the Collaborative Research Centre Sonderforschungsbereich FB386, German research foundation (DFG). We are very grateful for this support and thanks are due to Luc Lemaire (Free University, Brussels) for coordinating the EMS summer school proposals. We would also like to thank all anonymous referees for reading the chapters and helping in improving their presentation.

On behalf of the steering group,
B. Finkenstädt (Warwick), L. Held (Munich), and V. Isham (London)

Participants

Sofia Åberg, Lund University (Sweden),
Forecasting radar precipitation using image warping.

Paul Anderson, University of Warwick (United Kingdom),
How tangled is nature?

John Aston, Academia Sinica (Taiwan),
Wavelet based spatial and temporal neuroimage analysis.

Enrica Bellone, University College London (United Kingdom),
Identification and modelling of rain event sequences.

Susanne Breitner, Ludwig-Maximilians-University Munich (Germany),
Modelling the association between air pollution and cardiorespiratory
symptoms.

Petruta Caragea, Iowa State University (USA),
Alternative estimation of spatial parameters for large data sets.

Ignacio Cascos, Public University of Navarre (Spain),
Integral trimming.

Stella David, University of Augsburg (Germany),
Multivariate K-function.

Péter Elek, Eötvös Loránd University (Hungary),
Modelling extreme water discharges: comparison of a GARCH-type model
with conventional hydrologic models.

Georgia Escaramis, University of Barcelona (Spain),
Techniques to estimate confidence intervals of risks in disease mapping.

Francesco Fedele, University of Vermont (USA),
Successive wave crests in Gaussian seas.

Laura Ferracuti, University of Perugia (Italy),
MCEM estimation for multivariate geostatistical non-Gaussian models.

Edith Gabriel, University of Montpellier II (France),
Estimating and testing zones of abrupt change for spatial data.

Axel Gandy, University of Ulm (Germany),
On goodness of fit tests for Aalen's additive risk model.

Susanne Gschlößl, Munich University of Technology (Germany),
Modelling spatial claim frequencies using a conditional autoregressive prior.

Elizabeth Heron, University of Warwick (United Kingdom),
Spatial modelling of cracks in bone cement.

Mathias Hofmann and Volker Schmid, Ludwig-Maximilians-University
Munich (Germany),
A stochastic model for surveillance of infectious diseases.

Michael Höhle, Ludwig-Maximilians-University Munich (Germany),
Towards multivariate surveillance methods.

Rosaria Ignaccolo, University of Turin (Italy),
Model testing for spatial correlated data.

Janine Illian, University of Abertay at Dundee (United Kingdom),
Ecological communities – spatial point process modelling of highly complex
systems.

José Ernesto Jardim, IPIMAR - National Institute for Agriculture and
Fisheries Research (Portugal),
Designing bottom trawl surveys with geostatistical simulation.

Thomas Kneib, Ludwig-Maximilians-University Munich (Germany),
A general mixed model approach for spatio-temporal regression data.

Stephan Kötzer, University of Aarhus (Denmark),
Lower-tail behaviour in Wicksell's corpuscle problem.

Nadja Leith, University College London (United Kingdom),
Uncertainties in statistical downscaling.

Raquel Menezes, Lancaster University (United Kingdom),
Assessing spatial dependency under preferential sampling.

Tomasz Niedzielski, University of Wroclaw (Poland),
Testing for the Markov property within sedimentary rocks: case study from
the Roztoka-Mokrzeszw Graben (SW Poland).

Irene Oliveira, University of Tràs-os-Montes e Alto Douro (Portugal),
Hilbert CCA — A new methodology to study time series processes in
multivariate analysis.

Bianca Ortuani, University of Milan (Italy),
A geostatistical application in inverse problem for highly heterogeneous porous
media.

Zbynek Pawlas, Charles University in Prague (Czech Republic),
Weak convergence of empirical distribution functions in germ-grain models.

Nathalie Peyrard, INRA — Institut National de la Recherche Agronomique (France),
The pair and Bethe approximations for the contact process on a graph.

Alexandre Pintore, University of Oxford (United Kingdom),
Rank-reduced methods for non-stationary covariance functions.

Philipp Pluch, University of Klagenfurt (Austria),
A Bayesian way out of the dilemma of underestimation of prediction errors in kriging.

Michalea Prokesova, Charles University in Prague (Czech Republic),
Statistics for locally scaled point processes.

Jakob Gulddahl Rasmussen, Aalborg University (Denmark),
Approximate simulation of Hawkes Processes.

Fabio Rigat, EURANDOM (The Netherlands),
Bayesian predictive survival trees.

Abel Rodriguez, Duke University (USA),
Bayesian structural alignment of proteins.

Ida Scheel, University of Oslo (Norway),
Spread of infectious agents in salmon farming.

Jana Sillmann, Max Planck Institute for Meteorology (Germany),
Weather extremes in climate change simulations.

Sibylle Sturtz, University of Dortmund (Germany),
Poisson-Gamma models as a tool for analysing different spatial patterns in the presence of covariates.

Thordis Linda Thorarinsdottir, University of Aarhus (Denmark),
Spatio-temporal model for fMRI data.

Ingunn Fride Tvete, University of Oslo (Norway),
Examining patterns in the spawning of Pollock by the Alaska West coast.

Jan van de Kassteele, Wageningen University (The Netherlands),
A model for external drift kriging with uncertain covariates applied to air quality measurements and dispersion model output.

Hanne Wist, Norwegian University of Science and Technology (Norway),
Speeding up block-MCMC algorithms for hierarcial GMRF models.

Contributors

Ole E. Barndorff-Nielsen
The T.N. Thiele Centre
Department of Mathematical
Sciences
University of Aarhus
Aarhus, Denmark

Enrica Bellone
Department of Statistical
Science
University College London
London, United Kingdom

Richard E. Chandler
Department of Statistical Science
University College London
London, United Kingdom

Peter J. Diggle
Department of Mathematics
and Statistics
Lancaster University
Lancaster, United Kingdom

Montserrat Fuentes
Department of Statistics
North Carolina State University
Raleigh, North Carolina

Marc G. Genton
Department of Statistics
Texas A&M University
College Station, Texas

Tilmann Gneiting
Department of Statistics
University of Washington
Seattle, Washington

Peter Guttorp
Department of Statistics
University of Washington
Seattle, Washington

David Higdon
Los Alamos National Laboratory
Statistical Sciences Group
Los Alamos, New Mexico

Valerie Isham
Department of Statistical Science
University College London
London, United Kingdom

Eva B. Vedel Jensen
The T.N. Thiele Centre
Department of Mathematical
Sciences
University of Aarhus
Aarhus, Denmark

Kristjana Ýr Jónsdóttir,
The T.N. Thiele Centre
Department of Mathematical
Sciences
University of Aarhus
Aarhus, Denmark

Paul Northrop
Department of Statistical
Science
University College London
London, United Kingdom

Paul D. Sampson
Department of Statistics
University of
Washington
Seattle, Washington

Jürgen Schmiegel
The T.N. Thiele Centre
Department of Mathematical
Sciences
University of Aarhus
Aarhus, Denmark

Chi Yang
Department of Statistical Science
University College London
London, United Kingdom

Contents

CHAPTER 1

Spatio-Temporal Point Processes: Methods and Applications

Peter J. Diggle

Contents

1.1 Introduction

This chapter is concerned with the analysis of data whose basic format is $(x_i, t_i) : i = 1, \ldots, n$, where each x_i denotes the location and t_i the corresponding time of occurrence of an *event* of interest. We shall assume that the data form a complete record of all events which occur within a pre-specified spatial region A and a prespecified time-interval, $(0, T)$. We call a data-set of this kind a *spatio-temporal point pattern,* and the underlying stochastic model for the data a *spatio-temporal point process.*

1.1.1 Motivating examples

1.1.1.1 Amacrine cells in the retina of a rabbit

One general approach to analysing spatio-temporal point process data is to extend existing methods for purely spatial data by considering the time of occurrence as a distinguishing feature, or *mark,* attached to each event. Before giving an example of this, we give an even simpler example of a marked spatial

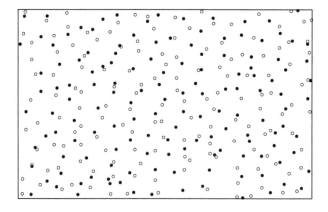

Figure 1.1 Amacrine cells in the retina of a rabbit. On and off cells are shown as open and closed circles, respectively. The rectangular region on which the cells are observed has dimension 1060 by 662 μm.

point pattern, in which the events are of just two qualitatively different types. Each event in Figure 1.1 represents the location of an amacrine cell in the retina of a rabbit. These cells play a fundamental role in mammalian vision. One type transmits information when a light goes *on*; the other type similarly transmits information when a light goes *off*. The data consist of the locations of 152 on cells and 142 off cells in a rectangular region of dimension 1060 by 662 μm.

The primary goal for the analysis of these data is to discriminate between two competing developmental hypotheses. The first hypothesis is that the pattern forms initially in two separate layers, corresponding to their pre-determined functionality; the second is that the pattern forms initially in a single, undifferentiated layer with function determined at a later developmental stage. One way to formalise this in statistical terms is to ask whether the two component patterns are statistically independent. Approximate independence would favour the first hypothesis. As we shall discuss in Section 1.2, this statement is a slight over-simplification but it provides a sensible starting point for an analysis of the data.

Our description and later analysis of these data are based on material in Diggle et al. (2005a). For a general discussion of the biological background, see Hughes (1985).

1.1.1.2 Bovine tuberculosis in Cornwall, U.K.

Our second example concerns the spatio-temporal distribution of reported cases of bovine tuberculosis (BTB) in the county of Cornwall, U.K., over the years 1991 to 2002. Individual cases are identified from annual inspections of farm herds; hence the effective time-resolution of the data is 1 year.

The prevalence of BTB has been increasing during the 12-year period covered by the data, but the observed annual counts exaggerate this effect because the scale of the annual inspection programme has also increased. Each recorded case is classified genetically, using the method of spoligotyping (Durr et al., 2000). The main scientific interest in these data lies not so much in the overall spatio-temporal distribution of the disease, but rather in the degree of spatial segregation amongst the different spoligotypes, and whether this spatial segregation is or is not stable over time. If the predominant mode of transmission is through local cross-infection, we might expect to find a stable pattern of spatial segregation, in which locally predominant spoligotypes persist over time; whereas if the disease is spread primarily by the introduction of animals from remote locations which are bought and sold at market, the resulting pattern of spatial segregation should be less stable over time (Diggle et al., 2005c).

Figure 1.2 shows the spatial distributions of cases corresponding to each of the four most common spoligotypes. The visual impression is one of strong spatial segregation, with each of the four types predominating in particular sub-regions.

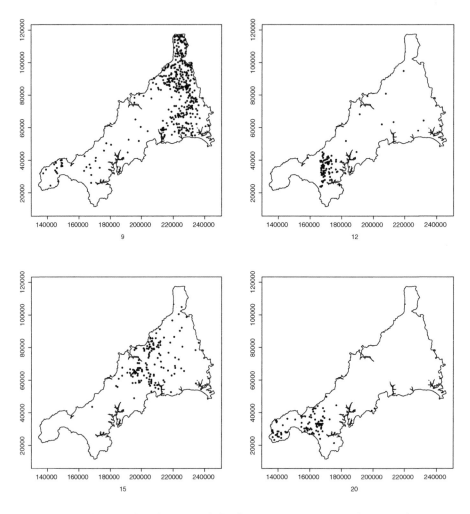

Figure 1.2 Spatial distributions of the four most common spoligotype data over the 14 years. Top row: spoligotype 9 (left) and spoligotype 12 (right). Bottom row: spoligotype 15 (left) and spoligotype 20 (right).

1.1.1.3 Gastroenteric disease in Hampshire, U.K.

Our third example concerns the spatio-temporal distribution of gastroenteric disease in the county of Hampshire, U.K., over the years 2001 and 2002. The data are derived from calls to NHS Direct, a 24-hour, 7-day phone-in service operating within the U.K. National Health Service. Each call to NHS Direct generates a data-record which includes the caller's post-code, the date of the call and a symptom code (Cooper et al., 2003). Figure 1.3 shows the locations of the 7167 calls from patients resident in Hampshire whose assigned symptom code corresponded to acute, non-specific gastroenteric disease. The spatial distribution of cases largely reflects that of the population of Hampshire, with strong concentrations in the large cities of Southampton and Portsmouth, and

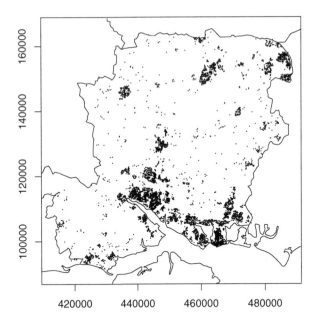

Figure 1.3 Locations of 7167 incident cases of non-specific gastroenteric disease in Hampshire, 1 January 2001 to 31 December 2002.

smaller concentrations in other towns and villages. Inspection of a dynamic display of the space-time coordinates of the cases suggests the kind of pattern typical of an endemic disease, in which cases can occur at any point in the study region at any time during the 2-year period. Occasional outbreaks of gastroenteric disease, which arise as a result of multiple infections from a common source, should result in anomalous spatially and temporally localised concentrations of cases.

The data were collected as part of the AEGISS project (Diggle et al., 2003), whose overall aim was to improve the timeliness of the disease surveillance systems currently used in the U.K. The specific statistical aims for the analysis of the data are to establish the normal pattern of spatial and temporal variation in the distribution of reported cases, and hence to develop a method of real-time surveillance to identify as quickly as possible any anomalous incidence patterns which might signal the onset of an outbreak requiring some form of public health intervention.

1.1.1.4 The U.K. 2001 epidemic of foot-and-mouth disease

Foot-and-mouth disease (FMD) is a highly infectious viral disease of farm livestock. The virus can be spread directly between animals over short distances in contaminated airborne droplets, and indirectly over longer distances, for example via the movement of contaminated material. The U.K. experienced a

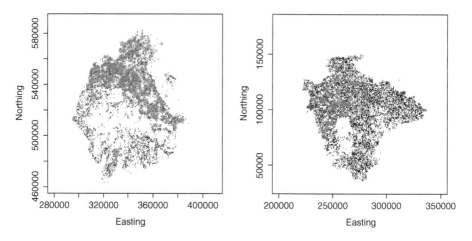

Figure 1.4 (SEE COLOR INSERT FOLLOWING PAGE 142) Locations of at-risk farms (black) and FMD case-farms (red) in Cumbria (left-hand panel) and in Devon (right-hand panel).

major FMD epidemic in 2001, which resulted in the slaughter of more than 6 million animals. Its estimated total cost to the U.K. economy was around £8 billion (U.K. National Audit Office, 2002). The epidemic affected 44 counties, and was particularly severe in the counties of Cumbria, in the north-west of England, and Devon, in the south-west. Figure 1.4 shows the spatial distributions of all farms in Cumbria and Devon which were at risk at the start of the epidemic, and of the farms which experienced the disease. In sharp contrast to the data on gastroenteric disease in Hampshire, the case-farms are strongly concentrated in sub-regions within each of the two counties. Dynamic plotting of the space-time locations of case-farms confirms the typical pattern of a highly infectious, epidemic disease. The predominant pattern is of transmission between near-neighbouring farms, but there are also a few, apparently spontaneous outbreaks of the disease far from any previously infected farms.

The main control strategies used during the epidemic involved the preemptive slaughter of animal holdings at farms thought to be at high risk of acquiring, and subsequently spreading, the disease. Factors which could affect whether a farm is at high risk include, most obviously, its proximity to infected farms, but also recorded characteristics such as the size and species composition of its holding. One objective in analysing these data is to formulate and fit a model for the dynamics of the disease which incorporates these effects. A model of this kind could then provide information on what forms of control strategy would be likely to prove effective in any future epidemic.

1.1.2 Chapter outline

In Section 1.2 we give a brief review of statistical methods for spatial point patterns, illustrated by an analysis of the amacrine cell data shown in Figure 1.1. We refer the reader to Diggle (2003) or Møller and Waagepetersen

(2003) for more detailed accounts of the methodology, and to Diggle et al. (2005a) for a full account of the data-analysis.

In Section 1.3 we discuss strategies for analysing spatio-temporal point process data. We argue that an important distinction in practice is between data for which the individual events (x_i, t_i) occur in a space-time continuum, and data for which the time-scale is either naturally discrete, or is made so by recording only the aggregate spatial pattern of events over a sequence of discrete time-periods. Our motivating examples include instances of each of these scenarios. Other scenarios which we do not consider further are when the locations are coarsely discretised by assigning each event to one of a number of sub-regions which form a partition of A. Methods for the analysis of spatially discrete data are typically based on Markov random field models. An early, classic reference is Besag (1974). Book-length treatments include Cressie (1991), Banerjee et al. (2003), and Rue and Held (2005).

In later sections we describe some of the available models and methods through their application to our motivating examples. This emphasis on specific examples is to some extent a reflection of the author's opinion that generic methods for analysing spatio-temporal data-sets have not yet become well established; certainly, they are less well established than is the case for purely spatial data. Nevertheless, in the final section of the chapter we will attempt to draw some general conclusions which go beyond the specific examples considered, and can in that sense be regarded as pointers towards an emerging general methodology.

1.2 Statistical methods for spatial point processes

1.2.1 Descriptors of pattern: spatial regularity, complete spatial randomness, and spatial aggregation

A convenient, and conventional, starting point for the analysis of a spatial point pattern is to apply one or more tests of the hypothesis of *complete spatial randomness* (CSR), under which the data are a realisation of a homogeneous Poisson process. A homogeneous Poisson process is a point process that satisfies two conditions: the number of events in any planar region A follows a Poisson distribution with mean $\lambda|A|$, where $|\cdot|$ denotes area and the constant λ is the *intensity*, or mean number of events per unit area; and the numbers of events in disjoint regions are independent. It follows that, conditional on the number of events in any region A, the locations of the events form an independent random sample from the uniform distribution on A (see, for example, Diggle, 2003, Section 4.4). Hence, CSR embraces two quite different properties: a uniform marginal distribution of events over the region A; and independence of events. We emphasise that this is only a starting point, and that the hypothesis of CSR is rarely of any scientific interest. Rather, CSR is a dividing hypothesis (Cox, 1977), a test of which leads to a qualitative classification of an observed pattern as regular, approximately random or aggregated.

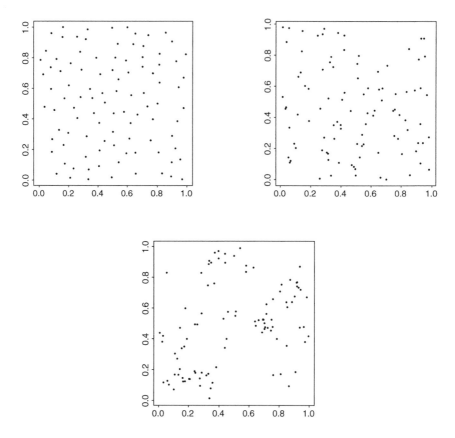

Figure 1.5 Examples of a regular (upper-left panel), a completely random (upper-right panel), and an aggregated (lower panel) spatial point pattern.

We do not attempt a precise mathematical definition of the descriptions "regular" and "aggregated." Roughly speaking, a regular pattern is one in which events are more evenly spaced throughout A than would be expected under CSR, and typically arises through some form of inhibitory dependence between events. Conversely, an aggregated pattern is one in which events tend to occur in closely spaced groups. Patterns of this type can arise as a consequence of marginal non-uniformity, or a form of attractive dependence, or both. In general, as shown by Bartlett (1964), it is not possible to distinguish empirically between underlying hypotheses of non-uniformity and dependence using the information presented by a single observed pattern. Figure 1.5 shows an example of a regular, a completely random, and an aggregated spatial point pattern. The contrasts amongst the three are clear.

1.2.2 Functional summary statistics

Tests of CSR which are constructed from functional summary statistics of an observed pattern are useful for two reasons: when CSR is conclusively rejected, their behaviour gives clues as to the kind of model which might provide a reasonable fit to the data; and they may suggest preliminary estimates of model parameters. Two widely used ways of constructing functional summaries are through nearest neighbour and second-moment properties. Third and higher-order moment summaries are easily defined, but appear to be rarely (possibly too rarely) used in data-analysis; an exception is Peebles and Groth (1975). They do feature, for example, in the theoretical analysis of ecological models, as discussed in Murrell, Dieckmann and Law (2004), and undoubtedly offer potential insights which are not captured by second-moment properties.

Two nearest neighbour summaries are the distribution functions of X, the distance from an arbitrary origin of measurement to the nearest event of the process, and of Y, the distance from an arbitrary event of the process to the nearest other event. We denote these by $F(x)$ and $G(y)$, respectively. The empirical counterpart of $F(x)$ typically uses the distances, d_i say, from each of m points in a regular lattice arrangement to the nearest event, leading to the estimate $\tilde{F}(x) = m^{-1} \sum I(d_i \leq x)$ where $I(\cdot)$ is the indicator function. Similarly, if e_i is the distance from each of n events to its nearest neighbour, then $\tilde{G}(y) = n^{-1} \sum I(e_i \leq y)$. Edge-corrected versions of these simple estimators are sometimes preferred, and are necessary if we wish to compare empirical estimates with the corresponding theoretical properties of a stationary point process.

Derivations, and further discussion, of results in the remainder of this section can be found, for example, in Diggle (2003, Chapter 4).

Under CSR, $F(x) = G(x) = 1 - \exp(-\lambda \pi x^2)$, where λ is the *intensity*, or mean number of events per unit area. Typically, in a regular pattern $G(x) < F(x)$, whereas in an aggregated pattern $G(x) > F(x)$.

To describe the second-moment properties of a spatial point process, we need some additional notation. Let dx denote an infinitesimal neighbourhood of the point x, and $N(dx)$ the number of events in dx. Then, the *intensity function* of the process is

$$\lambda(x) = \lim_{|dx| \to 0} \left\{ \frac{E[N(dx)]}{|dx|} \right\}.$$

Similarly, the *second-moment intensity function* is

$$\lambda_2(x, y) = \lim_{\substack{|dx| \to 0 \\ |dy| \to 0}} \left\{ \frac{E[N(dx)N(dy)]}{|dx||dy|} \right\},$$

and the *covariance density* is

$$\gamma(x, y) = \lambda_2(x, y) - \lambda(x)\lambda(y).$$

The process is stationary and isotropic if its statistical properties do not change under translation and rotation, respectively. If we now assume that the process is stationary and isotropic, the intensity function reduces to a constant, λ, equal to the expected number of events per unit area. Also, the second-moment intensity reduces to a function of distance, $\lambda_2(x, y) = \lambda_2(r)$ where $r = ||x - y||$ is the distance between x and y, and the covariance density is $\gamma(r) = \lambda_2(r) - \lambda^2$. In this case, the scaled quantity $\rho(r) = \lambda_2(r)/\lambda^2$ is called, somewhat misleadingly, the *pair correlation function*. For a homogeneous Poisson process, $g(r) = 1$ for all r.

A more tangible interpretation of the pair correlation function is obtained if we integrate over a disc of radius s. This gives the reduced second-moment measure, or *K-function*,

$$K(s) = 2\pi \int_0^s \rho(r) r \, dr. \tag{1.1}$$

Ripley (1976, 1977) introduced the K-function as a tool for data-analysis. One of its advantages over the pair correlation function is that it can be interpreted as a scaled expectation of an observable quantity. Specifically, let $E(s)$ denote the expected number of further events within distance s of an arbitrary event. Then,

$$K(s) = \lambda^{-1} E(s). \tag{1.2}$$

The result (1.2) leads to several useful insights. First, it suggests a method of estimating $K(s)$ directly by the method of moments, without the need for any smoothing; this is especially useful for relatively small data-sets. Second, it explains why $K(s)$ is a good descriptor of spatial pattern. For a completely random pattern, events are positioned independently; hence $E(s) = \lambda \pi s^2$ and $K(s) = \pi s^2$. This gives a benchmark against which to assess departures from CSR. For aggregated patterns, $K(s)$ is relatively large at small distances s because each event typically forms part of a "cluster" of mutually close events. Conversely, for regular patterns, $K(s)$ is relatively small at small distances s because each event tends to be surrounded by empty space. Another useful property is that $K(s)$ is invariant to random thinning, i.e., retention or deletion of events according to a series of independent Bernoulli trials. This follows immediately from (1.2), which expresses $K(s)$ as the ratio of two quantities, both of which vary by the same constant of proportionality under random thinning.

We use the following edge-corrected method of moments estimator proposed originally by Ripley (1976, 1977). For data $x_i \in A : i = 1, \ldots, n$, a natural estimator for $E(s)$ is

$$\tilde{E}(s) = n^{-1} \sum_{i=1}^{n} \sum_{j \neq i} I(r_{ij} \leq s), \tag{1.3}$$

where $r_{ij} = ||x_i - x_j||$. Except for very small values of s, this estimator suffers from substantial negative bias because events outside A are not recorded in the data. A remedy is to replace the simple count in (1.3) by a sum of weights w_{ij}, where w_{ij}^{-1} is the proportion of the circumference of the circle with centre

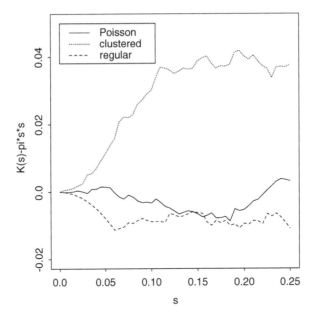

Figure 1.6 Estimates $\hat{K}(s) - \pi s^2$ for a regular (dashed line), a completely random (Poisson process, solid line), and an aggregated or clustered (dotted line) point pattern.

x_i and radius r_{ij} which lies within A. Finally, we estimate λ by $(n-1)/|A|$, where $|A|$ denotes the area of A, to give

$$\hat{K}(s) = |A|\{n(n-1)\}^{-1} \sum_{i=1}^{n} \sum_{j \neq i} w_{ij} I(r_{ij} \leq s). \qquad (1.4)$$

Ripley used $n/|A|$ to estimate λ. Our preference for $(n-1)/|A|$ has a slightly arcane theoretical justification which is discussed in Chetwynd and Diggle (1998) but this is clearly of no great consequence when n is large.

Figure 1.6 shows estimates $\hat{K}(s) - \pi s^2$ for each of the three point patterns shown in Figure 1.5. Subtraction of the CSR benchmark, $K(s) = \pi s^2$, emphasises departures from CSR, in effect acting as a magnifying glass applied to the estimate $\hat{K}(s)$.

Multivariate extensions of the K-function and its estimator were proposed by Lotwick and Silverman (1982). For a stationary, isotropic process let λ_j : $j = 1, \ldots, m$ denote the intensity of type j events. Define functions $K_{ij}(s) = \lambda_j^{-1} E_{ij}(s)$, where $E_{ij}(s)$ is the expected number of further type j events within distance s of an arbitrary type i event. Note that $K_{ij}(s) = K_{ji}(s)$. Although this equality is not obvious from the above definitions, it follows immediately from the multivariate analogue of our earlier definition (1.1) of $K(s)$ as an integrated version of the pair correlation function. However, direct extension of (1.4) to the multivariate case leads to two different estimates $\tilde{K}_{ij}(s)$ and

$\tilde{K}_{ji}(s)$ which, following Lotwick and Silverman (1982), we can combine to give the single estimate

$$\hat{K}_{ij}(s) = \{n_i \tilde{K}_{ij}(s) + n_j \tilde{K}_{ji}(s)\}/(n_i + n_j). \tag{1.5}$$

Two useful benchmark results for multivariate K-functions are:

1. If type i and type j events form independent processes, then $K_{ij}(s) = \pi s^2$;
2. If type i and type j events form a random labelling of a univariate process with K-function $K(s)$, then $K_{ii}(s) = K_{jj}(s) = K_{ij}(s) = K(s)$.

1.2.3 Functional summary statistics for the amacrines data

Figure 1.7 shows estimates $\hat{K}_{ij}(s) - \pi s^2$ for the amacrine cell data. Our interpretation of the three estimates is as follows. First, the near-equality of $\hat{K}_{11}(s)$ and $\hat{K}_{22}(s)$ suggests that the underlying biological process may be the same for both types of cell. Informally, the difference between $\hat{K}_{11}(s)$ and $\hat{K}_{22}(s)$ gives an upper bound to the size of the sampling fluctuations in the estimates. Second, both estimates show a strong inhibitory effect, with no two cells of the same type occurring within a distance of around 30 μm. Third, the

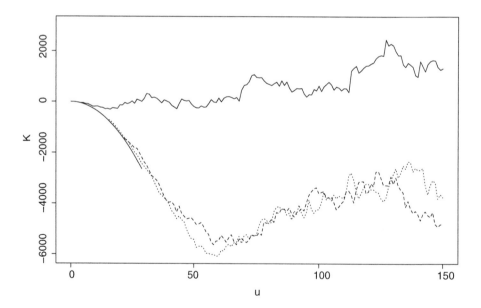

Figure 1.7 Estimates of the K-functions for the amacrine cell data. Each plotted function is $\hat{K}(s) - \pi s^2$. The dashed line corresponds to $\hat{K}_{11}(s)$ (on cells), the dotted line to $\hat{K}_{22}(s)$ (off cells), and the solid line to $\hat{K}_{12}(s)$. The parabola $-\pi s^2$ is also shown as a solid line.

estimate $\hat{K}_{12}(s)$ fluctuates around a value close to zero at small distances s, suggesting that the two component patterns are approximately independent. More specifically, $\hat{K}_{12}(s)$ does not show the strong inhibitory effect exhibited by both $\hat{K}_{11}(s)$ and $\hat{K}_{22}(s)$.

Collectively, these results are consistent with the first of the two developmental hypotheses for these data, namely that the component patterns of on and off cells form initially in two separate layers which later fuse to form the mature retina. Specifically, the separate layer hypothesis would imply statistical independence between the two component patterns; hence $K_{12}(s) = \pi s^2$. In fact, as we discuss below, the component patterns cannot strictly be independent because of the physical space required by each cell body. The data are clearly not compatible with random labelling of an initially undifferentiated pattern, as this would require all three estimated K-functions to be equal to within sampling variation. Furthermore, it is difficult to imagine how any biologically plausible labelling process could preserve strict inhibition between any two cells of the same type without imposing a similar constraint on two cells of opposite type. Hence, the analysis summarised in Figure 1.7 strongly favours the separate layer hypothesis.

1.2.4 Likelihood-based methods

Classical maximum likelihood estimation is straightforward for Poisson processes, but notoriously intractable for other point process models. Two more tractable alternatives are maximum pseudo-likelihood and Monte Carlo maximum likelihood. Both are particularly well suited to estimation in a class of models known as pairwise interaction point processes, and it is in this context that we discuss them here.

A third variant of likelihood-based estimation uses a *partial likelihood*. This method is best known in the context of survival analysis (Cox, 1972, 1975). We describe its adaptation to spatio-temporal point processes in Section 1.3.2.2.

1.2.4.1 Pairwise interaction point processes

Pairwise interaction processes form a sub-class of Markov point processes (Ripley and Kelly, 1977). They are defined by their likelihood ratio, $f(\cdot)$, with respect to a Poisson process of unit intensity. Hence, if $\chi = \{x_1, \ldots, x_n\}$ denotes a configuration of n points in a spatial region A, then $f(\chi)$ measures in an intuitive sense how much more likely is the configuration χ than it would be as a realisation of a Poisson process of unit intensity. For a pairwise interaction process, we need to specify a parameter β which governs the mean number of events per unit area and an *interaction function* $h(r)$, where r denotes distance. Intuitively, $h(r)$ is related to the likelihood that the model will generate pairs of events separated by a distance r, in the sense that the likelihood for a particular configuration of events depends on the product of $h(||x_i - x_j||)$ over all distinct pairs of events. Hence, for example, a value $h(r) = 0$ for all $r < \delta$ would imply that no two events can be separated by a distance less than δ. The likelihood ratio $f(\chi)$ for the resulting pairwise interaction point process is

$$f(\chi) = c(\beta, h)\beta^n \prod_{j<i} h(||x_i - x_j||), \tag{1.6}$$

where $c(\beta, h)$ is a normalising constant which is generally intractable. Note that in a homogeneous Poisson process, the number of points in A follows a Poisson distribution with mean proportional to $|A|$ and, conditional on the number of points in A, their locations form an independent random sample from the uniform distribution on A. It follows that a homogeneous Poisson process is a special case of a pairwise interaction process in which $h(u) = 1$ for all u, and β is the intensity. More generally, in (1.6) the parameter β is directly related to, but not necessarily equal to, the intensity. Provided that the specified form of $h(\cdot)$ is legitimate, values of $h(r)$ less than or greater than 1 correspond to processes that generate regular or aggregated patterns, respectively.

A sufficient condition for legitimacy is that $h(r) \leq 1$ for all r, as this guarantees a finite intensity for the resulting point process. It also leads to point patterns whose character is inhibitory, meaning that close pairs of events are relatively unlikely by comparison with a Poisson process of the same intensity. Pairwise interaction point process of this kind are widely used for modelling regular spatial point patterns.

Specifications in which $h(u) > 1$ are more problematic. An intuitive explanation for this is that if, over a range of distances r, the interaction function takes values $h(r) \geq h_0$, where $h_0 > 1$, then the product term on the right-hand side of (1.6) can be as large as $h_0^{n(n-1)/2}$, and this cannot be balanced by adjusting the value of β in (1.6). Hence, the likelihood increases without limit as $n \to \infty$. This perhaps explains why, even if we are prepared to consider n as fixed, pairwise interaction processes with $h(r) > 1$ tend to generate unrealistically strong spatial aggregation, with large clusters of near-coincident events. For a rigorous discussion of the properties of pairwise interaction processes with $h(r) > 1$, see Gates and Westcott (1986).

1.2.4.2 Maximum pseudo-likelihood

The method of maximum pseudo-likelihood was originally proposed by Besag (1975, 1978) as a method for real-valued, spatially discrete processes. Besag et al. (1982) derived a point process version by considering a limit of binary-valued processes on a lattice, as the lattice spacing tends to zero. For a finite-dimensional probability distribution, the pseudo-likelihood is the product of the full conditional distributions, i.e., the conditional distributions of each Y_i given the values of all other Y_j. Hence, if $Y = (Y_1, \ldots, Y_n)$ has joint probability density $f(y)$, then the pseudo-likelihood is, in an obvious notation, $\prod_{i=1}^{n} f(y_i|y_j, j \neq i)$.

For a point process, the pseudo-likelihood uses conditional intensities in place of the full conditional distributions. In particular, for a Markov point process with likelihood ratio $f(\cdot)$, the conditional intensity for an arbitrary point u given the observed configuration X on $A - \{u\}$ is

$$\lambda(u; X) = \begin{cases} f(X \cup \{u\})/f(X) : u \notin X \\ f(X)/f(X - \{u\}) : u \in X \end{cases}$$

and the log-pseudo-likelihood is

$$\sum_{i=1}^{n} \log \lambda(x_i; X) - \int_A \lambda(u; X) du.$$

1.2.4.3 Monte Carlo maximum likelihood

Monte Carlo maximum likelihood estimation, as described here, was proposed by Geyer and Thompson (1992). Geyer (1999) and Møller and Waagepetersen (2003) discuss the method in the context of point process models including, but not restricted to, pairwise interaction point processes.

Conditional on the number of events in a specified region A, the likelihood for a pairwise interaction point process can be written in principle as

$$\ell(\theta) = c(\theta) f(X; \theta), \tag{1.7}$$

where $X = \{x_1, \ldots, x_n\}$ is the observed configuration of the n events in A. For most models of interest, the normalising constant $c(\theta)$ in (1.7) is intractable. However, note that

$$c(\theta)^{-1} = \int_X f(X; \theta) dX$$
$$= \int_X f(X; \theta) \times \frac{c(\theta_0)}{c(\theta_0)} \times \frac{f(X; \theta_0)}{f(X; \theta_0)},$$

for any value of θ_0. If we now define $r(X; \theta, \theta_0) = f(X; \theta)/f(X; \theta_0)$, then we can write

$$c(\theta)^{-1} = c(\theta_0)^{-1} \int_X r(X; \theta, \theta_0) c(\theta_0) f(X; \theta) dX$$
$$= c(\theta_0)^{-1} \mathrm{E}_{\theta_0}[r(X; \theta, \theta_0)] dX,$$

where $\mathrm{E}_{\theta_0}[\cdot]$ denotes expectation with respect to the distribution of X when $\theta = \theta_0$. This in turn allows us to re-express the likelihood (1.7) as

$$\ell(\theta) = c(\theta_0) f(X; \theta) / \mathrm{E}_{\theta_0}[r(X; \theta, \theta_0)]. \tag{1.8}$$

It follows from (1.8) that for any fixed value θ_0, the maximum likelihood estimator $\hat{\theta}$ maximises

$$L_{\theta_0}(\theta) = \log f(X; \theta) - \log \mathrm{E}_{\theta_0}[r(X; \theta, \theta_0)]. \tag{1.9}$$

Now choose any value θ_0, simulate realisations $X_j : j = 1, \ldots, s$ with $\theta = \theta_0$, and define

$$L_{\theta_0, s}(\theta) = \log f(X; \theta) - \log s^{-1} \sum_{j=1}^{s} [r(X_j; \theta, \theta_0)]. \qquad (1.10)$$

Then, the value $\hat{\theta}_{MC}$ which maximises (1.10) is a *Monte Carlo maximum likelihood estimator* (MCMLE) for θ. Note the indefinite article. A Monte Carlo log-likelihood $L_{\theta_0, s}(\theta)$ is typically a smooth function of θ and can easily be maximised numerically, but it is also a function of θ_0, s, and the simulated realisations X_j. In practice, we would want to choose s sufficiently large that the Monte Carlo variation introduced by using a sample average in place of the expectation on the right-hand side of (1.9) is negligible. However, for given s the behaviour of the MCMLE is critically dependent on the choice of θ_0, the ideal being to choose θ_0 equal to $\hat{\theta}$, in which case the Monte Carlo variation in (1.10) is zero at $\theta = \hat{\theta}$. More generally, obtaining a sufficiently accurate Monte Carlo approximation to the intractable expectation in (1.9) raises a number of practical issues which, in the author's experience, make it difficult to automate the procedure.

1.2.5 Bivariate pairwise interaction point processes

The family of pairwise interaction point processes can readily be extended to the multivariate case by specifying a set of interaction functions, one for each possible pair of types of event. In the bivariate case, and again treating the numbers of events of each type as fixed, we denote the data by a pair of configurations, $X_1 = \{x_{1i} : i = 1, \ldots, n_1\}$ and $X_2 = \{x_{2i} : i = 1, \ldots, n_2\}$.

In a bivariate pairwise interaction process, the joint density for X_1 and X_2 is $f(X_1, X_2) \propto P_{11} P_{22} P_{12}$, where

$$P_{11} = \prod_{i=2}^{n_1} \prod_{j=1}^{i-1} h_{11}(\|x_{1i} - x_{1j}\|),$$

$$P_{22} = \prod_{i=2}^{n_2} \prod_{j=1}^{i-1} h_{22}(\|x_{2i} - x_{2j}\|),$$

and

$$P_{12} = \prod_{i=1}^{n_1} \prod_{j=1}^{n_2} h_{12}(\|x_{1i} - x_{2j}\|).$$

As in the univariate case, a sufficient condition for the legitimacy of the model is that $0 \leq h_{ij}(r) \leq 1$ for all r. However, and in contrast to the univariate case, a model of this kind can easily generate spatially aggregated

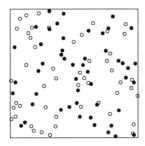 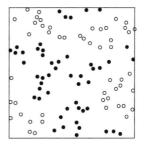

Figure 1.8 Simulated realisations of bivariate pairwise interaction point processes each with 50 events of either type on the unit square and simple inhibitory interaction functions. In both panels, the minimum permissible distance between any two events of the same type is 0.025. In the left-hand panel, the two component patterns are independent. In the right-hand panel, the minimum permissible distance between any two events of opposite types is 0.1.

component patterns. Figure 1.8 shows an example from Diggle et al. (2005a), in which marginal aggregation is induced by specifying a strongly inhibitory interaction between events of opposite type.

1.2.6 Likelihood-based analysis of the amacrine cell data

The analysis in Section 1.2.3 suggests that a suitable model for the amacrine data might be a bivariate pairwise interaction process with strongly inhibitory marginal properties and approximate independence between the two components.

Our first stage in fitting a model of this kind is to use maximum pseudo-likelihood estimation in conjunction with a piece-wise constant specification of $h(u)$ to identify a candidate model for the interaction within each component pattern. We then use Monte Carlo maximum likelihood to fit a suitable parametric model to each component. Figure 1.9 shows the result, together with a Monte Carlo maximum likelihood estimate using the parametric model

$$h(u; \theta) = \begin{cases} 0 : u \leq \delta \\ 1 - \exp[-\{(u - \delta)/\phi\}^\alpha] : u > \delta \end{cases}. \tag{1.11}$$

The fit adopts common parameter values for the two types of cell, on the basis of a Monte Carlo likelihood ratio test under the assumption that the two component processes are independent; for details, see Diggle et al. (2005a). In fitting the parametric model, we used a fixed value $\delta = 10\,\mu\text{m}$, representing the approximate physical size of each cell body, and estimated the remaining parameters as $\hat{\phi} = 49.08$ and $\hat{\alpha} = 2.92$.

For the bivariate analysis, we use the same parametric form (1.11) for the three interaction functions $h_{11}(r)$, $h_{22}(r)$, and $h_{12}(r)$. For ϕ tending to zero,

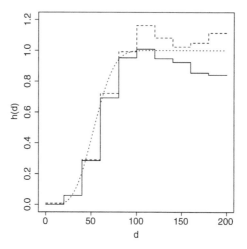

Figure 1.9 Non-parametric maximum pseudo-likelihood estimates of the pairwise interaction functions for on cells (solid line) and for off cells (dashed line), together with parametric fit assuming common parameter values for both types of cell (dotted line). Each plotted function shows $\hat{h}(r)$.

the model for $h_{12}(r)$ reduces to a simple inhibitory form, $h_{12}(r) = 0$ for $r < \delta_{12}$ and $h_{12}(r) = 1$ otherwise, with independence of the two components as the special case $\delta_{12} = 0$. Independence is strictly impossible because no two cells can occupy the same location, but it is reasonable to treat δ_{12} as a parameter to be estimated because the two types of cell are located at slightly different depths within the retina. Diggle et al. (2005a) conclude that a bivariate model with a simple inhibitory $h_{12}(r)$, and $\hat{\delta}_{12} = 4.9$ gives a reasonable fit to the data.

1.3 Strategies for the analysis of spatio-temporal point patterns

Many of the tools used to analyse spatial point process data can be extended to the spatio-temporal setting. Functional summaries based on low-order moments can be extended in the obvious way by considering configurations of events at specified spatial and temporal positions.

For example, Diggle et al. (1995) defined a spatio-temporal K-function $K(s,t)$ such that $\lambda K(s,t)$ is the expected number of further events within distance s and time t of an arbitrary event of the process. Bhopal et al. (1992) used an estimate of $K(s,t)$ to analyse the spatio-temporal distribution of apparently sporadic cases of Legionnaires' disease. However, the spatio-temporal setting opens up other modelling and analysis strategies which take more explicit account of the directional character of time, and the consequently richer opportunities for scientific inference.

1.3.1 Strategies for discrete-time data

As noted earlier, discrete-time spatio-temporal point process data can arise in two ways: either the underlying process genuinely operates in discrete-time or an underlying continuous-time process is observed at a discrete sequence of time-points. A hypothetical example of the former would be the yearly sequence of spatial point distributions formed by the natural regeneration of an annual plant community. The Cornwall BTB data described in Section 1.1.1.2 are an example of the latter.

1.3.1.1 Transition models

For genuinely discrete-time processes, a natural strategy is to build a transition model to describe the changes between successive times. In symbolic notation, if $\mathcal{P}_t$ denotes the spatial point process at time t, a transition model for the joint distribution of $\mathcal{P} = \{\mathcal{P}_1, \mathcal{P}_2, \ldots, \mathcal{P}_t\}$ takes the form

$$[\mathcal{P}] = [\mathcal{P}_1][\mathcal{P}_2|\mathcal{P}_1] \ldots [\mathcal{P}_t|\mathcal{P}_{t-1}, \ldots \mathcal{P}_1]. \tag{1.12}$$

A convenient working assumption would be that the process is Markov in time, and this may have some mechanistic justification when times correspond to successive generations, as in our hypothetical example.

1.3.1.2 A transition model for spatial aggregation

A standard example of a spatial point process model which leads to spatially aggregated patterns is a Neyman-Scott clustering process (Neyman and Scott, 1958), defined as follows. *Parent* events form a homogeneous Poisson process with intensity ρ events per unit area. The parents then generate numbers of *offspring* as an independent random sample from the Poisson distribution with mean μ. The positions of the offspring relative to their parents are an independent random sample from the bivariate normal distribution with mean zero and variance matrix $\sigma^2 I$, where I denotes the identity matrix. The observed point pattern is then taken to be the superposition of the offspring from all parents.

An obvious way to turn the Neyman-Scott process into a transition model is to let the offspring of one generation become the parents for the next generation. Kingman (1977) discusses a model of this kind in questioning whether the basic Neyman-Scott formulation can arise as the equilibrium distribution of a spatio-temporal process. Note in particular that the spatio-temporal process defined in this way may die out after a finite number of generations. Figure 1.10 shows the result of a simulation on the unit square with periodic boundary conditions, i.e., events are generated on a torus, which is then unwrapped to form the unit square region A. The model parameters are $\rho = 100$,

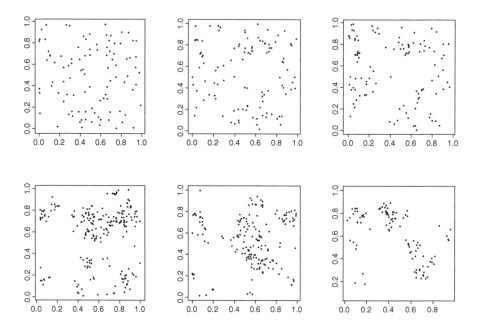

Figure 1.10 Simulated realisation of a transition model for spatial aggregation. The first row shows the first three generations of the process, starting from a completely random spatial distribution. The second row shows the 50th, 70th, and 90th generations.

$\sigma = 0.025$, and $\mu = 1$. Setting $\mu = 1$ implies that the mean number of events in each generation is ρ, but with a variance which increases from one generation to the next; if N_t denotes the number of events in the t-th generation, then $E[N_t] = \rho$ for all t and $\mathrm{Var}(N_t) = t\rho$.

The simulation shows how the spatial aggregation in the resulting patterns tends to increase with successive generations, although it is not clear that this informal description can be expressed rigorously, not least because on any finite region the process is certain eventually to become extinct.

1.3.1.3 Marked point process models

When time is artificially discrete as a consequence of the data-recording process, a principled approach would be to formulate a continuous-time model and to deduce from the model the statistical properties of the observed, discrete-time data. A more pragmatic strategy is to analyse the data as a marked spatial point process, treating time as an ordered categorical mark attached to the spatial location of each event. From this point of view, methods for the analysis of multivariate spatial point processes, which are marked point processes with categorical marks, can be applied directly and the discussion of multivariate point processes in Section 1.2 is relevant. However, the way in which the results of any such analysis are interpreted should, so far as is possible, take account of the natural ordering of time.

This pragmatic strategy is unlikely to deliver a model with a direct, mechanistic interpretation but is a useful approach for descriptive analysis. In Section 1.4 we illustrate the approach with an analysis of the Cornwall BTB data, as reported in Diggle et al. (2005c).

1.3.2 Analysis strategies for continuous-time data

Even when continuous-time data are available, we shall preserve a distinction between empirical and mechanistic modelling. An empirical model aspires to provide a good descriptive fit to the data, but does not necessarily admit a context-specific scientific interpretation. A mechanistic model is more ambitious, embodying features which relate directly to the underlying science. To some extent, this is a false dichotomy. On the one hand, a good empirical model will include parameters that are interpretable in ways relevant to the scientific context and, minimally, should furnish an answer to a scientifically interesting question. On the other hand, even a mechanistic model will be at best an idealised, and quite possibly a crude, approximation to the truth. From a statistical perspective, a simple but well-identified model may be more valuable than an over-complicated model incorporating more parameters than can reasonably be estimated from the available data.

1.3.2.1 Empirical modelling: log-Gaussian spatio-temporal Cox processes

A Cox process, introduced in one time dimension by Cox (1955), is a Poisson process with a varying intensity which is itself a stochastic process. For our purposes, we need a model for a non-negative valued spatio-temporal stochastic process $\Lambda(x,t)$. Then, conditionally on $\Lambda(x,t)$, our point process is a Poisson process with intensity $\Lambda(x,t)$. This implies that, again conditionally on $\Lambda(x,t)$, the number of events in any spatio-temporal region, say $A \times (0,T)$, is Poisson-distributed with mean

$$\mu_{A,T} = \int_0^T \int_A \Lambda(x,t) \, dx dt,$$

and the locations and times of the events are an independent random sample from the distribution on $A \times (0,T)$ with probability density proportional to $\Lambda(x,t)$.

By far the most tractable class of real-valued spatio-temporal stochastic processes is the Gaussian process, $S(x,t)$ say, for which the joint distribution of $S(x_i, t_i)$ for any set of points (x_i, t_i) is multivariate normal. A log-Gaussian Cox process is a Cox process whose intensity is of the form $\Lambda(x,t) = \exp\{S(x,t)\}$, where $S(x,t)$ is a Gaussian process (Møller et al., 1998).

The properties of a log-Gaussian Cox process are determined by the mean and covariance structure of $S(x,t)$. Note first that any spatial and/or temporal variation in the mean of $S(x,t)$ translates into a multiplicative, deterministic component to $\Lambda(x,t)$; hence we can always re-express our model as $\Lambda(x,t) = \lambda(x,t)\exp\{S(x,t)\}$ where the mean of $S(x,t)$ is a constant. In the stationary case, a convenient parameterisation is to set $E[S(x,t)] = -0.5\sigma^2$, where $\sigma^2 = \text{Var}\{S(x,t)\}$. This gives $E[\exp\{S(x,t)\}] = 1$, and hence $\lambda(x,t)$ is the unconditional space-time intensity, or mean number of events per unit time per unit area in an infinitesimal neighbourhood of the point (x,t). In the remainder of this section we assume that $\lambda(x,t) = 1$ and focus on the specification of the stochastic component $\exp\{S(x,t)\}$.

In general, if $S(x,t)$ has mean $-0.5\sigma^2$ and covariance function

$$\text{Cov}\{S(x,t), S(x',t')\} = \sigma^2\rho(x,x',t,t'),$$

then the covariance function of $\exp\{S(x,t)\}$ is

$$\gamma(x,x',t,t') = \exp\{\sigma^2\rho(x,x',t,t')\} - 1,$$

and $\gamma(\cdot)$ is also the covariance density of the Cox process (see Brix and Diggle, 2001, but note also the correction in Brix and Diggle, 2003). In the stationary case, $\gamma(x,x',t,t') = \gamma(u,v)$, where $u = ||x - x'||$ and $v = |t - t'|$.

Brix and Møller (2001) consider a sub-class of spatio-temporal log-Gaussian Cox processes in which $S(x,t) = S(x) + g(t)$ where $g(t)$ is a deterministic function. One interpretation of this sub-class is that $S(x)$ represents spatial environmental variation which does not vary over time, and $\exp\{g(t)\}$ represents a time-varying birth-rate for new events. A natural extension would be to replace $g(t)$ by a stationary stochastic process $G(t)$, which would give an additive decomposition of the correlation function of $S(x,t)$ as

$$\rho(u,v) = \{\sigma_S^2\rho_S(u) + \sigma_G^2\rho_G(v)\}/(\sigma_S^2 + \sigma_G^2).$$

Brix and Diggle (2001) develop an approach to spatio-temporal prediction using a separable spatio-temporal correlation function

$$\rho(u,v) = r(u)\exp(-v/\beta). \tag{1.13}$$

In (1.13), $r(u)$ is any valid spatial correlation function, whilst the exponential term reflects the underlying Markov-in-time structure of the model for $S(x,t)$, which they derive as follows.

First, consider a discretisation of continuous two-dimensional space into a fine grid, say of size M by N, and write S_t for the MN-element vector of values of $S(x,t)$ at the grid-points. Now, assume that S_t evolves over time according to the stochastic differential equation

$$dS_t = (A - BS_t)dt + dU_t \tag{1.14}$$

where A is an MN-element vector, B is a non-singular $MN \times MN$ matrix, and U_t is a discrete-space approximation of spatial Brownian motion. In the stationary case, (1.14) corresponds to a Gaussian process $S(x,t)$ with spatio-temporal covariance structure

$$\text{Cov}\{S(x,t), S(x-u, t-v)\} = \sigma^2 r(u) \exp(-vB).$$

Brix and Diggle focus on the special case in which $B = \beta^{-1}I$ for some scalar $\beta > 0$, in which case $S(x,t)$ has variance σ^2 and separable spatio-temporal correlation function given by (1.13).

Separability of the spatial and temporal correlation properties is a reasonable working assumption for an empirical, descriptive model, but may be too inflexible for some applications. For example, it implies that for any single location, x_0 say, the conditional distribution of $S(x_0, t)$ given the *whole* of the process $S(x, t')$ for some $t' < t$ depends only on $S(x_0, t')$. Gneiting (2002) reviews the relevant literature and proposes a general class of stationary, non-separable spatio-temporal covariance functions.

Within the log-Gaussian Cox process framework, model-specification corresponds exactly to the problem of specifying a model for a spatio-temporal Gaussian process. See, for example, Chapters 4 and 6 of this volume.

1.3.2.2 Mechanistic modelling: conditional intensity and a partial likelihood

Accepting that the distinction between what we have chosen to call empirical and mechanistic models is not sharp, a mechanistic model is one that seeks to explain how the evolution of the process depends on its past history, in a way which can be interpreted in terms of underlying scientific mechanisms. A natural way to specify a model of this kind is through its conditional intensity function. Let $\mathcal{H}_t$ denote the accumulated history of the process, i.e., the complete set of locations and times of events occurring up to time t. Then, the conditional intensity function, $\lambda(x, t | \mathcal{H}_t)$, represents the conditional intensity for an event at location x and time t, given $\mathcal{H}_t$. This assumes, amongst other things, that multiple, coincident events cannot occur; for a rigorous discussion, see for example Daley and Vere-Jones (1988, Chapter 2).

A defining property of a Poisson process is that its conditional intensity function is equal to its unconditional intensity; in other words, the future of the process is stochastically independent of its past. A more interesting example of a conditional intensity function is the following, which bears some resemblance to the discrete-time transition model illustrated in Section 1.3.1.2. Each event of the process at time zero subsequently produces offspring according to an inhomogeneous temporal Poisson process with intensity $\alpha(u)$, realised independently for different events. As in the earlier discrete-time example, the positions of the offspring relative to their parents are an independent random sample from the bivariate normal distribution with mean zero and variance matrix $\sigma^2 I$. Each offspring, independently, then follows the same rules as their parent: it produces offspring according to an inhomogeneous Poisson process with intensity $\alpha(u - t)$, where t denotes its birth-time, and each offspring is spatially dispersed relative to its parent according to a bivariate normal distribution with mean zero and variance $\sigma^2 I$. The events of the process are the resulting collection of locations x and birth-times t. If we order the events of the process so that $t_i < t_{i+1}$ for all i, then the history at time t is $\mathcal{H}_t = \{(x_i, t_i) : i = 1, \ldots, N_t\}$, where N_t is the number of events to have occurred by time t. Writing $f(x) = (2\pi\sigma^2)^{-1} \exp\{-x'x/(2\sigma^2)\}$, the conditional intensity function is

$$\lambda(x, t)|\mathcal{H}_t) = \sum_{i=1}^{N_t} \alpha(t - t_i) f(x - x_i).$$

The number of offspring produced by any event of this process is Poisson-distributed, with mean

$$\mu = \int_0^\infty \alpha(u) \, du,$$

which we therefore assume to be finite. The number of events as a function of time, N_t, forms a simple branching process, and eventual extinction is certain if $\mu \leq 1$. Otherwise, the probability of extinction depends both on μ and on the initial conditions at time $t = 0$. The spatial character of the process varies considerably, according to both the detailed model specification and the initial conditions. However, the cumulative spatial distribution of events occurring up to time t tends to become progressively more strongly aggregated as t increases, because of the combined effects of the successive clustering of groups of offspring around their respective parents, together with the extinction of some lines of descent.

For data $(x_i, t_i) \in A \times (0, T) : i = 1, \ldots, n$, with $t_1 < t_2 < \cdots < t_n$, the log-likelihood associated with any point process specified through its conditional intensity function can be written as

$$L(\theta) = \sum_{i=1}^n \log \lambda(x_i, t_i|\mathcal{H}_{t_i}) - \int_0^T \int_A \lambda(x, t|\mathcal{H}_t) \, dx dt. \tag{1.15}$$

See, for example, Daley and Vere-Jones (1988, Chapter 13). Two major obstacles to the use of (1.15) in practice are that the form of the conditional intensity may itself be intractable, and that even when the conditional intensity is available, as in our example above, direct evaluation of the integral term in (1.15) is seldom straightforward. Monte Carlo methods are becoming more widely available for problems of this kind (Geyer, 1999; Møller and Waagepetersen, 2003). However, in practice these methods often need careful tuning to each application and the associated cost of developing and running reliable code can be an obstacle to their routine use.

As an alternative, computationally simpler approach to inference for models which are defined through their conditional intensity, Diggle (2005) proposed a partial likelihood, which is obtained by conditioning on the locations x_i and times t_i and considering the resulting log-likelihood for the observed time-ordering of the events $1, \ldots, n$. Now let

$$p_i = \lambda(x_i, t_i | \mathcal{H}_{t_i}) / \sum_{j=i}^{n} \lambda(x_j, t_i | \mathcal{H}_{t_i}). \tag{1.16}$$

Then, the partial log-likelihood is

$$L_p(\theta) = \sum_{i=1}^{n} \log p_i. \tag{1.17}$$

This method is a direct adaptation to the space-time setting of the seminal proposal in Cox (1972) for proportional hazards modelling of survival data; see also Møller and Sorensen (1994) or Lawson and Leimich (2000). As discussed in Cox (1975), estimates obtained by maximising the partial likelihood inherit the general asymptotic properties of maximum likelihood estimators, although their use may entail a loss of efficiency by comparison with full maximum likelihood estimation. The exact conditions under which the method gives consistent estimation for spatial point process models have yet to be established. Also, some parameters of the original model may be unidentifiable from the partial likelihood. The loss of identifiability can be advantageous if the non-identified parameters are nuisance parameters. This often applies, for example, in the proportional hazards model for survival data (Cox, 1972), where it is helpful that the baseline hazard function can be left unspecified. Otherwise, and again as exemplified by the proportional hazards model for survival data, other methods of estimation are needed to recover the unidentified parameters; see, for example, Andersen et al. (1992).

1.4 Bovine tuberculosis: non-parametric smoothing methods for estimating spatial segregation

Recall that the data for this application are the locations, genotypes, and year of detection of all known cases of bovine tuberculosis (BTB) amongst farm animals in the county of Cornwall, U.K. Figure 1.2 showed maps of the locations for each of the four most common genotypes, collapsed over time. These maps give a clear impression of spatial segregation, by which we mean that the county can be partitioned approximately into sub-regions where one or other of the four genotypes predominates. More formally, suppose that the pattern of cases in Figure 1.2 is generated by a multivariate Poisson process with intensities $\lambda_k(x)$ corresponding to cases of genotype $k = 1, \ldots, 4$. Then, the probability that a case at location x will be of type j, conditional on there being a case of one of the four types at x, is

$$p_j(x) = \lambda_j(x) / \sum_{k=1}^{4} \lambda_k(x).$$

We say that the pattern is *unsegregated* if $p_j(x) = p_j$ for all j and all x. At the opposite extreme, the pattern is *completely segregated* if, at each x, $p_j(x) = 1$ for one of $j = 1, \ldots, 4$. Estimating and mapping the type-specific probabilities $p_j(x)$ allows an assessment of intermediate degrees of spatial segregation.

A specific question posed by these data is whether the pattern of spatial segregation is stable over time, as this would point towards constant re-infection by the locally predominant genotype, rather than to introduced infections transmitted through cattle bought and sold at market. We present here an analysis taken from Diggle et al. (2005c) which is intended to answer this question. Our aim is to estimate possibly time-varying type-specific probability surfaces $p_{jt}(x)$, the conditional probabilities that a case at location x in time-period t will be of type j.

In this spatio-temporal setting, we assume that cases form a discrete-time sequence of multivariate Poisson processes with intensity functions $\lambda_{jt}(x)$. Although this is not strictly consistent with the infectious nature of the disease, it provides a reasonable working model within which we can answer the specific question of interest. The Poisson assumption lends itself more naturally to the general approach described in Sections 1.2 and 1.3.1.3, where we consider the time-dimension as a qualitative mark attached to the spatial location of each event, rather than to a transition modelling approach as described in Sections 1.3.1.1 and 1.3.1.2. The latter would be more natural if the goal of the analysis was to model the transmission of the disease between farms, but this is ruled out because of the very coarse time-resolution of the annual inspection regime. We therefore approach the analysis of the data as a problem in non-parametric intensity estimation for a multivariate Poisson process.

The spatial distribution of farms at risk is not uniform over the county although it is, to a good approximation, constant over the 12-year time period covered by the data. Also, as noted earlier, variation in the temporal intensity of the disease is confounded with variation in the extent of the annual testing programme. We therefore factorise $\lambda_{jt}(x)$ as

$$\lambda_{jt}(x) = \lambda_0(x)\mu_0(x,t)\rho_{jt}(x), \qquad (1.18)$$

where $\lambda_0(x)$ represents the spatial intensity of farms, $\mu_0(x,t)$ the spatio-temporal intensity of recorded cases of unspecified type, and the functions $\rho_{jt}(x)$ represent the spatially, temporally, and genotypically varying risks which are the quantities of interest. The type-specific probabilities associated with (1.18) are

$$p_{jt}(x) = \lambda_{jt}(x)/\sum_{k=1}^{4}\lambda_{kt}(x) = \rho_{jt}(x)/\sum_{k=1}^{4}\rho_{kt}(x), \qquad (1.19)$$

which implies that spatio-temporal variations in the pattern of segregation can be estimated without our knowing either the spatial distribution of farms (although this is available if required) or, more importantly (because it is not available), the spatio-temporal variation in the extent of the inspection programme. Note that to estimate non-specific spatial variation in risk, we would need to make the additional assumption that $\mu_0(x,t) = \mu_0(t)$.

We estimate the type-specific probability surfaces using a simple kernel regression method. Let n_t denote the number of recorded cases in time-period t. Write the data for time-period t in the form of a set of non-specific case-locations $x_{it} : i = 1, \ldots, n_t$ and associated labels Y_{it} giving the genotype of each case. Then, our kernel regression estimator of $p_{jt}(x)$ is

$$\hat{p}_{jt}(x) = \sum_{i=1}^{n_t} w_{ij}(x) I(Y_{it} = j), \tag{1.20}$$

where

$$w_{ij}(x) = w_j(x - x_{it}) / \sum_{k=1}^{n_t} w_j(x - x_{kt}),$$

$w_j(x) = w(x/h_j)/h_j^2$, and

$$w(x) = \exp(-||x||^2/2). \tag{1.21}$$

The Gaussian kernel function (1.21) could be replaced by any other non-negative-valued weighting function. The choice of kernel function is usually less important than the choice of the band-width constants, h_j, which determine the amount of smoothing applied to the data in estimating the $p_{jt}(x)$. Using a common set of band-widths across time-periods makes for ease of interpretation. We shall also use a common band-width, h, for all four components of the p-surface, as this ensures that $\sum_{j=1}^m \hat{p}_j(x) = 1$ for every location x.

To choose the value of h, we proceed as follows. Conditioning on the case-locations, the log-likelihood for the labels Y_{it} under the Poisson process model has the multinomial form

$$L(h) = \sum_{t=1}^m \sum_{i=1}^{n_t} \sum_{j=1}^m I(Y_{it} = j) \log p_{jt}(x_{it}; h).$$

Maximising $L(h)$ would give the degenerate solution $\hat{h} = 0$. To avoid this, we use a cross-validated form of the log-likelihood,

$$L_c(h_1, \ldots, h_4) = \sum_{t=1}^r \sum_{i=1}^{n_t} \sum_{j=1}^m I(Y_{it} = j) \log \hat{p}_{jt}^{(it)}(x_{it}),$$

where $\hat{p}_{jt}^{(it)}(x_i)$ denotes the kernel estimator (1.20), but based on all of the data except (x_{it}, Y_{it}).

The data from the years prior to 1997 are too sparse to allow non-parametric estimation of the $p_{jt}(x)$. We therefore applied the kernel regression method to data from the years 1997 to 2002. To maintain sufficient numbers of cases per genotype per time-period, we also combined data from successive years to give $r = 3$ discrete 2-year time-periods.

The null hypothesis of interest is that $p_{jt}(x) = p_j(x)$ for $j = 1, \ldots, 4$ and $t = 1, \ldots, r$. An *ad hoc* statistic for a Monte Carlo test of this hypothesis is

$$T^p = \sum_{j=1}^{4} \sum_{x \in X} \sum_{t=1}^{r} \left(\hat{p}_{jt}(x) - \bar{p}_j(x) \right)^2, \tag{1.22}$$

where $X = \{x_i^t : i = 1, 2, \ldots, n_t; t = 1, 2, \ldots, s\}$, $\hat{p}_{jt}(x)$ is the estimated type-specific probability surface for type j in time-period t and $\bar{p}_j(x) = r^{-1} \sum_{t=1}^{r} \hat{p}_{jt}(x)$.

The band-width which maximises the cross-validated log-likelihood is $h = 9647$ metres. A Monte Carlo test based on the statistic (1.22) using 999 simulations gave an attained significance level of 0.015; hence the null hypothesis is rejected at the conventional 5% level. However, the changes in the segregation pattern over time appear to be somewhat subtle. Figures 1.11, 1.12, and 1.13 show the estimated type-specific probability surfaces over the three consecutive time periods. The general pattern is of an increase in the extent of spatial segregation over time. Thus, genotype 15 becomes progressively

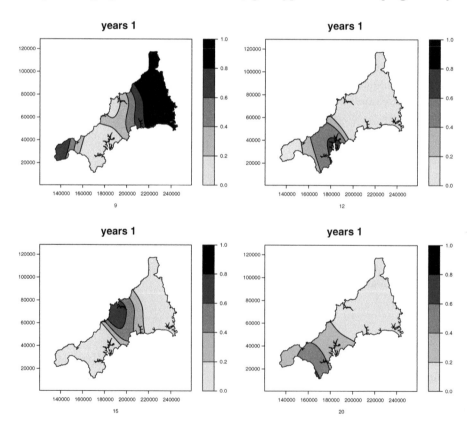

Figure 1.11 Kernel regression estimates of type-specific probability surfaces for cases in years 1997–1998.

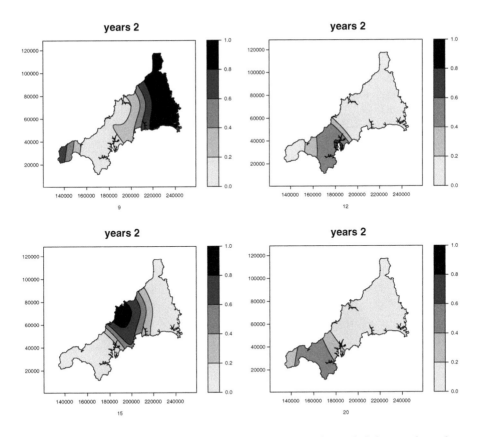

Figure 1.12 Kernel regression estimates of type-specific probability surfaces for cases in years 1999–2000.

more dominant in north-central Cornwall, genotype 20 has established near-dominance in the far west by 2001/02, and spoligotype 9 is dominant in the east of Cornwall, but within a territory which becomes more confined to an area close to the eastern boundary as time proceeds. Finally, genotype 15 shows an apparently stable spatial distribution over the three time periods.

In summary, our findings from our analysis of the BTB data are that there is very strong spatial segregation amongst the four most common genotypes, and that the pattern of spatial segregation is broadly consistent over time, albeit with subtle but statistically significant differences between successive 2-year periods.

Further methodological developments in progress include replacing the non-parametric kernel regression methodology by a hierarchical stochastic model, in which the spatio-temporal intensities are determined by a multivariate, discrete-time log-Gaussian process; hence $\lambda_j(x,t) = \exp\{S_j(x,t)\}$ where $\{S_1(x,t), \ldots, S_4(x,t)\}$ is a quadri-variate Gaussian process. The hierarchical stochastic modelling framework is better able to deal with the relatively

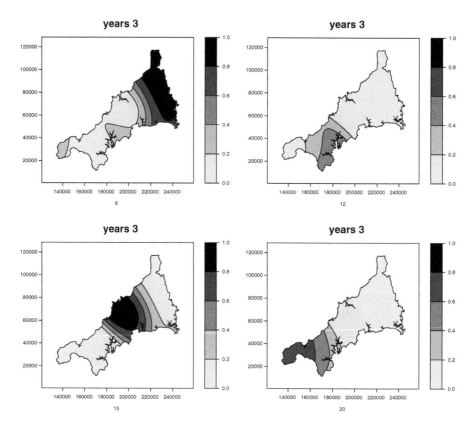

Figure 1.13 Kernel regression estimates of type-specific probability surfaces for cases in years 2001–2002.

small numbers of cases of each genotype in each time-period by exploiting the assumed statistical dependencies between genotypes and over time.

1.5 Real-time surveillance for gastroenteric disease: log-Gaussian Cox process modelling

Our second application has previously been reported in greater detail by Diggle et al. (2003) and by Diggle et al. (2005b). It concerns the development of a real-time surveillance system for non-specific gastroenteric disease in the county of Hampshire, U.K., using the data described in Section 1.1.1.3.

The methodological problem is to build a descriptive model of the normal pattern of spatio-temporal variation in the distribution of incident cases, and to use the model to identify unusual, spatially and temporally localised, departures from this pattern, which we call "anomalies." The wider aim is that statistical evidence of current anomalies in the spatio-temporal distribution of incident cases can then be combined with other forms of evidence, for example, reports from pathology laboratory analyses of faecal samples, to trigger an earlier response to an emerging problem than is typically achieved by current surveillance systems. For further discussion, see Diggle et al. (2003).

Our descriptive model is a log-Gaussian Cox process of the kind proposed by Brix and Diggle (2001) and discussed in Section 1.3.2.1. Our model needs to allow for both spatial and temporal heterogeneity in the rate of calls to NHS Direct. The heterogeneity arises through a combination of factors including spatial variation in the density of the population at risk and aspects of the pattern of usage of NHS Direct by different sectors of the community; for example, there is a clear day-of-the-week effect due to the relative inaccessibility of other medical service on weekends, and anecdotal evidence suggests that NHS Direct is used disproportionately more often by young families than by the elderly. We assume that these spatial and temporal effects operate independently, whereas the spatially and temporally localised anomalies which we wish to detect are governed by a spatio-temporally correlated stochastic process. Hence, within the framework of log-Gaussian Cox processes, we postulate a spatio-temporal intensity function for incident cases of the form

$$\lambda(x,t) = \lambda_0(x)\mu_0(t)\exp\{S(x,t)\}, \qquad (1.23)$$

where $S(x,t)$ is a stationary, spatio-temporal Gaussian process with expectation $E[S(x,t)] = -0.5\sigma^2$, variance $\text{Var}\{S(x,t)\} = \sigma^2$, and separable correlation function $\text{Corr}\{S(x,t), S(x-u,t-v)\} = \rho(u,v) = r(u)\exp(-v/\beta)$. As noted earlier, this specification guarantees that $E[\exp\{S(x,t)\}] = 1$ for all x and t. We now add the constraint that $\lambda_0(x)$ integrates to 1 over the study-region. The function $\mu_0(t)$ then represents the time-varying total intensity, or mean number per unit time, of incident cases over the whole county, whilst $\lambda_0(x)$ becomes a probability density function for the spatial distribution of incident cases, averaged over time. The spatial correlation function $r(u)$ is in principle arbitrary, but we have found that a simple exponential, $r(u) = \exp(-u/\phi)$, gives a reasonable fit to the Hampshire data.

To fit the model, we need to estimate the functions $\lambda_0(x)$ and $\mu_0(t)$, and the additional parameters σ^2, ϕ, and β which specify the Gaussian process $S(x,t)$. We consider each of these estimation problems in turn.

For the spatial density $\lambda_0(x)$, it is hard to envisage a suitable parametric model. Also, we cannot assume that the spatial distribution of the relevant population, namely users of NHS Direct, matches that of the overall population distribution over the county of Hampshire; hence we cannot use census information to estimate $\lambda_0(x)$. We therefore use a non-parametric kernel density estimation method. This is very similar to the kernel regression method discussed in Section 1.4 but adapted to the density estimation setting. Using the Gaussian kernel (1.21), and band-width h, the kernel estimate of $\lambda_0(x)$ is

$$\hat{\lambda}_0(x) = n^{-1}\sum_{i=1}^{n} h^{-2}w\{h^{-1}(x-x_i)\}, \qquad (1.24)$$

where $x_i : i = 1, \ldots, n$ are the case-locations. Because of the very severe variations in population density across the county, any choice of a fixed bandwidth h is liable to be an unsatisfactory compromise between the relatively large and small band-widths which would be appropriate in the more rural and urban areas, respectively. We therefore follow a suggestion in Silverman (1986), in which we first construct a fixed band-width pilot estimator $\tilde{\lambda}_0(x_i)$ using (1.24) with a subjectively chosen band-width h_0, then calculate $\tilde{g}$, the geometric mean of $\tilde{\lambda}_0(x_i) : i = 1, \ldots, n$, and use a locally adaptive kernel estimator

$$\hat{\lambda}_0(x) = n^{-1} \sum h_i^{-2} \phi\{(x - x_i)/h_i\}$$

with $h_i = h_0\{\tilde{\lambda}_0(x_i)/\tilde{g}\}^{-0.5}$. This has the required effect of applying a larger band-width in the more rural areas, where $\tilde{\lambda}_0(x)$ is relatively small. Figure 1.14 shows the resulting estimate $\hat{\lambda}_0(x)$.

To estimate the mean number of cases per day, $\mu_0(t)$, a parametric approach is more reasonable. We use a Poisson log-linear regression model incorporating

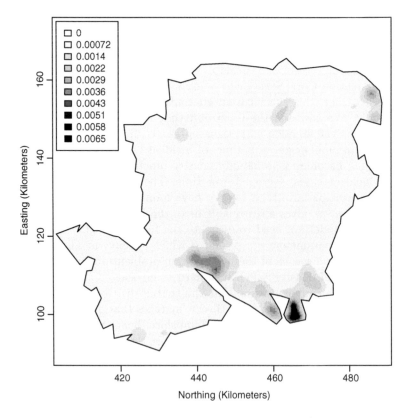

Figure 1.14 Locally adaptive kernel estimate of $\lambda_0(x)$ for the Hampshire gastroenteric disease data.

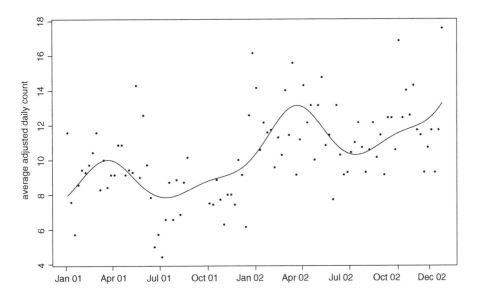

Figure 1.15 Fitted log-linear model for the mean number of gastroenteric disease cases $\hat{\mu}_0(t)$ per day adjusted for day-of-the-week effects (solid line), and observed numbers of cases per day averaged over successive weeks (solid dots).

day-of-the-week effects as a 7-level factor, time-of-year effects as a sine-cosine wave at frequency $\omega = 2\pi/365$ plus its first harmonic, and a linear trend to reflect progressive uptake in the usage of NHS Direct during the period covered by the data. Hence,

$$\log \mu_0(t) = \delta_{d(t)} + \alpha_1 \cos(\omega t) + \eta_1 \sin(\omega t) + \alpha_2 \cos(2\omega t) + \eta_2 \sin(2\omega t) + \gamma t, \tag{1.25}$$

where $d(t)$ codes the day of the week. The Poisson formulation does not account for the extra-Poisson variation which, as anticipated, the data exhibit, but nevertheless produces consistent estimates on the assumption that (1.25) is a correct specification for $\mu_0(t)$. Figure 1.15 compares the resulting estimate, centred on the average of the seven daily intercepts $\delta_{d(t)}$, with the observed numbers of calls per day, averaged over each week. The spring peak in incidence is a well-known feature of this group of diseases.

To estimate the parameters of the stochastic component of the model, $S(x, t)$, we have used a simple method-of-moments approach, based on matching empirical and theoretical second-moment properties of the data and model, respectively. We are currently developing an implementation of a Monte Carlo maximum likelihood method, based on material in Møller and Waagepetersen (2003). A partial justification for the method-of-moments approach is that the main goal of the analysis is real-time spatial prediction, whose precision is limited by the relatively low daily incidence of cases, whereas parameter estimation draws on the complete set of data; hence, efficiency of parameter estimation is not crucial.

Consider first the parameters of the spatial covariance structure of $S(x, t)$, namely,

$$\text{Cov}\{S(x, t), S(x - u, t)\} = \sigma^2 \exp(-|u|/\phi).$$

The corresponding spatial pair correlation function is $g(u) = \exp\{\sigma^2 \exp(-|u|/\phi)\}$, and the estimation method consists of minimising the criterion

$$\int_0^{u_0} [\{\log \hat{g}(u)\} - \{\log g(u)\}]^2 du, \qquad (1.26)$$

where $u_0 = 2$ km is chosen subjectively as the apparent range of the spatial correlation, and $\hat{g}(u)$ is a non-parametric estimate of the pair correlation function. We use a time-averaged kernel estimator,

$$\hat{g}(u) = T^{-1} \sum_{t=1}^{T} \hat{g}_t(u)$$

where

$$\hat{g}_t(u) = \frac{1}{2\pi u |A|} \sum_{i=1}^{n_t} \sum_{i \neq j} \frac{K_h(u - \| x_i - x_j \|) w_{ij}}{\lambda_t(x_i) \lambda_t(x_j)}. \qquad (1.27)$$

In (1.27), n_t is the number of incident cases on day $t = 1, \ldots, T$, each of the summations over $i \neq j$ refers to pairs of events occurring on the same day, A is the study area, w_{ij} is Ripley's (1977) edge-correction as used in (1.4), $\lambda_t(x) = \hat{\lambda}_0(x)\hat{\mu}_0(t)$, and

$$K_h(u) = \begin{cases} 0.75h^{-1}(1 - u^2/h^2) : -h \leq u \leq h \\ \qquad\qquad\qquad : \text{otherwise.} \end{cases}$$

To estimate the temporal correlation parameter β, we again match empirical and theoretical second-moment properties, as follows. Let N_t denote the numbers of incident cases on day t. For our model, the time-variation in $\mu_0(t)$ makes the covariance structure of N_t non-stationary. We obtain

$$\text{Cov}(N_t, N_{t-v}) = \mu_0(t)\mathbf{1}(v = 0) + \{\mu_0(t)\mu_0(t - v)\}$$
$$\times \left\{ \int_W \int_W \lambda_0(x_1)\lambda_0(x_2)e^{\sigma^2 e^{(-v/\beta)}e^{(-u/\phi)}} dx_1 dx_2 - 1 \right\} \qquad (1.28)$$

Note that the expression for $\text{Cov}(N_t, N_{t-v})$ given by Brix and Diggle (2001) is incorrect. To estimate β we minimise

$$\sum_{v=1}^{v_0} \sum_{t=v+1}^{n} \{\hat{C}(t, v) - C(t, v; \beta)\}^2,$$

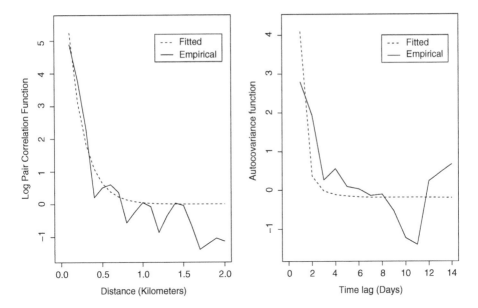

Figure 1.16 Empirical (solid line) and fitted (dashed line) second-moment properties for the Hampshire gastroenteric disease data. The left-hand panel shows the logarithm of the pair correlation function, the right-hand panel the temporal autocovariance function.

where $v_0 = 14$ days, $\hat{C}(t, v) = N_t N_{t-v} - \hat{\mu}_0(t)\hat{\mu}_0(t - v)$, and $C(t, v; \theta) = \text{Cov}(N_t, N_{t-v})$ as defined in (1.28) but plugging in the previously estimated values for σ^2 and ϕ.

The resulting estimates are shown in Figure 1.16. Note in particular that the scales on which the spatial and temporal dependence decay are broadly consistent with the known character of the diseases in question, which are spread by direct contact between infected individuals and generally have latent periods on the order of a few days.

We now use the fitted model for spatial prediction, as follows. First, and with the same justification as given above for the use of potentially inefficient methods of parameter estimation, we use plug-in values of the estimated model parameters, thereby ignoring the effects of parameter uncertainty on the predictive distributions of interest. We then use a Metropolis-adjusted Langevin algorithm, as described in Brix and Diggle (2001), to generate samples from the conditional distribution of $S(x, t)$ given data up to time t. For display purposes, we choose a critical threshold value $c > 1$ and map predictive exceedance probabilities,

$$p_t(x) = \text{P}(\exp\{S(x, t)\} > c | \text{data}). \tag{1.29}$$

We would argue that mapping $p_t(x)$ is more relevant than mapping predictive estimates, $\hat{S}(x, t)$. The latter are highly variable because of the relatively low

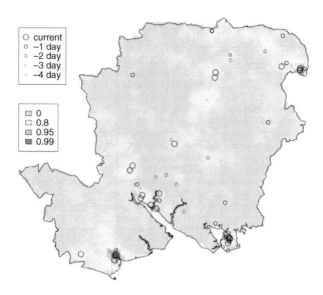

Figure 1.17 (SEE COLOR INSERT) Predictive exceedance probabilities $p_t(x)$ for the Hampshire gastroenteric disease data on a day in March 2003, using the threshold value $c = 2$. Incident cases over five consecutive days are shown as circles of diminishing size to correspond to the progressive discounting of past data in constructing the predictive distribution of $S(x, t)$.

daily incidence and would be liable to over-interpretation. We suggest that the value of c should be chosen by the public health practitioner to represent a multiplicative increase in the local daily incidence which, if verified, would be regarded as practically significant. A high predictive probability that this threshold has been exceeded at a particular time and place would then suggest that some kind of follow-up action may be required, for example to ascertain whether recent cases in the immediate vicinity might have a common cause or are more likely to be unrelated chance occurrences.

Figure 1.17 shows an example of the resulting map of $p_t(x)$ for one day in March 2003, using the threshold value $c = 2$. Note that the colour scale for $p_t(x)$ is continuous, but non-linear so that only predictive probabilities close to 1 show up as orange or red. We also emphasise that the value $c = 2$ was chosen for illustrative purposes and is much smaller than would be associated, in this application, with a genuine outbreak of public health concern.

As noted above, one way to refine the methodology associated with this application would be to use Monte Carlo maximum likelihood estimation for the covariance parameters of $S(x, t)$. A brief outline of this method follows. We approximate the continuous spatio-temporal domain $A \times (0, T)$ by a fine lattice, and denote by X and S the point process data and the latent Gaussian process in this discretised domain. The likelihood for the model, $L(\theta; X)$ say, is the marginal distribution of X, which is intractable, whereas the joint distribution of X and S can be written explicitly as the product of the conditional distribution of X given S and the marginal distribution of S, both of which have known standard forms—Poisson and Gaussian, respectively. It follows that for any fixed value θ_0,

$$
\begin{aligned}
L(\theta; X) &= \int f(S, X; \theta) dS \\
&= \int f(S, X; \theta) \times \frac{L(\theta_0; X)}{L(\theta_0; X)} \times \frac{f(S, X; \theta_0)}{f(S, X; \theta_0)} dS \\
&= L(\theta_0; X) \int \frac{f(S, X; \theta)}{f(S, X; \theta_0)} f(S|X; \theta_0) dS.
\end{aligned}
$$

Hence, the likelihood ratio between θ and θ_0 is

$$
L(\theta; X)/L(\theta_0; X) = \mathrm{E}_{\theta_0} \left[\frac{f(S, X; \theta)}{f(S, X; \theta_0)} \right], \tag{1.30}
$$

where $\mathrm{E}_{\theta_0}[\cdot]$ denotes expectation with respect to the conditional distribution of S given X at $\theta = \theta_0$. In (1.30), we now replace the theoretical expectation by a sample mean over Monte Carlo simulations, using the same Metropolis-adjusted Langevin algorithm as used earlier to simulate samples from the predictive distribution of S given X. Note also that writing $S = a(\theta) + B(\theta)Z$, where Z is a vector of independent standard Gaussian variates, gives $f(S, X; \theta) = f(X|Z; \theta) f(Z)$, in which case the ratio within the expectation term on the right-hand side of (1.30) reduces to $f(X|Z; \theta)/f(X|Z; \theta_0)$.

Another possible refinement of the methodology would be to use a stochastic model of daily incidence in place of the deterministic $\mu_0(t)$. In particular, the rising trend evident during the 2-year period covered by the data which were used to develop the model was not sustained thereafter. More generally, patterns in the usage of NHS Direct are liable to fluctuate in response to changes in policy-related factors within the NHS system whose combined effects are hard to capture in a deterministic model. A pragmatic strategy would be the following. Postulate a temporal log-Gaussian Cox process for the daily incident counts, N_t, as a series of conditionally independent Poisson counts with conditional expectations $\exp(\alpha_t + M_t)$, where α_t encodes day-of-the-week effects and M_t is a latent Gaussian process. Use this model to generate real-time predictions $\hat{\alpha}_t + \hat{M}_t$, which can then be plugged into the spatial prediction algorithm in place of the log-linear regression estimates $\hat{\mu}_0(t)$.

1.6 Foot-and-mouth disease: mechanistic modelling and partial likelihood analysis

To illustrate the partial likelihood method described in Section 1.1.3.2.2, we analyse the foot-and-mouth data using a mechanistic model proposed by Keeling et al. (2001). The analysis reported here is taken from Diggle (2005).

The model assumes that the rate of transmission of infection between an infected farm i and a susceptible farm j is given by

$$\lambda_{ij}(t) = \lambda_0(t)A_iB_jf(||x_i - x_j||)I_{ij}(t), \tag{1.31}$$

where $\lambda_0(t)$ is a baseline rate of infection; A_i and B_j encode characteristics of the farms in question which affect their infectivity and susceptibility, respectively; $f(\cdot)$ is a *transmission kernel* which models the spread of infection as a function of distance; and $I_{ij}(t)$ is an indicator that farm i is infective and farm j susceptible at time t. The data include, as applicable for each animal-holding farm, the dates on which the disease was reported and on which the stock was culled. To allow for reporting delays, we assume that a farm is reported as a case τ days after it becomes infective. A farm remains infective unless and until its stock is culled. A farm is susceptible if it is not infective and has not had its stock culled.

In the analysis reported here, we assume that $\tau = 5$ days and model the infectivity and susceptibility factors as

$$A_i = (\alpha n_{1i} + n_{2i}) \tag{1.32}$$

and

$$B_j = (\beta n_{1j} + n_{2j}), \tag{1.33}$$

where n_{1i} and n_{2i} are the numbers of cows and sheep held by farm i at the start of the epidemic. The parameters α and β therefore represent the relative infectiousness and susceptibility, respectively, of cows to sheep. For the transmission kernel, we assume that

$$f(u) = \exp\{-(u/\phi)^\kappa\} + \rho. \tag{1.34}$$

Keeling et al. (2001) did not specify a functional form for $f(\cdot)$, but (1.34) captures the qualitative features of the results which they reported in graphical form. In particular, they reported a sharper-than-exponential decay with distance, which in (1.34) would correspond to $\kappa < 1$, whilst ρ in (1.34) represents the contribution to the epidemic from apparently spontaneous cases which occur remotely from any previous case.

For any farm k, we define $\lambda_k(t) = \sum_j \lambda_{jk}(t)$, from which we obtain the conditional intensities

$$\lambda(x_i, t_i | \mathcal{H}_{t_i}) = \lambda_i(t_i) / \sum_k \lambda_k(t_i).$$

The partial log-likelihood follows by substitution of the conditional intensities into (1.16) and (1.17). To maximise the partial log-likelihood we use the Nelder-Mead simplex algorithm (Nelder and Mead, 1965) as implemented in the R function `optim()`, which provides a numerical estimate of the Hessian matrix.

In the model for the transmission kernel, the parameters κ and ρ are poorly identified because the cases which appear to correspond to long-range transmission are few in number and can be explained empirically either by including a small, positive value of ρ or by adjusting the value of κ. Because ρ corresponds formally to what is known to be a real effect, namely the indirect spread of infection via the movement of farm equipment and staff, we retain ρ as a positive-valued parameter to be estimated, but fix $\kappa = 0.5$ to correspond to the observation in Keeling et al. (2001) that the transmission kernel is more sharply peaked than exponential.

We first investigated whether the data in Cumbria and Devon support the assumption of a common set of parameters in the two counties. The likelihood ratio test statistic for common versus separate parameters is 2.98 on 4 degrees of freedom; hence $p = 0.56$ and we therefore accepted the hypothesis of common parameter values. We then obtained common parameter estimates $(\hat{\alpha}, \hat{\beta}, \hat{\phi}, \hat{\rho}) = (4.92, 30.68, 0.39, 9.9 \times 10^{-5})$. For all practical purposes, $\hat{\rho} \approx 0$, although a likelihood ratio test formally rejects $\rho = 0$ because the likelihood is sensitive to the precise probabilities which the model assigns to rare events.

One question of specific interest is whether the infectivities and susceptibilities for individual farms, A_i and B_j, are linear or sub-linear in the numbers of animals. To investigate this, we extend (1.32) and (1.33) to $A_i = (\alpha n_{1i}^{\gamma} + n_{2i}^{\gamma})$ and $B_j = (\beta n_{1j}^{\gamma} + n_{2j}^{\gamma})$, respectively, where γ is an additional parameter to be estimated. Fitting this five-parameter model results in a large increase in the maximised log-likelihood, from -6196.3 to -5861.4.

Another possible extension of the model would be to include farm-level covariates by defining $A_i = (\alpha n_{1i}^{\gamma} + n_{2i}^{\gamma}) \exp(z_i' \delta)$, where z_i is a vector of covariates for farm i, with a similar expression for the susceptibilities B_j. The z_i might, for example, codify management practices or other measured characteristics of individual farms which could affect their propensity to transmit, or succumb to, the disease. By way of illustration, we consider adding a log-linear effect of farm area to the model. The likelihood ratio statistic for the covariate effect is 3.26 on 1 degree of freedom, corresponding to $p = 0.07$. However, we can expect this test to be rather weak because the observed distribution of farm area is extremely skewed, and the few farms with large areas will therefore have high leverage.

Estimates for the five-parameter model are shown in Table 1.1, together with approximate 95% confidence limits deduced from the numerical estimate

Table 1.1 Parameter estimation for the five-parameter model fitted to combined data from Cumbria and Devon

Parameter	Estimate	95% Confidence Interval	
α	1.42	1.13	1.78
β	36.17	0.19	92.92
ϕ	0.41	0.36	0.48
ρ	1.3×10^{-4}	8.5×10^{-5}	2.1×10^{-4}
γ	0.13	0.09	0.21

of the Hessian matrix. Optimisation was conducted on the log-scale for all parameters, which is why the confidence limits are not symmetric about the point estimates. Estimated correlations amongst the parameter estimates are all small, the largest being 0.25 between $\log\phi$ and $\log\rho$. The results in Table 1.1 indicate a strongly sub-linear dependence of infectivity and susceptibility on the numbers of animals. Note that under the weak form of dependence implied by the estimate of γ, the estimate of β is very imprecise.

These results are qualitatively similar to those reported in Keeling et al. (2001), although they considered only the case $\gamma = 1$. They reported point estimates $\tilde{\alpha} = 1.61$ and $\tilde{\beta} = 15.2$. They did not specify a parametric model for the transmission kernel but their Figure 1B shows similar behaviour to our fitted model, decaying from 1 at $u = 0$ to approximately 0.1 at $u = 1$ km, compared with our $\hat{f}(1) = 0.21$.

Finally, we use a simple adaptation of the Nelson-Aalen estimator (Andersen et al., 1992, Chapter 4) to obtain a non-parametric estimate of the cumulative base-line hazard,

$$\hat{\Lambda}_0(t) = \int_0^t \hat{\lambda}_0(u) du.$$

We re-write (1.31) as $\lambda_{ij}(t) = \lambda_0(t)\rho_{ij}(t)$ and define $\rho(t) = \sum_i \sum_j \rho_{ij}(t)$. The Nelson-Aalen estimator is now given by

$$\hat{\Lambda}_0(t) = \int_0^t \hat{\rho}(u)^{-1} dN(u)$$

$$= \sum_{i:t_i \leq t} \hat{\rho}(t_i)^{-1}, \tag{1.35}$$

where $\hat{\rho}(t)$ is the parametric estimate of $\rho(t)$ implied by the fitted model. Figure 1.18 shows the Nelson-Aalen estimates obtained from the Cumbria and Devon data. The generally lower estimates for Devon are consistent with the lower overall prevalence of the disease (137 cases out of 8182 at-risk farms in Devon, 657 cases out of 5090 at-risk farms in Cumbria). Both estimates are approximately linear over the first 2 to 3 months, by which time the epidemic in Devon has almost run its course. The slope of the Cumbria estimate increases thereafter. This does not necessarily imply a failure of the culling strategies being applied since the model already takes account of their effects, but rather suggests that external environmental effects, for example the increase in animal movements outdoors in spring and summer, may have promoted an increase in the virulence of the disease process.

This analysis of the foot-and-mouth data demonstrates the feasibility of using the partial likelihood approach to answer a variety of questions relevant to an understanding of the underlying disease process. The method of fitting is quite flexible, and it would be straightforward to extend the model in various ways, for example by including additional farm-level covariates.

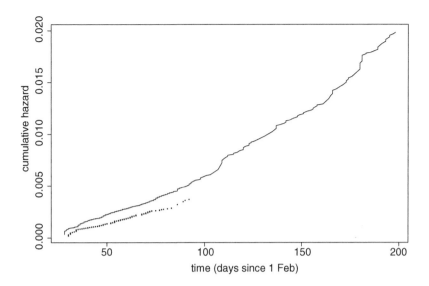

Figure 1.18 Estimated cumulative baseline hazards in Cumbria (solid line) and in Devon (dotted line) for the five-parameter model.

A fuller analysis of the data will be reported separately. More generally, the partial likelihood provides a useful method for analysing a range of spatio-temporal point process models which are specified via their conditional intensity function. The method is based on a generally accepted principle of inference with known asymptotic properties, whilst being computationally straightforward and therefore well suited to routine use.

1.7 Discussion

In this chapter we have tried to indicate the scope for simple adaptation of existing methods, and for development of new methods, to analyse spatio-temporal point process data. Data-sets of this kind are becoming more widely available, perhaps reflecting the increased ability in many fields of science to capture such data routinely. More fundamentally, interesting scientific questions concerning the dynamic spatio-temporal behaviour of natural systems cannot easily be addressed using purely spatial or temporal information.

In the author's opinion, simple adaptations of existing spatial statistical methods, as illustrated by our analysis of the bovine tuberculosis data, will continue to be valuable for descriptive analysis and are more or less forced on us when the data are coarsely discretised in time. But when data with a fine resolution in both the spatial and temporal dimensions are available, models which explicitly recognise the directional nature of time by conditioning future behaviour on past outcomes are likely to be more insightful.

Our analyses of the gastroenteric disease data and of the foot-and-mouth data illustrate this general philosophy. For the gastroenteric disease application, our model for the latent stochastic process $S(x, t)$ is, essentially, a multivariate time series model incorporating a qualitatively sensible, if admittedly also computationally convenient, Markov dependence structure in time. The cross-correlation structure of the model is then chosen so as to have a parsimonious spatial interpretation whilst giving a reasonable empirical fit to the data. In contrast, for the foot-and-mouth data, we use a previously proposed mechanistic model in which the conditioning on past events is used explicitly to quantify the current risk of transmission of the disease from an infective to a susceptible farm.

As in other areas of statistics, the development of computationally intensive Monte Carlo methods of inference has greatly enhanced our ability to fit relatively complex and realistic models. However, it is still all too easy to find combinations of data and model for which fitting by Monte Carlo methods is computationally infeasible. Also, the requirement to tune Monte Carlo algorithms to each non-standard application imposes very real constraints on the statistician's ability to compare a range of candidate models within a reasonable time-scale whilst ensuring that Markov-chain Monte Carlo algorithms have converged to their equilibrium distributions and that inferences about parameter combinations of interest are appropriately insensitive to pragmatic choices of multivariate priors. For these reasons, we see a place for methods of inference such as the partial likelihood method proposed in Diggle (2005) and applied here to the foot-and-mouth data, which are capable of routine implementation whilst still being based on generally accepted statistical principles.

Acknowledgments

This work was supported by the U.K. Engineering and Physical Sciences Research Council through the award of a Senior Fellowship to Peter Diggle (Grant number GR/S48059/01) and by the U.S.A. National Institute of Environmental Health Sciences through Grant number 1 R01 ES012054, Statistical Methods for Environmental Epidemiology.

I thank Prof. Abbie Hughes and Dr. Elzbieta Wieniawa Narkiewicz for providing the amacrine cell data, and Laura Green, Matt Keeling, Graham Medley, Sir David Cox, and Hugues Lassalle for very helpful discussions concerning the foot-and-mouth data.

The application to the bovine tuberculosis data was supported by the U.K. Department for Environment, Food and Rural Affairs (Grant number SE3020). Roger Sainsbury of the State Veterinary Service helped collect the data. Jackie Inwald of the Department of Bacterial Diseases, VLA Weybridge, carried out the spoligotyping.

The application to gastroenteric disease surveillance was supported by a grant from the Food Standards Agency, U.K., and from the National Health Service Executive Research and Knowledge Management Directorate.

References

Andersen, P.K., Borgan, O., Gill, R.D., and Keiding, N. (1992). *Statistical Models Based on Counting Processes*. New York: Springer.

Banerjee, S., Carlin, B.P., and Gelfand, A.E. (2003). *Hierarchical Modelling and Analysis for Spatial Data*. London: CRC Press.

Bartlett, M.S. (1964). The spectral analysis of two-dimensional point processes. *Biometrika*, 51, 299–311.

Besag, J. (1974). Spatial interaction and the statistical analysis of lattice systems (with Discussion). *Journal of the Royal Statistical Society*, B 36, 192–225.

Besag, J.E. (1975). Statistical analysis of non-lattice data. *The Statistician*, 24, 179–195.

Besag, J.E. (1978). Some methods of statistical analysis for spatial data. *Bulletin of the International Statistical Institute*, 47, Book 2, 77–92.

Besag, J., Milne, R., and Zachary, S. (1982). Point process limits of lattice processes. *Journal of Applied Probability*, 19, 210–216.

Bhopal, R.S., Diggle, P.J., and Rowlingson, B.S. (1992). Pinpointing clusters of apparently sporadic cases of legionnaires' disease. *British Medical Journal*, 304, 1022–1027.

Brix, A. and Diggle, P.J. (2001). Spatio-temporal prediction for log-Gaussian Cox processes. *Journal of the Royal Statistical Society*, B 63, 823–841.

Brix, A. and Diggle, P.J. (2003). Corrigendum: Spatio-temporal prediction for log-Gaussian Cox processes. *Journal of the Royal Statistical Society*, B 65, 946.

Brix, A. and Møller, J. (2001). Space-time multi-type log Gaussian Cox processes with a view to modelling weed data. *Scandinavian Journal of Statistics*, 28, 471–488.

Chetwynd, A.G. and Diggle, P.J. (1998). On estimating the reduced second moment measure of a stationary spatial point process. *Australian and New Zealand Journal of Statistics*, 40, 11–15.

Cooper, D.L., Smith, G.E., O'Brien, S.J., Hollyoak, V.A., and Baker, M. (2003). What can analysis of calls to NHS Direct tell us about the epidemiology of gastrointestinal infections in the community? *Journal of Infection*, 46, 101–105.

Cox, D.R. (1955). Some statistical methods related with series of events (with discussion). *Journal of the Royal Statistical Society*, B 17, 129–157.

Cox, D.R. (1972). Regression models and life tables (with Discussion). *Journal of the Royal Statistical Society*, B 34, 187–220.

Cox, D.R. (1975). Partial likelihood. *Biometrika*, 62, 269–275.

Cox, D.R. (1977). The role of significance tests. *Scandinavian Journal of Statistics*, 4, 49–71.

Cressie, N.A.C. (1991). *Statistics for Spatial Data*. New York: Wiley.

Daley, D.J. and Vere-Jones, D. (1988). *An Introduction to the Theory of Point Processes*. New York: Springer.

Diggle, P.J. (2003). *Statistical Analysis of Spatial Point Patterns (second edition)*. London: Arnold.

Diggle, P.J. (2005). A partial likelihood for spatio-temporal point processes. Johns Hopkins University, Department of Biostatistics Working Papers, Working Paper 75. http://www.bepress.com/jhubiostat/paper75.

Diggle, P.J., Chetwynd, A.G., Haggkvist, R., and Morris, S. (1995). Second-order analysis of space-time clustering. *Statistical Methods in Medical Research*, 4, 124–136.

Diggle, P.J., Eglen, S.J., and Troy, J.B. (2005a). Modelling the bivariate spatial distribution of amacrine cells. In *Case Studies in Spatial Point Processes*, Ed. A. Baddeley, J. Mateu, D. Stoyan. New York: Springer (to appear).

Diggle, P.J., Knorr-Held, L., Rowlingson, B., Su, T., Hawtin, P., and Bryant, T. (2003). Towards on-line spatial surveillance. In *Monitoring the Health of Populations: Statistical Methods for Public Health Surveillance.*, Ed. R. Brookmeyer and D. Stroup. Oxford: Oxford University Press.

Diggle, P.J., Rowlingson, B., and Su, T.-L. (2005b). Point process methodology for on-line spatio-temporal disease surveillance. *Environmetrics*, 16, 423–434.

Diggle, P.J., Zheng, P., and Durr, P. (2005c). Non-parametric estimation of spatial segregation in a multivariate point process. *Applied Statistics*, 54, 645–658.

Durr, P.A., Hewinson, R.G., and Clifton-Hadley, R.S. (2000). Molecular epidemiology of bovine tuberculosis. I. *Mycobacterium bovis* genotyping. *Revue Scientifique et Technique de l'Office Internationale des Epizooties*, 19, 675–688.

Gates, D.J. and Westcott, M. (1986). Clustering estimates in spatial point distributions with unstable potentials. *Annals of the Institute of Statistical Mathematics*, 38 A, 55–67.

Geyer, C. (1999). Likelihood inference for spatial point processes. In *Stochastic Geometry: Likelihood and Computation*, Ed O.E. Barndorff-Nielsen, W.S. Kendall, and M.N.M. van Lieshout, 79–140.

Geyer, C.J. and Thompson, E.A. (1992). Constrained Monte Carlo maximum likelihood for dependent data (with Discussion). *Journal of the Royal Statistical Society*, B 54, 657–699.

Gneiting, T. (2002). Nonseparable, stationary covariance functions for space-time data. *Journal of the American Statistical Association*, 97, 590–600.

Hughes, A. (1985) New perspectives in retinal organisation. *Progress in Retinal Research*, 4, 243–314.

Keeling, M.J., Woolhouse, M.E.J., Shaw, D.J., Matthews, L., Chase-Topping, M., Haydon, D.T., Cornell, S.J., Kappey, J., Wilesmith, J., and Grenfell, B.T. (2001). Dynamics of the 2001 U.K. foot and mouth epidemic: stochastic dispersal in a heterogeneous landscape. *Science*, 294, 813–817.

Kingman, J.F.C. (1977). Remarks on the spatial distribution of a reproducing population. *Journal of Applied Probability*, 14, 577–583.

Lawson, A.B. and Leimich, P. (2000). Approaches to space-time modelling of infectious disease behaviour. *Mathematical Medicine and Biology*, 17, 1–13.

Lotwick, H.W. and Silverman, B.W. (1982). Methods for analysing spatial processes of several types of points. *Journal of the Royal Statistical Society,* B 44, 406–413.

Møller, J. and Sorensen, M. (1994). Statistical analysis of a spatial birth-and-death process model with a view to modelling linear dune fields. *Scandinavian Journal of Statistics*, 21, 1–19.

Møller, J. and Waagepetersen, R.P. (2003). *Statistical Inference and Simulation for Spatial Point Processes*. London: Chapman & Hall.

Møller, J., Syversveen, A., and Waagepetersen, R. (1998). Log Gaussian Cox processes. *Scandinavian Journal of Statistics*, 25, 451–482.

Murrell, D.J., Dieckmann, U., and Law, R. (2004). On moment closures for population dynamics in continuous space. *Journal of Theoretical Biology*, 229, 421–432.

Nelder, J.A. and Mead, R. (1965). A Simplex algorithm for function minimisation. *Computer Journal*, 7, 308–313.

Neyman, J. and Scott, E.L. (1958). Statistical approach to problems of cosmology. *Journal of the Royal Statistical Society*, Series B 20, 1–43.

Peebles, P.J.E. and Groth, E.J. (1975). Statistical analysis of catalogs of extragalactic objects. V. Three point correlation function for the galaxy distribution in the Zwicky catalog. *Astrophysical Journal*, 196, 1–11.

Ripley, B.D. (1976). The second-order analysis of stationary point processes. *Journal of Applied Probability*, 13, 255–266.

Ripley, B.D. (1977). Modelling spatial patterns (with discussion). *Journal of the Royal Statistical Society* B 39, 172–212.

Ripley, B.D. and Kelly, F.P. (1977). Markov point processes. *Journal of the London Mathematical Society*, 15, 188–192.

Rue, H. and Held, L. (2005). *Gaussian Markov Random Fields: Theory and Applications*. London: CRC Press.

Silverman, B.W. (1986). *Density Estimation for Statistics and Data Analysis*. London: Chapman & Hall.

U.K. National Audit Office (2002). The 2001 Outbreak of Foot and Mouth Disease. Report by the Comptroller and Auditor General, HC 939, Session 2001–2002. London: The Stationery Office.

CHAPTER 2

Spatio-Temporal Modelling — with a View to Biological Growth

Eva B. Vedel Jensen, Kristjana Ýr Jónsdóttir, Jürgen Schmiegel, and Ole E. Barndorff-Nielsen

Contents

2.1 Introduction

Modelling of biological growth patterns is a field of mathematical biology that has attracted much attention in recent years, see e.g. Chaplain et al. (1999) and Capasso et al. (2002). The biological systems modelled are diverse and comprise growth of plant populations, year rings of trees, capillary networks, bacteria colonies, and tumours. This chapter deals with spatio-temporal models for such random growing objects, using spatio-temporal point processes or the theory of Lévy bases. For both types of models, the Poisson process will play a key role, either as a reference process or more directly in the model construction.

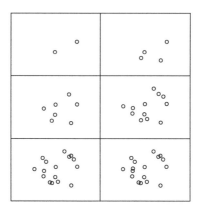

Figure 2.1 The development of a particular type of weed plant (Trifolium, clover) at six different time points. See Brix (1998, 1999) and Brix and Møller (2001).

The first main group of models to be discussed is based on spatio-temporal point processes. We let $Z = \{(t_i, \xi_i)\}$ be a spatio-temporal point process on $S = \mathbb{R}_+ \times \mathcal{X}$ where $\mathcal{X}$ is a bounded subset of $\mathbb{R}^d$ with positive volume $|\mathcal{X}|$. The object at time t is given by

$$X_t = \{\xi_i \in \mathcal{X} : (t_i, \xi_i) \in Z, t_i \leq t\}.$$

Note that $X_{t'} \subseteq X_t$ for $t' \leq t$. The object X_t is called the cumulative spatial point process at time t. Figure 2.1 shows an example of a growth pattern that may be modelled using this framework. The data come from an experiment on a Danish barley field and show the development in a subregion of the field of a particular type of weed plant (Trifolium, clover) on six different dates. Such point patterns show clustering compared to a Poisson pattern and have been the inspiration for developing important new models of Cox type; cf. Brix (1998, 1999), Brix and Diggle (2001), Brix and Møller (2001), Møller (2003) and references therein. The points ξ_i may also be used as centres of 'cells' $\Xi(\xi_i) \subset \mathbb{R}^d$, modelled as random compact sets; see Molchanov (2005). An important early example of such a model, describing tumour growth, can be found in Cressie and Hulting (1992), see also Cressie and Laslett (1987) and Cressie (1991a,b).

A second main group of models describes how the boundary of a full-dimensional object expands in time. We will mainly discuss growth models based on Lévy bases, cf. Barndorff-Nielsen et al. (2003), Barndorff-Nielsen and Schmiegel (2004) and references therein. One example of such a model describes the growth of a star-shaped object, using its radial function. In the planar case, the radial function of an object Y_t at time t gives the distance $R_t(\phi)$ from a reference point z to the boundary of Y_t in direction $\phi \in [-\pi, \pi)$.

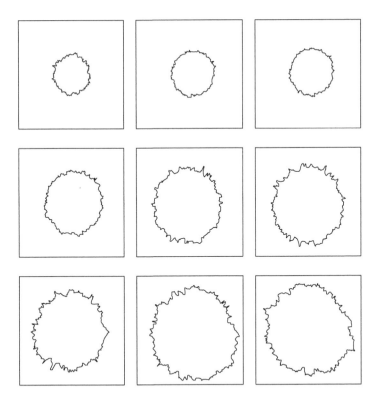

Figure 2.2 Contours of a brain tumour cell island at nine different time points. See Brú et al. (1998).

For such objects, we study model specifications of the type

$$R_t(\phi) = \exp\left(\int_{A_t(\phi)} f_t(\xi, \phi) Z(\mathrm{d}\xi)\right),$$

where Z is a factorizable or a normal Lévy basis, $A_t(\phi)$ is an ambit set (a concept introduced in Barndorff-Nielsen and Schmiegel (2004)), and $f_t(\xi, \phi)$ is a deterministic weight function. As we shall see, it is possible for such a model to derive an explicit expression for

$$\mathrm{Cov}(R_t(\phi), R_{t'}(\phi'))$$

in terms of the three components of the model. Figure 2.2 shows an example of a growth pattern that may be modelled using this framework. These data are part of a larger data set that has been discussed in Brú et al. (1998). Notice that the boundary of the growing tumour cell island is very irregular.

In Section 2.2, models based on spatio-temporal point processes are presented while models based on Lévy bases are dealt with in Section 2.3. Basic results for spatio-temporal point processes are briefly reviewed in Appendix A.

2.2 Models based on spatio-temporal point processes

We will start by giving a short and fairly self-contained introduction to the theory of spatio-temporal point processes. For a more detailed treatment, cf. Daley and Vere-Jones (2002). Expositions with emphasis on purely spatial point processes can be found in Stoyan et al. (1995), Van Lieshout (2000), Diggle (2003), and Møller and Waagepetersen (2003).

2.2.1 Setup

Let $Z = \{(t_i, \xi_i)\}$ be a spatio-temporal point process on $S = \mathbb{R}_+ \times \mathcal{X}$. We assume that the projections of Z on $\mathcal{X}$ and $\mathbb{R}_+$ are both simple point processes (no multiple points). In the following we let

$$Z_t = \{(t_i, \xi_i) \in Z : t_i \le t\}$$

be the restriction of Z to $S_t = (0, t] \times \mathcal{X}$. Note that since Z is locally finite and S_t is bounded, Z_t is a finite random subset of S_t. The corresponding cumulative spatial processes are denoted

$$X = \{\xi_i : (t_i, \xi_i) \in Z\}, \quad X_t = \{\xi_i : (t_i, \xi_i) \in Z_t\}.$$

Note that X and X_t are the projections of Z and Z_t, respectively, on $\mathcal{X}$; see also Figure 2.3.

Since the projection of Z on $\mathbb{R}_+$ is simple, the temporal part of the process gives a natural ordering of the points that does not exist in general for a spatial

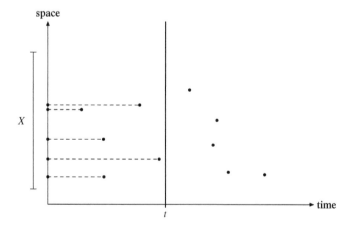

Figure 2.3 An illustration of the setup. The points constitute the spatio-temporal point process Z and the dashed lines indicate the projections on $\mathcal{X}$ of points arrived before or at time t. The projected points in $\mathcal{X}$ constitute X_t.

point process. This feature will be used at various places in the following. Unless stated otherwise, the numbering of the points of Z is such that

$$t_1 < t_2 < \cdots < t_n < \cdots.$$

It will be assumed that Z_t has a density g_{Z_t} with respect to the unit rate Poisson point process on S_t. Because of the natural ordering of the time axis, there are also alternative and perhaps more natural ways of specifying the distribution of Z. Thus, the process can be defined by two families of conditional probability densities,

$$\{p_n(t \mid t_{(n-1)}, \xi_{(n-1)}) : n \in \mathbb{N}\} \tag{2.1}$$

and

$$\{f_n(\xi \mid t_{(n-1)}, \xi_{(n-1)}, t_n) : n \in \mathbb{N}\}, \tag{2.2}$$

with respect to the Lebesgue measure on $\mathbb{R}$ and $\mathbb{R}^d$, respectively. Here and in what follows we use the short notation $t_{(n)}, \xi_{(n)}$ for

$$(t_1, \xi_1), \ldots, (t_n, \xi_n).$$

The density $p_n(\cdot \mid t_{(n-1)}, \xi_{(n-1)})$ describes the distribution of the n-th time point given the history of the whole process up to time t_{n-1}, whereas the density $f_n(\cdot \mid t_{(n-1)}, \xi_{(n-1)}, t_n)$ describes the distribution of the spatial point at time t_n given the history up to time t_{n-1} and the arrival time of the n-th point. The density $p_n(\cdot \mid t_{(n-1)}, \xi_{(n-1)})$ has support (t_{n-1}, ∞) while the density $f_n(\cdot \mid t_{(n-1)}, \xi_{(n-1)}, t_n)$ has support $\mathcal{X}$.

In Appendix A, it is shown for a general spatio-temporal point process how the density of the process Z_t can be expressed in terms of conditional densities. A proof of this well-known result can be found in Daley and Vere-Jones (2002) on conditional intensities and likelihoods for marked point processes. An alternative proof may be found in Appendix A. Here, we just present the result. The density of Z_t with respect to the unit rate Poisson point process on S_t can be expressed as

$$g_{Z_t}(z) = \exp(t|\mathcal{X}|) \prod_{i=1}^{n} [p_i(t_i \mid t_{(i-1)}, \xi_{(i-1)}) f_i(\xi_i \mid t_{(i-1)}, \xi_{(i-1)}, t_i)]$$
$$\times S_{n+1}(t \mid t_{(n)}, \xi_{(n)}),$$

where

$$z = \{(t_1, \xi_1), \ldots, (t_n, \xi_n)\}, \quad t_1 < \cdots < t_n.$$

Here,

$$S_{n+1}(t \mid t_{(n)}, \xi_{(n)}) = \int_{t}^{\infty} p_{n+1}(u \mid t_{(n)}, \xi_{(n)}) du, \quad t > t_n,$$

is the survival function of $p_{n+1}(\cdot \mid t_{(n)}, \xi_{(n)})$.

For spatio-temporal point processes it is of particular interest to study the conditional intensities. For a sequence $\{(t_i, \xi_i)\}$ with

$$0 = t_0 < t_1 < \cdots < t_n < \cdots,$$

the conditional intensity is

$$\lambda^\star(t, \xi) = \lambda_g(t) f^\star(\xi \mid t), \quad \text{if } t_{n-1} < t \leq t_n,$$

where

$$\lambda_g(t) = \frac{p_n(t \mid t_{(n-1)}, \xi_{(n-1)})}{S_n(t \mid t_{(n-1)}, \xi_{(n-1)})}, \quad \text{if } t_{n-1} < t \leq t_n,$$

$$f^\star(\xi \mid t) = f_n(\xi \mid t_{(n-1)}, \xi_{(n-1)}, t), \quad \text{if } t_{n-1} < t \leq t_n.$$

The conditional intensity $\lambda^\star$ has a simple intuitive interpretation. Thus,

$$\lambda^\star(t, \xi) dt d\xi$$

can be interpreted as the conditional probability of observing a point at (t, ξ) given the previous history of the process and a waiting time for the n-th point at least up until t. It can be shown that the density of Z_t can be written as

$$g_{Z_t}(z) = \exp\left(-\int_{S_t} [\lambda^\star(s, \xi) - 1] ds d\xi\right) \prod_{i=1}^n \lambda^\star(t_i, \xi_i), \qquad (2.3)$$

where

$$z = \{(t_1, \xi_1), \ldots, (t_n, \xi_n)\}, \quad t_1 < \cdots < t_n.$$

For further details, see Appendix A.

2.2.2 The Poisson case

The process Z is Poisson with intensity function λ if the number $Z(A)$ of points falling in any Borel subset $A \in \mathcal{B}(S)$ is Poisson distributed with parameter

$$\int_A \lambda(s, \xi) ds d\xi.$$

For a Poisson process, $Z(A_1)$ and $Z(A_2)$ are independent if A_1 and A_2 are disjoint. If Z is Poisson, the conditional intensity function $\lambda^\star$ is equal to the unconditional intensity function λ, and the density of Z_t with respect to the unit rate Poisson point process on S_t is given by

$$g_{Z_t}(z) = \exp\left(-\int_{S_t} [\lambda(s, \xi) - 1] d\xi ds\right) \prod_{i=1}^n \lambda(t_i, \xi_i).$$

The distribution of the cumulative spatial process X_t is also Poisson with intensity function

$$\lambda^t(\xi) = \int_0^t \lambda(s,\xi) ds.$$

If the intensity function can be written as $\lambda(t,\xi) = \lambda_1(t)\lambda_2(\xi)$, then

$$\lambda^t(\xi) = a(t)\lambda_2(\xi),$$

where

$$a(t) = \int_0^t \lambda_1(s) ds.$$

Thus, if the intensity function is of product form, the cumulative point pattern at time t is a scaled version of a spatial template Poisson point process with intensity function $\lambda_2(\xi)$.

In the Poisson case, the conditional densities (2.1) and (2.2) are

$$p_n(t \mid t_{(n-1)}, \xi_{(n-1)}) = \lambda_g(t) \exp\left(-\int_{t_{n-1}}^t \lambda_g(s) ds\right), \quad t > t_{n-1},$$

and

$$f_n(\xi \mid t_{(n-1)}, \xi_{(n-1)}, t_n) = \frac{\lambda(t_n, \xi)}{\lambda_g(t_n)}, \quad \xi \in \mathcal{X},$$

where

$$\lambda_g(t) = \int_{\mathcal{X}} \lambda(t,\xi) d\xi.$$

These results hold under the assumption that

$$\int_t^\infty \lambda_g(s) ds = \infty \quad \text{for all } t \geq 0.$$

The conditional densities can be given simple interpretations. Thus, let T_n be the arrival time of the n-th point. Then, given $(t_{(n-1)}, \xi_{(n-1)})$,

$$\int_{t_{n-1}}^{T_n} \lambda_g(s) ds$$

is exponentially distributed with parameter 1. The density of the position Ξ_n of the n-th point, given $(t_{(n-1)}, \xi_{(n-1)}, t_n)$, is proportional to $\lambda(t_n, \cdot)$.

Poisson point processes play a fundamental role in stochastic geometry; see Stoyan et al. (1995) and references therein. In relation to growth modelling, an important model class is the class of Boolean models. In a purely spatial context, a Boolean model is defined by

$$Y = \cup_i \Xi(\eta_i),$$

where $\{\eta_i\}$ is a Poisson point process on $\mathcal{X}$ and $\{\Xi(\eta_i)\}$ is a sequence of independent and identically distributed random sets, independent of the points $\{\eta_i\}$ and centred at the points of the point process. In particular, Boolean models have been used in the modelling of tumour growth by Cressie and coworkers; see Cressie and Laslett (1987), Cressie (1991a,b) and Cressie and Hulting (1992). Their model can be described as a sequence of Boolean models such that the tumour Y_t at time t is a union of balls placed at uniform random positions inside the tumour Y_{t-1} at time $t-1$. Formally this means

$$Y_t = \cup\{B_d(\xi_i, R) : \xi_i \in Y_{t-1}\},$$

where $\{\xi_i\}$ is a homogeneous Poisson point process in $\mathcal{X}$. They also consider more regular point patterns than Poisson and, in the planar case, rectangular cells instead of circular cells in the growth process and establish the relation to interacting particle systems. Suitable transformations are introduced in order to be able to model a higher growth rate at parts of the boundary with higher curvature.

More recently, Deijfen (2003) has studied a random object Y_t defined as a connected union of randomly sized balls constructed from a spatio-temporal Poisson point process. It is shown that the asymptotic shape of the object is spherical. Other stochastic models of tumour growth also have asymptotic spherical growth; cf. Richardson (1973), Schurger (1979), Bramson and Griffeath (1980), Bramson and Griffeath (1981) and Durret and Liggett (1981).

The model suggested in Cressie and Hulting (1992) has been the inspiration for the discrete Markov growth model studied recently in Jónsdóttir and Jensen (2005). The increments follow a cyclic Gaussian p-order process on the circle; see also Hobolth et al. (2003).

2.2.3 Cox processes

A spatio-temporal Cox process Z on S is a spatio-temporal Poisson point process with a *random* intensity function Λ; cf. Cox (1955), Møller and Waagepetersen (2003) and references therein. Such a process exhibits clustering between the points. The intensity function of a spatio-temporal Cox process is given by $\lambda(t, \xi) = \mathbb{E}\Lambda(t, \xi)$ and the pair correlation function by

$$\rho((t, \xi), (s, \eta)) = \frac{\mathbb{E}(\Lambda(t, \xi)\Lambda(s, \eta))}{\mathbb{E}\Lambda(t, \xi)\mathbb{E}\Lambda(s, \eta)}.$$

Note that

$$\rho((t, \xi), (s, \eta))\mathbb{E}\Lambda(t, \xi)\mathbb{E}\Lambda(s, \eta)dtd\xi dsd\eta$$

can be interpreted as the probability that Z contains points simultaneously in two infinitesimal regions around (t, ξ) and (s, η).

It is clear that Z_t is a spatio-temporal Cox process on S_t driven by the restriction Λ_t of Λ to S_t. The cumulative spatial process X_t is a Cox process on $\mathcal{X}$ driven by

$$\Lambda^t(\xi) = \int_0^t \Lambda(s, \xi)\,\mathrm{d}s.$$

The intensity function of X_t is $\lambda^t(\xi) = \mathbb{E}\Lambda^t(\xi)$. It can be shown that for $t' \leq t$, $X_{t'}$ has the same distribution as a process obtained by independent thinning of the points in X_t with retention probability for a point located at $\xi \in \mathcal{X}$ given by

$$p_{t',t}(\xi) = \frac{\lambda^{t'}(\xi)}{\lambda^t(\xi)}.$$

A spatio-temporal Cox process with log-Gaussian intensity function has been used with success in the analysis of weed data of the type shown in Figure 2.1; see Møller et al. (1998) or Brix and Møller (2001). More specifically, the model considered is of the following form

$$\log \Lambda(t, \xi) = m(t, \xi) + W(\xi),$$

where m is a mean function satisfying

$$m(t', \xi) \leq m(t, \xi) \quad \text{for } t' \leq t, \xi \in \mathcal{X},$$

and W is a zero mean Gaussian process on $\mathcal{X}$. In fact, a bivariate model of Cox type is considered for the modelling of two types of weed plants. An extension of this approach, involving an Ornstein-Uhlenbeck stochastic differential equation, is discussed in Brix and Diggle (2001).

Another important class of spatio-temporal Cox processes is the class of spatio-temporal shot noise Cox processes. As an example, consider a Cox process Z driven by

$$\Lambda(t, \xi) = \sum_{(u,c,\gamma) \in \Phi} \gamma k((u, c), (t, \xi)),$$

where $k((u, c), \cdot)$ is a kernel (i.e., a probability density on S) and Φ is a Poisson point process on $S \times \mathbb{R}_+$. A comprehensive treatment of the purely spatial case can be found in Møller (2003). See also the recent paper by Møller and Torrisi (2005). The process can be viewed as a cluster process since

$$Z|\Phi \sim \cup_{(u,c,\gamma) \in \Phi} U_{u,c,\gamma},$$

where $U_{u,c,\gamma}$, $(u, c, \gamma) \in \Phi$, are independent spatio-temporal Poisson processes with intensity functions

$$\gamma k((u, c), (\cdot, \cdot)).$$

The cumulative spatial process X_t is also a cluster process of Cox-type since

$$X_t | \Phi \sim \cup_{(u,c,\gamma) \in \Phi} V_{u,c,\gamma},$$

where $V_{u,c,\gamma}$, $(u, c, \gamma) \in \Phi$, are independent Poisson point processes on $\mathcal{X}$ with intensity function

$$\gamma \int_0^t k((u,c),(s,\cdot)) \mathrm{d}s.$$

Another important model class is that of shot noise G Cox processes, also suggested for the analysis of the weeds data; cf. Brix (1998, 1999) and Brix and Chadoeuf (2002). These processes are defined through the class of G-measures, which originates from the so-called G-family of probability distributions, extending Gamma and inverse Gaussian distributions.

2.2.4 Markov point processes

In recent years, Markov models for inhomogeneous spatial point processes have been studied quite intensively; see Stoyan and Stoyan (1998), Baddeley et al. (2000), Jensen and Nielsen (2000), Hahn et al. (2003) and references therein. The majority of inhomogeneous models have been constructed by introducing inhomogeneity into a homogeneous Markov point process X, defined on a bounded subset $\mathcal{X}$ of $\mathbb{R}^d$. In this section, we will discuss extensions of these inhomogeneous point process models to a spatio-temporal framework. Extensions can be constructed in an *empirical* fashion, writing down a density for the spatio-temporal point pattern or in a *mechanistic* fashion, using conditional intensities.

We start by giving a short review of recently suggested inhomogeneous spatial Markov point processes. It should be noted that Markov models are primarily appropriate for describing inhibition between the points.

2.2.4.1 Inhomogeneous spatial point processes

In principle, any given homogeneous template point process can be turned into an inhomogeneous point process by independent thinning with a retention probability $p(\xi)$ that depends on the location $\xi \in \mathcal{X}$. As Baddeley et al. (2000) show, second-order functions such as Ripley's K-function can be defined for thinned point processes such that they are identical to the corresponding second-order functions of the original process. However, thinning changes the interaction structure. Thus, if a very regular point process is subjected to inhomogeneous thinning, regions of low intensity seem to exhibit almost no interaction and look similar to a realization of a Poisson process while regions with high intensity will show the original very regular pattern.

Another method that is applicable to any template process is to generate inhomogeneity by a nonlinear transformation of the spatial coordinates. Jensen and Nielsen (2000) prove that the process resulting from transformation of a Markov point process is again Markov. Transformation does in general not preserve (local) isotropy of the template process. See also Jensen and Nielsen (2004).

Ogata and Tanemura (1986) and Stoyan and Stoyan (1998) suggest to introduce inhomogeneity into Markov or Gibbs models by location-dependent first-order interactions. As an example, consider a Strauss template X on $\mathcal{X}$ with parameters $\beta > 0$, $\gamma \in [0, 1]$, and $R > 0$, which is defined by a density

$$f_X(x) \propto \beta^{n(x)} \gamma^{s(x)}, \quad s(x) = \sum_{\{\eta, \xi\} \subseteq x} \mathbf{1}_{[0,R]}(\|\eta - \xi\|), \tag{2.4}$$

with respect to the unit rate Poisson process on $\mathcal{X}$. In (2.4), the sum is over all pairs of different points in x. The resulting inhomogeneous process has density

$$f_X(x) \propto \prod_{\eta \in x} \beta(\eta) \gamma^{s(x)}, \quad s(x) = \sum_{\{\eta, \xi\} \subseteq x} \mathbf{1}_{[0,R]}(\|\eta - \xi\|) \tag{2.5}$$

with respect to the unit rate Poisson process on $\mathcal{X}$. For such an inhomogeneous process, the degree of regularity in the resulting process depends on the intensity, as in the case of thinning described above.

An approach that preserves locally the geometry of the template model, in particular the degree of regularity and also isotropy, was introduced in Hahn et al. (2003). It can be applied to models that are specified by a density with respect to the unit rate Poisson process. The idea of the approach is that a location-dependent scale factor $c(\xi) > 0$ changes the local specification of the model such that in a neighbourhood of any point $\xi \in \mathcal{X}$, the inhomogeneous process behaves like the template process scaled by the factor $c(\xi)$. This is achieved by defining the locally scaled process X_c by a density $f_{X_c}^{(c)}$ with respect to an inhomogeneous Poisson process of rate $c(\xi)^{-d}$. The density $f_{X_c}^{(c)}$ is obtained (up to a normalizing constant) from the template density f_X by replacing all k-dimensional volume measures ν^k, $k = 0, 1, \ldots, d$, that occur in the definition of f_X by their locally scaled counterparts ν_c^k, where $\nu_c^k(A) := \int_A c(u)^{-k} \nu^k(du)$ for all $A \in \mathcal{B}_d$. Note that according to previous notation, $\nu^d(\cdot) = |\cdot|$.

A locally scaled version of the Strauss process has thereby the density

$$f_{X_c}^{(c)}(x) \propto \beta^{n(x)} \gamma^{s_c(x)}, \quad s_c(x) = \sum_{\{\eta, \xi\} \subseteq x} \mathbf{1}_{[0,R]}(\nu_c^1([\eta, \xi])), \tag{2.6}$$

where

$$\nu_c^1([\eta, \xi]) := \int_{[\eta, \xi]} c(u)^{-1} \nu^1(du)$$

is the locally scaled length of the segment $[\eta, \xi]$. This modification applies to all Markov point processes where the higher-order interaction is a function of pairwise distances. The resulting inhomogeneous point process is again Markov, now with respect to the neighbour relation

$$\eta \sim \xi \quad \Longleftrightarrow \quad \nu_c^1([\eta, \xi]) \leq R.$$

Since evaluation of the integral in the locally scaled length measure may be computationally expensive in the general case, the scaled distance of two points may be approximated by

$$\nu_c^1([\eta, \xi]) \approx \frac{\|\eta - \xi\|}{(c(\eta) + c(\xi))/2}. \tag{2.7}$$

Using (2.7) in (2.6), and adjusting the first-order term in (2.6), we get the density f_{X_c} of X_c with respect to the unit rate Poisson process as

$$f_{X_c}(x) \propto \beta^{n(x)} \gamma^{s_c(x)} \prod_{\eta \in x} c(\eta)^{-d}, \quad s_c(x) = \sum_{\{\eta, \xi\} \subseteq x} 1_{[0, \frac{c(\eta) + c(\xi)}{2} R]}(\|\eta - \xi\|).$$
$$\tag{2.8}$$

As shown in Hahn et al. (2003), if the scaling function is slowly varying compared to the interaction radius, the local intensity in a point ξ of such a locally scaled process is in good approximation proportional to $c(\xi)^{-d}$. Figures 2.4 and 2.5 show realizations of locally scaled Strauss processes. Notice that locally these processes look like a scaled version of a homogeneous Strauss process. The statistical analysis of local scaling models is discussed in Prokešová et al. (2005).

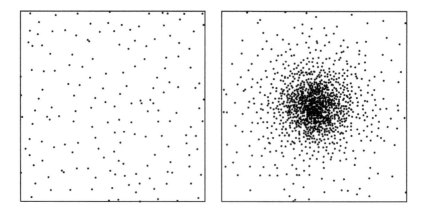

Figure 2.4 (Left) Homogeneous template Strauss process X on $[-1, 1]^2$ with parameters $\beta = 200$, $\gamma = 0.1$, $R = 0.1$. (Right) Inhomogeneous Strauss process X_c with $c(\xi) = \|\xi\|^2 + 0.1$. (Reprinted from Hahn et al. (2003) with permission.)

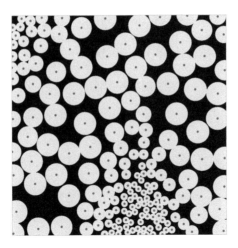

Figure 2.5 Inhomogeneous Strauss process obtained by local scaling. The value of the interaction parameter is $\gamma = 0$, corresponding to a hardcore model. To each point ξ, a circular disc of radius $\frac{c(\xi)}{2} R$ is attached.

2.2.4.2 Spatio-temporal extensions

How can we construct spatio-temporal extensions of these inhomogeneous Markov point processes? An *empirical* approach is to write an expression for the density g_{Z_t} of Z_t with respect to the unit rate Poisson point process on S_t. If the interactions are purely spatial, an obvious suggestion is

$$g_{Z_t}(z) \propto \prod_{i=1}^{n} \lambda(t_i, \xi_i) \cdot \prod_{y \subseteq_2 x} \varphi(y),$$

where

$$z = \{(t_1, \xi_1), \dots, (t_n, \xi_n)\},$$
$$x = \{\xi_1, \dots, \xi_n\},$$

and $\subseteq_2$ indicates that only subsets with at least two elements are considered. If

$$\lambda(t, \xi) = \lambda_1(t)\lambda_2(\xi),$$

then it can be shown that the density of X_t is of the form

$$f_{X_t}(x) \propto a(t)^{n(x)} \prod_{\xi \in x} \lambda_2(\xi) \prod_{y \subseteq_2 x} \varphi(y),$$

where

$$a(t) = \int_0^t \lambda_1(s)\mathrm{d}s.$$

Furthermore, $X_{t'}$ can be obtained from X_t by independent thinning, $t' \leq t$.

Inspired by these results, we may consider backwards temporal thinning of a spatial Markov point process X with intensity function λ, say. Let the resulting spatio-temporal point process be denoted by

$$Z = \{(T_\xi, \xi) : \xi \in X\}.$$

If, conditionally on X,

$$T_\xi, \xi \in X,$$

are independent and T_ξ has density p_ξ, then the cumulative process X_t has intensity function

$$\lambda^t(\xi) = \lambda(\xi) \int_0^t p_\xi(s)\mathrm{d}s.$$

Furthermore, for all $t' \leq t$, $X_{t'}$ can be obtained from X_t by independent thinning, with retention probability for a point located at ξ given by

$$p_{t',t}(\xi) = \int_0^{t'} p_\xi(s)\mathrm{d}s \Big/ \int_0^t p_\xi(s)\mathrm{d}s.$$

Note that the special K-function defined in Baddeley et al. (2000) will be the same for all processes X_t. Note also that the thinning, backwards in time, implies that X_t may look Poisson-like for small t.

Below, we study thinning of a locally scaled Strauss process.

Example 2.2.1 *Temporal thinning of a locally scaled Strauss process.* Let $c_1 : \mathbb{R}_+ \to \mathbb{R}$ and $c_2 : \mathbb{R}^d \to \mathbb{R}$ be positive and bounded local scaling functions for time and space, respectively. Let the density of X be a locally scaled Strauss process

$$f_X(x) \propto \beta^{n(x)} \gamma^{s_{c_2}(x)} \prod_{\eta \in x} c_2(\eta)^{-d},$$

where

$$s_{c_2}(x) = \sum_{\{\eta, \xi\} \subseteq x} \mathbf{1}_{[0,R]}\left(\nu_{c_2}^1([\eta, \xi])\right).$$

A birth time at ξ is distributed with a density which does not depend on ξ:

$$p_\xi(t) \propto c_1(t)^{-1}, \tag{2.9}$$

if $t \in [0, T]$, and $p_\xi(t) = 0$, otherwise.

Figure 2.6 shows the result of a simulation of a temporal thinning of such a locally scaled Strauss process on $[-1, 1]^2$ with $\beta = 100$, $\gamma = 0.01$, $R = 0.1$, and local scaling function $c_2(\xi) = 0.2 + 4\|\xi\|^2$. The birth times have density given in (2.9) with $c_1(t) = 0.2 + 0.05t$ and $T = 12$. The figure shows the corresponding cumulative point patterns $X_{t'}$ for $t' = 2, 4, 8, 12$. Note that for small t', $X_{t'}$ appears Poisson-like.

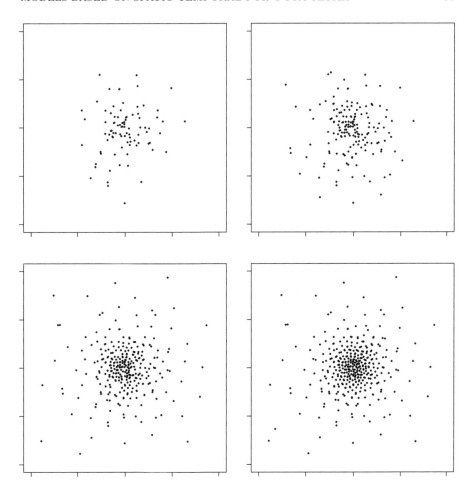

Figure 2.6 Result of a simulation of a backwards thinning of a locally scaled Strauss process on $[-1,1]^2$. For details, see Example 2.2.1.

Another possibility is specification of a spatio-temporal point process model in terms of conditional intensities; see e.g., Hawkes (1971) or Schoenberg et al. (2002) and references therein. This approach could be characterized as *mechanistic*. Here, the form of the conditional intensity may be motivated by the form of the Papangelou conditional intensity for a purely spatial point process. A local scaling example is given below.

Example 2.2.2 Let $c_1 : \mathbb{R}_+ \to \mathbb{R}$ and $c_2 : \mathbb{R}^d \to \mathbb{R}$ be positive and bounded scaling functions for time and space, respectively. We define the spatio-temporal point process Z by its conditional intensities,

$$\lambda^\star(t, \xi) = \frac{\beta \gamma^{s_{c_2}(\xi | \xi_{(n-1)})}}{c_1(t) c_2(\xi)^d}, \quad t_{n-1} < t \leq t_n, \ \xi \in \mathcal{X},$$

where

$$s_{c_2}(\xi \mid \xi_{(n-1)}) = \sum_{i=1}^{n-1} \mathbf{1}_{[0,R]}\left(\nu_{c_2}^1([\xi, \xi_i])\right).$$

In this case, the density of Z_t is of the following form:

$$g_{Z_t}(z) = \exp\left(-\sum_{i=1}^{n}\int_{t_{i-1}}^{t_i}\int_{\mathcal{X}}\left(\frac{\beta\gamma^{s_{c_2}(\xi|\xi_{(i-1)})}}{c_1(t)c_2(\xi)^d} - 1\right)dtd\xi\right)\prod_{i=1}^{n}\frac{\beta\gamma^{s_{c_2}(\xi_i|\xi_{(i-1)})}}{c_1(t_i)c_2(\xi_i)^d}$$

$$= \exp\left(-\sum_{i=1}^{n}\int_{t_{i-1}}^{t_i}\int_{\mathcal{X}}\left(\frac{\beta\gamma^{s_{c_2}(\xi|\xi_{(i-1)})}}{c_1(t)c_2(\xi)^d} - 1\right)dtd\xi\right)\beta^{n(z)}\gamma^{\sum_{i=1}^{n}s_{c_2}(\xi_i|\xi_{(i-1)})}$$

$$\times \prod_{i=1}^{n}\frac{1}{c_1(t_i)c_2(\xi_i)^d}.$$

Figure 2.7 shows the result of a simulation on $[-1,1]^2$ of this type of spatio-temporal extension of the locally scaled Strauss process, with $\beta = 3.61$,

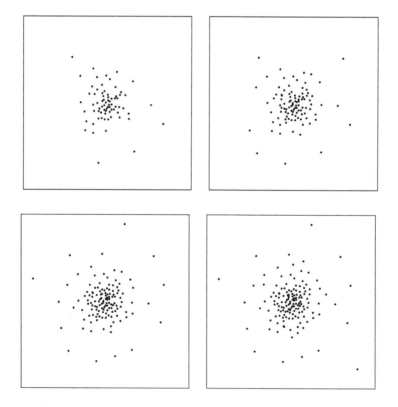

Figure 2.7 Result of a simulation on $[-1,1]^2$ of the spatio-temporal extension of the locally scaled Strauss process described in Example 2.2.2. For details, see the text.

$\gamma = 0.01$, $R = 0.1$, $c_1(t) = 0.2 + 0.05t$, and $c_2(\xi) = 0.2 + 4\|\xi\|^2$. The parameter values are thereby identical to those chosen in the previous example, except for the value of β which has been chosen so that the number of points are expected to be the same in the two examples.

2.3 Lévy-based growth models

In this section, we will discuss an alternative approach to the modelling of random growing objects. We will concentrate on the case of random planar star-shaped objects. The model describes how the boundary of the growing object expands in time. The basic notion of this approach is that of Lévy bases.

2.3.1 Setup

This subsection provides a very brief overview of the theory of Lévy bases, in particular, the theory of integration with respect to a Lévy basis. For a more detailed account, see Barndorff-Nielsen and Schmiegel (2004) and references therein. We will use the following notation for the cumulant function of a random variable X:

$$C\{\lambda \ddagger X\} = \log \mathbb{E}(e^{i\lambda X}).$$

Let $(\mathcal{R}, \mathcal{A})$ be a measurable space. In what follows, we will assume that $\mathcal{R}$ is a Borel subset of $\mathbb{R}^n$. Let $\mathcal{A} = \mathcal{B}(\mathcal{R})$ be the σ-algebra of Borel subsets of $\mathcal{R}$ and let $\mathcal{A}_b$ denote the class of bounded elements of $\mathcal{A}$. A collection of random variables $Z = \{Z(A) : A \in \mathcal{A}\}$ or $Z = \{Z(A) : A \in \mathcal{A}_b\}$ is said to be an *independently scattered random measure*, if for every sequence $\{A_n\}$ of disjoint sets in $\mathcal{A}$, respectively $\mathcal{A}_b$, the random variables $Z(A_n)$ are independent and $Z(\bigcup A_n) = \sum Z(A_n)$ a.s., where in case $Z = \{Z(A) : A \in \mathcal{A}_b\}$ we furthermore require $\cup A_n \in \mathcal{A}_b$. If, in addition, $Z(A)$ is *infinitely divisible* for all $A \in \mathcal{A}$, Z will be called a *Lévy basis*. (The need to distinguish between the two cases $\mathcal{A}$ and $\mathcal{A}_b$ comes from the possible difficulty in controlling the countable sum $\sum Z(A_n)$ if Z can take both positive and negative values. When that is the case, Z should strictly speaking be referred to as an independently scattered signed random measure.) When Z is a Lévy basis, the cumulant function of $Z(A)$ can be written as

$$C\{\lambda \ddagger Z(A)\} = i\lambda a(A) - \frac{1}{2}\lambda^2 b(A) + \int_{\mathbb{R}} (e^{i\lambda x} - 1 - i\lambda x 1_{[-1,1]}(x))U(\mathrm{d}x, A),$$

(2.10)

where a is a signed measure on $\mathcal{A}$ or $\mathcal{A}_b$, b is a positive measure on $\mathcal{A}$ or $\mathcal{A}_b$, $U(\mathrm{d}x, A)$ is a Lévy measure on $\mathbb{R}$ for fixed A and a measure on $\mathcal{A}$ or $\mathcal{A}_b$ for fixed $\mathrm{d}x$. The Lévy basis is said to have characteristic triplet (a, b, U) and the measure U is referred to as the generalized Lévy measure.

The cumulant function (2.10) can also be expressed in an infinitesimal form:

$$C\{\lambda \ddagger Z(\mathrm{d}\xi)\} = i\lambda a(\mathrm{d}\xi) - \frac{1}{2}\lambda^2 b(\mathrm{d}\xi) + \int_{\mathbb{R}} (e^{i\lambda x} - 1 - i\lambda x \mathbf{1}_{[-1,1]}(x))U(\mathrm{d}x, \mathrm{d}\xi).$$

Without essential loss of generality we can assume that the measure U factorizes as

$$U(\mathrm{d}x, \mathrm{d}\xi) = V(\mathrm{d}x, \xi)\mu(\mathrm{d}\xi),$$

where μ is some measure on $\mathcal{R}$ and $V(\mathrm{d}x, \xi)$ is a Lévy measure for fixed ξ. A Lévy basis is called *factorizable* if the Lévy measure $V(\cdot; \xi)$ does not depend on ξ. If a Lévy basis is factorizable, then one can write

$$C\{\lambda \ddagger Z(\mathrm{d}\xi)\} = i\lambda a(\mathrm{d}\xi) - \frac{1}{2}\lambda^2 b(\mathrm{d}\xi) + C\{\lambda \ddagger Z'\}\mu(\mathrm{d}\xi),$$

where Z' is an infinitely divisible random variable with cumulant function

$$C\{\lambda \ddagger Z'\} = \int_{\mathbb{R}} (e^{i\lambda x} - 1 - i\lambda x \mathbf{1}_{[-1,1]}(x))V(\mathrm{d}x).$$

We will now give examples of Lévy bases.

Example 2.3.1 *Poisson Lévy basis.* The Poisson basis has characteristic triplet $(\mu, 0, U)$, where $U(\mathrm{d}x, \mathrm{d}\xi) = \delta_1(\mathrm{d}x)\mu(\mathrm{d}\xi)$ and δ_1 is Dirac's delta function at 1, so Z is factorizable. Clearly the cumulant function of $Z(A)$ is

$$C\{\lambda \ddagger Z(A)\} = (e^{i\lambda} - 1)\mu(A).$$

We have $Z(A) \sim Pois(\mu(A))$. The Poisson point process was dealt with in more detail in Subsection 2.2.2.

Example 2.3.2 *Normal Lévy basis.* The normal Lévy basis has characteristic triplet $(\alpha\mu, \sigma^2\mu, 0)$ and the cumulant function is

$$C\{\lambda \ddagger Z(A)\} = i\lambda\alpha\mu(A) - \frac{1}{2}\lambda^2\sigma^2\mu(A).$$

Note that $Z(A) \sim N(\alpha\mu(A), \sigma^2\mu(A))$.

Example 2.3.3 *Gamma Lévy basis.* The gamma Lévy basis has characteristic triplet $(a, 0, U)$ where

$$a(\mathrm{d}\xi) = \frac{1 - e^{-\alpha(\xi)}}{\alpha(\xi)}\mu(\mathrm{d}\xi),$$

$$U(\mathrm{d}x, \mathrm{d}\xi) = \mathbf{1}_{\mathbb{R}_+}(x)x^{-1}e^{-\alpha(\xi)x}\mathrm{d}x\mu(\mathrm{d}\xi),$$

and $\alpha(\xi) > 0$. If $\alpha(\xi)$ does not depend on ξ, $\alpha(\xi) = \alpha$, say, $Z(A) \sim G(\mu(A), \alpha)$, $A \in \mathcal{A}$.

Example 2.3.4 *Inverse Gaussian Lévy basis.* When Z has characteristic triplet $(a, 0, U)$ where

$$a(\mathrm{d}\xi) = \frac{1}{\sqrt{2\pi}} \int_0^1 u^{-\frac{1}{2}} e^{-\frac{1}{2}\gamma^2(\xi)u} \mathrm{d}u \mu(\mathrm{d}\xi),$$

$$U(\mathrm{d}x, \mathrm{d}\xi) = \frac{1}{\sqrt{2\pi}} \mathbf{1}_{\mathbb{R}_+}(x) x^{-\frac{3}{2}} e^{-\frac{1}{2}\gamma^2(\xi)x} \mathrm{d}x \mu(\mathrm{d}\xi),$$

and $\gamma(\xi) > 0$, Z constitutes an inverse Gaussian Lévy basis. If $\gamma(\xi)$ does not depend on ξ, $\gamma(\xi) = \gamma$, say, then $Z(A) \sim IG(\mu(A), \gamma)$, $A \in \mathcal{A}$.

The Poisson, gamma, and inverse Gaussian Lévy bases are examples of the random G-measures introduced in Brix (1999). These measures are purely discrete and can be represented as

$$Z(A) = a_0(A) + \int_0^\infty y \tilde{Z}(A \times \mathrm{d}y),$$

where $\tilde{Z}$ is a Poisson measure on $\mathcal{R} \times [0, \infty)$. This result is an example of a Lévy-Ito representation.

The usefulness of the definitions above becomes clear in connection with the integration of measurable functions f with respect to a Lévy basis Z. We consider the integral of a measurable function f on $\mathcal{R}$ with respect to a factorizable Lévy basis Z. For simplicity we denote this integral by $f \bullet Z$. For the theory of integration with respect to independently scattered random measures, see Kallenberg (1989) and Kwapien and Woyczynski (1992). A key result for many calculations is (subject to minor regularity conditions)

$$C\{\lambda \ddagger f \bullet Z\} = i\lambda(f \bullet a) - \frac{1}{2}\lambda^2(f^2 \bullet b) + \int_{\mathcal{R}} C\{\lambda f(\xi) \ddagger Z'\} \mu(\mathrm{d}\xi). \quad (2.11)$$

Similarly, for the logarithm of the Laplace transform of $f \bullet Z$,

$$K\{\lambda \ddagger f \bullet Z\} = C\{-i\lambda \ddagger f \bullet Z\},$$

we have

$$K\{\lambda \ddagger f \bullet Z\} = \lambda(f \bullet a) + \frac{1}{2}\lambda^2(f^2 \bullet b) + \int_{\mathcal{R}} K\{\lambda f(\xi) \ddagger Z'\} \mu(\mathrm{d}\xi). \quad (2.12)$$

If Z is a normal Lévy basis, $Z(A) \sim N(\alpha\mu(A), \sigma^2\mu(A))$, we have

$$C\{\lambda \ddagger f \bullet Z\} = \int_{\mathcal{R}} C\{\lambda f(\xi) \ddagger Z'\} \mu(\mathrm{d}\xi), \quad\quad (2.13)$$

and

$$K\{\lambda \ddagger f \bullet Z\} = \int_{\mathcal{R}} K\{\lambda f(\xi) \ddagger Z'\} \mu(\mathrm{d}\xi), \quad\quad (2.14)$$

where $Z' \sim N(\alpha, \sigma^2)$. Note that (2.13) and (2.14) also hold for a factorizable Lévy basis with $a \equiv b \equiv 0$.

2.3.2 Lévy-based growth models

Let us consider a planar compact object with size and shape changing over time, where the object at time t is denoted by $Y_t \subset \mathbb{R}^2$. In the following we will assume that Y_t is star-shaped with respect to a point $z \in \mathbb{R}^2$ for all t. Then the boundary of the object Y_t can be determined by its radial function $R_t = \{R_t(\phi) : \phi \in [-\pi, \pi)\}$, where

$$R_t(\phi) = \max\{r : z + r(\cos\phi, \sin\phi) \in Y_t\}, \quad \phi \in [-\pi, \pi).$$

In the following we will let

$$\mathcal{R} = [-\pi, \pi) \times (0, \infty)$$

and $\mathcal{A}$ the Borel σ-algebra of $\mathcal{R}$. The idea behind the following definitions is based on the intuitive picture of an ambit set $A_t(\phi)$, associated to each point (ϕ, t), which defines the causal correlation cone. The ambit set satisfies

$$A_t(\phi) \subseteq [-\pi, \pi) \times [0, t].$$

Examples are shown in Figure 2.8. The radial process $R_t(\phi)$ is defined as the integral of some weight function over the attached ambit set, with respect to a positive factorizable Lévy basis or as the exponential of such an integral with respect to a factorizable or normal Lévy basis.

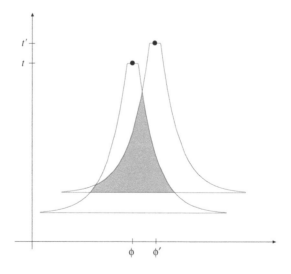

Figure 2.8 Examples of ambit sets $A_t(\phi)$ and $A_{t'}(\phi')$, respectively. Their intersection (shown hatched) determines the dependence structure of the growth process.

Definition 2.3.1 Let Z be a factorizable positive Lévy basis on $\mathcal{R}$. The field of radial functions $R = \{R_t(\phi)\}$ follows a *linear Lévy growth model* if

$$R_t(\phi) = \int_{A_t(\phi)} f_t(\xi; \phi) Z(\mathrm{d}\xi).$$

The ambit set $A_t(\phi) \in \mathcal{A}$ and the positive deterministic weight function $f_t(\xi; \phi)$, which is assumed to be suitable for the integral to exist, must be defined cyclically such that $R_t(\cdot)$ is cyclic.

Definition 2.3.2 Let Z be a factorizable or normal Lévy basis. The field of radial functions $R = \{R_t(\phi)\}$ follows an *exponential Lévy growth model* if

$$R_t(\phi) = \exp\left(\int_{A_t(\phi)} f_t(\xi; \phi) Z(\mathrm{d}\xi) \right).$$

The ambit set $A_t(\phi) \in \mathcal{A}$ and the deterministic weight function $f_t(\xi; \phi)$, assumed to be suitable for the integrals to exist, must be defined cyclically such that $R_t(\cdot)$ is cyclic.

There are many interesting problems to study within this model framework. Basically, it is the Lévy basis Z, the ambit sets $A_t(\phi)$, and the weight functions $f_t(\xi; \phi)$ which determine the growth dynamics. These three ingredients can be chosen arbitrarily so that a great variety of different growth dynamics can be obtained.

Below, we study linear and exponential growth models separately.

2.3.2.1 Linear Lévy growth models

Let us assume that R follows a linear Lévy growth model and that the Lévy basis has no Gaussian part; i.e., $b \equiv 0$. Using the key relation (2.12), we get that

$$\mathbb{E}(R_t(\phi)) = \int_{A_t(\phi)} f_t(\xi; \phi) a(\mathrm{d}\xi) + \mathbb{E}(Z') \int_{A_t(\phi)} f_t(\xi; \phi) \mu(\mathrm{d}\xi) \quad (2.15)$$

$$\mathbb{V}(R_t(\phi)) = \mathbb{V}(Z') \int_{A_t(\phi)} f_t(\xi; \phi)^2 \mu(\mathrm{d}\xi), \quad (2.16)$$

where $\mathbb{V}$ is the notation used for variance. Using (2.16) and the independence properties of a Lévy basis, we furthermore have

$$\mathrm{Cov}(R_t(\phi), R_{t'}(\phi')) = \mathbb{V}(Z') \int_{A_t(\phi) \cap A_{t'}(\phi')} f_t(\xi; \phi) f_{t'}(\xi; \phi') \mu(\mathrm{d}\xi). \quad (2.17)$$

The proof of (2.17) goes as follows. Let $A = A_t(\phi)$ and $A' = A_{t'}(\phi')$. Then,

$$
\begin{aligned}
\text{Cov}(R_t(\phi), R_{t'}(\phi')) &= \text{Cov}\left(\int_A f_t(\xi; \phi) Z(\mathrm{d}\xi), \int_{A'} f_{t'}(\xi; \phi') Z(\mathrm{d}\xi)\right) \\
&= \text{Cov}\left(\int_{A \cap A'} f_t(\xi; \phi) Z(\mathrm{d}\xi), \int_{A \cap A'} f_{t'}(\xi; \phi') Z(\mathrm{d}\xi)\right) \\
&= \frac{1}{2}\left[\mathbb{V}\left(\int_{A \cap A'} (f_t(\xi; \phi) + f_{t'}(\xi; \phi')) Z(\mathrm{d}\xi)\right)\right. \\
&\quad - \mathbb{V}\left(\int_{A \cap A'} f_t(\xi; \phi) Z(\mathrm{d}\xi)\right) \\
&\quad \left. - \mathbb{V}\left(\int_{A \cap A'} f_{t'}(\xi; \phi') Z(\mathrm{d}\xi)\right)\right].
\end{aligned}
$$

Modelling of a given covariance structure reduces to solving (2.17) for the weight-function f and the shape and size of the ambit sets $A_t(\phi)$. In practice this might be a complicated task, but for special applications it is possible. Equation (2.17) also provides some useful geometric interpretation of the covariance structure. This can most easily be seen for the simple case of a constant weight-function $f_t(\xi; \phi) \equiv f$ for all $\xi, (\phi, t) \in \mathcal{R}$. In this case, (2.17) reduces to

$$
\text{Cov}(R_t(\phi), R_{t'}(\phi')) \propto \mu(A_t(\phi) \cap A_{t'}(\phi')). \tag{2.18}
$$

For a constant weight function, the modelling of spatio-temporal covariances thus reduces to the problem of finding ambit sets $A_t(\phi)$ whose measure of overlap

$$
\mu(A_t(\phi) \cap A_{t'}(\phi'))
$$

fulfills (2.18) (see Figure 2.8 for an illustration). Note that only the measure of the overlap is involved and not the shape of the overlap.

In some growth examples it might be more natural to specify the model in terms of the time derivative of $R_t(\phi)$. For instance, as

$$
R_t'(\phi) = \int_{A_t(\phi)} f_t(\xi; \phi) Z(\mathrm{d}\xi).
$$

The induced model for $R_t(\phi)$ is again a linear Lévy growth model. We thus have

$$
\begin{aligned}
R_t(\phi) &= \int_0^t \int_{\mathcal{R}} \mathbf{1}_{A_s(\phi)}(\xi) f_s(\xi; \phi) Z(\mathrm{d}\xi) \mathrm{d}s \\
&= \int_{\mathcal{R}} \int_0^t \mathbf{1}_{A_s(\phi)}(\xi) f_s(\xi; \phi) \mathrm{d}s Z(\mathrm{d}\xi) \\
&= \int_{\bar{A}_t(\phi)} \bar{f}_t(\xi; \phi) Z(\mathrm{d}\xi), \tag{2.19}
\end{aligned}
$$

where

$$
\bar{A}_t(\phi) = \cup_{0 \le s \le t} A_s(\phi)
$$

and

$$\bar{f}_t(\xi;\phi) = \int_0^t \mathbf{1}_{A_s(\phi)}(\xi) f_s(\xi;\phi)\mathrm{d}s.$$

The representation (2.19) is, of course, not unique. If the ambit sets are of the form

$$A_t(\phi) = B_t \cap C_\phi,$$

where $B_t \subseteq [-\pi, \pi) \times [0, t]$, we may instead choose

$$\bar{A}_t(\phi) = C_\phi \cap ([-\pi, \pi) \times [0, t])$$

and

$$\bar{f}_t(\xi;\phi) = \int_0^t \mathbf{1}_{B_s}(\xi) f_s(\xi;\phi)\mathrm{d}s.$$

2.3.2.2 Exponential Lévy growth models

In the following we will assume that we have a factorizable Lévy basis with $b \equiv 0$ or a normal Lévy basis. Equation (2.14) allows us to calculate arbitrary n-point correlations in an exponential Lévy growth model. If we assume the correlations are finite, i.e., $\mathbb{E}\{R_{t_1}(\phi_1) \cdot \cdots \cdot R_{t_n}(\phi_n)\} < \infty$, we have

$$\frac{\mathbb{E}\{\prod_{j=1}^n R_{t_j}(\phi_j)\}}{\prod_{j=1}^n \mathbb{E}\{R_{t_j}(\phi_j)\}} = \frac{\exp\left\{\int_{\mathcal{R}} \mathrm{K}\left\{\sum_{j=1}^n f_{t_j}(\xi,\phi_j)\mathbf{1}_{A_{t_j}(\phi_j)}(\xi) \ddagger Z'\right\} \mu(\mathrm{d}\xi)\right\}}{\exp\left\{\sum_{j=1}^n \int_{\mathcal{R}} \mathrm{K}\left\{f_{t_j}(\xi,\phi_j)\mathbf{1}_{A_{t_j}(\phi_j)}(\xi) \ddagger Z'\right\} \mu(\mathrm{d}\xi)\right\}}.$$

(2.20)

In the proof of (2.20) the following reformulation can be used

$$\mathbb{E}\left\{\prod_{j=1}^n R_{t_j}(\phi_j)\right\} = \mathbb{E}\exp\left\{\int_{\mathcal{R}}\left[\sum_{j=1}^n f_{t_j}(\xi;\phi_j)\mathbf{1}_{A_{t_j}(\phi_j)}(\xi)\right] Z(\mathrm{d}\xi)\right\}$$

$$= \exp\left(\mathrm{K}\left\{1 \ddagger \int_{\mathcal{R}}\left[\sum_{j=1}^n f_{t_j}(\xi;\phi_j)\mathbf{1}_{A_{t_j}(\phi_j)}(\xi)\right] Z(\mathrm{d}\xi)\right\}\right).$$

For a constant weight function $f_t(\xi;\phi) \equiv f$ for all ξ, $(\phi, t) \in \mathcal{R}$, it follows that

$$\frac{\mathbb{E}\{R_{t_1}(\phi_1)R_{t_2}(\phi_2)\}}{\mathbb{E}\{R_{t_1}(\phi_1))\mathbb{E}(R_{t_2}(\phi_2)\}} = \exp\{\overline{\mathrm{K}}\,\mu(A_{t_1}(\phi_1) \cap A_{t_1}(\phi_2))\}$$

(2.21)

where $\overline{\mathrm{K}} = \mathrm{K}\{2f \ddagger Z'\} - 2\mathrm{K}\{f \ddagger Z'\}$.

In some cases it may be natural to formulate a linear model for the time derivative of $\ln R_t(\phi)$

$$(\ln R_t(\phi))' = \int_{A_t(\phi)} f_t(\xi; \phi) Z(\mathrm{d}\xi).$$

The resulting model for $R_t(\phi)$ is an exponential Lévy growth model

$$R_t(\phi) = \exp\left(\int_{\bar{A}_t(\phi)} \bar{f}_t(\xi; \phi) Z(\mathrm{d}\xi)\right),$$

where $\bar{A}_t(\phi)$ and $\bar{f}_t(\xi; \phi)$ are specified as in the previous subsection.

Figure 2.9a shows a simulation from an exponential Lévy growth model with

$$A_t(\phi) = \{(\theta, s) : t - T(t) \le s \le t, |\theta - \phi| \le \Theta(s)\},$$

$(\phi, t) \in \mathcal{R}$, $f_t(\xi; \phi) = f$, and Z a normal Lévy basis. Here, $T(t)$ and $\Theta(s)$ represent the temporal and spatial dependencies, respectively. The similarities

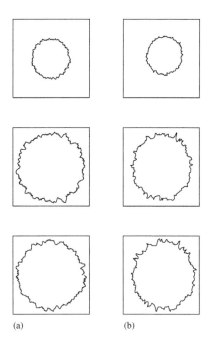

(a)   (b)

Figure 2.9 Comparison of a simulation of a log-normal model (a) with *in vitro* tumour growth (b) at times $t = 25, 45, 51$ (arbitrary units). Parameters of the simulation are $\mu = 0.11$, $\sigma = 0.01$, $T(t) = t/20$, $\Theta(\phi) = \pi/90$, and $f_t(\phi) = 1$. For details, see the text.

between the simulation and the observed *in vitro* growth pattern in Figure 2.9b are striking. A detailed analysis of these data will be presented elsewhere.

It still remains to fully explore the flexibility of the Lévy growth models. In particular, relations like (2.18) and (2.21) describe the correlation structure as a function of the measure of overlap between pairs of ambit sets. These relations can be used to create covariance models for processes defined on the circle. On the other hand, it will be interesting to investigate if a given covariance model can be obtained using the ambit set approach.

Acknowledgments

The authors thank Ute Hahn from the Department of Mathematics, University of Augsburg, for fruitful discussions on models based on spatio-temporal point processes. This work was supported in part by grants from the Danish Natural Science Research Council and the Carlsberg Foundation.

Appendix A: Conditional densities and conditional intensities

Let Π be the unit rate Poisson point process on S and Π_t the restriction of Π to S_t. We let Ω be the set of all locally finite subsets of S and Ω_t the set of finite subsets of S_t. On Ω_t, we use the σ-algebra $\mathcal{A}_t$ generated by

$$\{z \in \Omega_t : n(z \cap B) = k\}, \quad k \in \mathbb{N}_0, B \in \mathcal{B}_t,$$

where $\mathcal{B}_t$ is the Borel σ-algebra on S_t.

We will first state the following basic result for the Poisson point process.

Lemma A.1 Let π_t be the distribution of Π_t and $g_t : (\Omega_t, \mathcal{A}_t) \to (\mathbb{R}, \mathcal{B}(\mathbb{R}))$ be a Borel function. Then

$$\int_{\Omega_t} g_t(z) \pi_t(dz) \tag{2.22}$$

$$= \sum_{n=0}^{\infty} \exp(-t|\mathcal{X}|) \int_{\mathcal{X}} \cdots \int_{\mathcal{X}} \int_0^t \int_{t_1}^t \cdots \int_{t_{n-1}}^t g_t(\{(t_1, \xi_1), \ldots, (t_n, \xi_n)\})$$

$$\times dt_n \cdots dt_1 d\xi_n \cdots d\xi_1,$$

where $|\cdot|$ denotes the Lebesgue measure on $\mathbb{R}^d$.

Proof of Lemma A.1 For the restriction Π_t of the unit rate Poisson point process to S_t, the number N of points (t_i, ξ_i) in S_t is Poisson distributed with parameter $t|\mathcal{X}|$ and conditionally on $N = n$, Π_t is distributed as

$$\{(t_1, \xi_1), \ldots, (t_n, \xi_n)\},$$

where $(t_i, \xi_i), i = 1, \ldots, n$, are independent and uniform in S_t. It follows that

$$\int_{\Omega_t} g_t(z)\pi_t(dz)$$

$$= \sum_{n=0}^{\infty} \exp(-t|\mathcal{X}|)\frac{(t|\mathcal{X}|)^n}{n!} \int_{\mathcal{X}} \cdots \int_{\mathcal{X}} \int_0^t \cdots \int_0^t g(\{(t_1, \xi_1), \ldots, (t_n, \xi_n)\})$$

$$\times \frac{1}{(t|\mathcal{X}|)^n} dt_n \cdots dt_1 d\xi_n \cdots d\xi_1$$

$$= \sum_{n=0}^{\infty} \exp(-t|\mathcal{X}|) \int_{\mathcal{X}} \cdots \int_{\mathcal{X}} \int_0^t \int_{t_1}^t \cdots \int_{t_{n-1}}^t g(\{(t_1, \xi_1), \ldots, (t_n, \xi_n)\})$$

$$dt_n \cdots dt_1 d\xi_n \cdots d\xi_1.$$

If Z_t has density

$$g_{Z_t}(z), \quad z \in \Omega_t$$

with respect to the unit rate Poisson point process Π_t on S_t, then for $A \in \mathcal{A}_t$,

$$P(Z_t \in A) = \int_A g_{Z_t}(z)\pi_t(dz) = \int_{\Omega_t} 1[z \in A]g_{Z_t}(z)\pi_t(dz)$$

and Lemma A.1 can be used to calculate the integral.

The density of Z_t can be expressed in terms of the two families of conditional densities (2.1) and (2.2) presented in the main text; see the proposition below.

Proposition A.2 Let

$$g_n(t_{(n)}, \xi_{(n)}) = \prod_{i=1}^{n} p_i(t_i \mid t_{(i-1)}, \xi_{(i-1)})f_i(\xi_i \mid t_{(i-1)}, \xi_{(i-1)}, t_i)$$

be the density of the first n points of Z. Then, the density of Z_t with respect to Π_t is

$$g_{Z_t}(z) = \exp(t|\mathcal{X}|)g_n(t_{(n)}, \xi_{(n)})S_{n+1}(t \mid t_{(n)}, \xi_{(n)}), \qquad (2.23)$$

if $z \in \Omega_t$ is of the form

$$z = \{(t_1, \xi_1), \ldots, (t_n, \xi_n)\}, \quad t_1 < \cdots < t_n.$$

Here,

$$S_{n+1}(t \mid t_{(n)}, \xi_{(n)}) = \int_t^{\infty} p_{n+1}(u \mid t_{(n)}, \xi_{(n)})du, \quad t > t_n,$$

is the survival function of $p_{n+1}(\cdot \mid t_{(n)}, \xi_{(n)})$.

Proof of Proposition A.2 For $A \in \mathcal{A}_t$, we find

$$P(Z_t \in A)$$
$$= \sum_{n=0}^{\infty} P(Z_t \in A, n(Z_t) = n)$$
$$= \sum_{n=0}^{\infty} \int_{\mathbb{R}_+ \times \mathcal{X}} \cdots \int_{\mathbb{R}_+ \times \mathcal{X}} \mathbf{1}[\{(t_1,\xi_1),\ldots,(t_n,\xi_n)\} \in A]\mathbf{1}[t_{n+1} > t]$$
$$g_{n+1}(t_{(n+1)},\xi_{(n+1)})dt_{n+1}d\xi_{n+1}\cdots dt_1 d\xi_1$$
$$= \sum_{n=0}^{\infty} \int_{\mathcal{X}} \cdots \int_{\mathcal{X}} \int_0^t \int_{t_1}^t \cdots \int_{t_{n-1}}^t \mathbf{1}[\{(t_1,\xi_1),\ldots,(t_n,\xi_n)\} \in A]g_n(t_{(n)},\xi_{(n)})$$
$$S_{n+1}(t \mid t_{(n)},\xi_{(n)})dt_n \cdots dt_1 d\xi_n \cdots d\xi_1.$$

Now Lemma A.1 implies the result.

Another possibility is specification of the model in terms of the conditional intensities. For an increasing sequence

$$(t_1,\xi_1),\ldots,(t_n,\xi_n),\ldots, \quad t_1 < \cdots < t_n < \cdots, \tag{2.24}$$

we define the conditional intensity at (t,ξ) by

$$h_n(t,\xi \mid t_{(n-1)},\xi_{(n-1)}) = \frac{p_n(t \mid t_{(n-1)},\xi_{(n-1)})f_n(\xi \mid t_{(n-1)},\xi_{(n-1)},t)}{S_n(t \mid t_{(n-1)},\xi_{(n-1)})}, \tag{2.25}$$

for $t_{n-1} < t \leq t_n$ $(t_0 = 0)$. Note that the conditional intensity h_n is the product of the hazard function

$$\frac{p_n(t \mid t_{(n-1)},\xi_{(n-1)})}{S_n(t \mid t_{(n-1)},\xi_{(n-1)})}$$

for the n-th time point given the history $(t_{(n-1)},\xi_{(n-1)})$ and the density

$$f_n(\xi \mid t_{(n-1)},\xi_{(n-1)},t)$$

of the n-th spatial point given the history $(t_{(n-1)},\xi_{(n-1)})$ and the n-th time point t.

References

Baddeley, A. J., Møller, J., and Waagepetersen, R. P. (2000). Non- and semiparametric estimation of interaction in inhomogeneous point patterns. *Statistica Neerlandica*, 54:329–350.

Barndorff-Nielsen, O. E. and Schmiegel, J. (2004). Lévy based spatial-temporal modelling, with applications to turbulence. *Russian Math. Surveys*, 59(1):65–95.

Barndorff-Nielsen, O. E., Schmiegel, J., Eggers, H. C., and Greiner, M. (2003). A class of spatio-temporal and causal stochastic processes, with application to multiscaling and multifractality. Technical report, MaPhySto Research Report no. 33, University of Aarhus.

Bramson, M. and Griffeath, D. (1980). The asymptotic behavior of a probabilistic model for tumour growth. In Jager, W., Rost, H., and Tautu, P., editors, *Biological Growth and Spread*, number 38 in Springer Lecture Notes in Biomathematics, pages 165–172.

Bramson, M. and Griffeath, D. (1981). On the Williams–Bjerknes tumour growth model, I. *Ann. Prob.*, 9:173–185.

Brix, A. (1998). *Spatial and Spatio-temporal Models for Weed Abundance*. PhD thesis, Royal Veterinary and Agricultural University, Copenhagen.

Brix, A. (1999). Generalized gamma measures and shot-noise Cox processes. *Adv. Appl. Prob.*, 31:929–953.

Brix, A. and Chadoeuf, J. (2002). Spatio-temporal modelling of weeds by shot-noise G Cox processes. *Biom. J.*, 44:83–99.

Brix, A. and Diggle, P. J. (2001). Spatiotemporal prediction for log-Gaussian Cox processes. *J. R. Statist. Soc. B.*, 63:823–841.

Brix, A. and Møller, J. (2001). Space-time multitype log Gaussian Cox processes with a view to modelling weeds. *Scand. J. Statist.*, 28:471–488.

Brú, A., Pastor, J. M., Fernaud, I., Brú, I., Melle, S., and Berenguer, C. (1998). Super-rough dynamics on tumour growth. *Phys. Rev. Lett.*, 81:4008–4011.

Capasso, V., Agur, Z., Arino, O., Chaplain, M., Gyllenbarg, M., Heesterbeek, J., Kaufman, M., Krivan, V., Stevens, A., and Tracqui, P. (2002). Book of abstracts, 5th Conference of the European Society of Mathematical and Theoretical Biology. Milano.

Chaplain, M. A. J., Singh, G. D., and McLachlan, J. C. (1999). *On Growth and Form: Spatio-temporal Pattern Formation in Biology*. Wiley, Chichester.

Cox, D. R. (1955). Some statistical models connected with series of events. *J. R. Statist. Soc. B.*, 17:129–164.

Cressie, N. (1991a). Modelling growth with random sets. In Possolo, A. and Hayward, C. A., editors, *Spatial Statistics and Imaging*, pages 31–45. IMS Lecture Notes, Proceedings of the 1988 AMS-IMS-SIAM Joint Summer Research Conference.

Cressie, N. (1991b). *Statistics for Spatial Data*. Wiley, New York.

Cressie, N. and Hulting, F. L. (1992). A spatial statistical analysis of tumor growth. *J. Amer. Statist. Assoc.*, 87:272–283.

Cressie, N. and Laslett, G. M. (1987). Random set theory and problems of modelling. *SIAM Review*, 29:557–574.

Daley, D. J. and Vere-Jones, D. (2002). *An Introduction to the Theory of Point Processes. Volume I: Elementary Theory and Methods*. Springer, New York, second edition.

Deijfen, M. (2003). Asymptotic shape in a continuum growth model. *Adv. Appl. Prob. (SGSA)*, 35:303–318.

Diggle, P. J. (2003). *Statistical Analysis of Spatial Point Patterns*. Edward Arnold, London, second edition.

Durret, R. and Liggett, T. (1981). The shape of the limit set in Richardson's growth model. *Ann. Prob.*, 9:186–193.

Hahn, U., Jensen, E. B. V., van Lieshout, M. N. M., and Nielsen, L. S. (2003). Inhomogeneous point processes by location dependent scaling. *Adv. Appl. Prob. (SGSA)*, 35(2):319–336.

Hawkes, A. G. (1971). Point spectra of some mutually exciting point processes. *J. R. Statist. Soc. B*, 33:438–443.

Hobolth, A., Pedersen, J., and Jensen, E. B. V. (2003). A continuous parametric shape model. *Ann. Inst. Statist. Math.*, 55:227–242.

Jensen, E. B. V. and Nielsen, L. S. (2000). Inhomogeneous Markov point processes by transformation. *Bernoulli*, 6:761–782.

Jensen, E. B. V. and Nielsen, L. S. (2004). Statistical inference for transformation inhomogeneous point processes. *Scand. J. Statist.*, 31:131–142.

Jónsdóttir, K. Ý. and Jensen, E. B. V. (2005). Gaussian radial growth. *Image Anal. Stereol.*, 24:117–126.

Kallenberg, O. (1989). *Random Measures*. Akademie Verlag, Berlin, 4 edition.

Kwapien, S. and Woyczynski, W. A. (1992). *Random Series and Stochastic Integrals: Single and Multiple*. Birkhäuser, Basel.

Molchanov, I. S. (2005). *Theory of Random Sets*. Springer, London.

Møller, J. (2003). Shot noise Cox processes. *Adv. Appl. Prob. (SGSA)*, 35:614–640.

Møller, J., Syversveen, A. R., and Waagepetersen, R. P. (1998). Log Gaussian Cox processes. *Scand. J. Statist.*, 25:451–482.

Møller, J. and Torrisi, G. L. (2005). Generalised shot noise Cox processes. *Adv. Appl. Prob. (SGSA)*, 37:48–74.

Møller, J. and Waagepetersen, R. P. (2003). *Statistical Inference and Simulation for Spatial Point Processes*. Chapman and Hall/CRC, Boca Raton.

Ogata, Y. and Tanemura, M. (1986). Likelihood estimation of interaction potentials and external fields of inhomogeneous spatial point patterns. In Francis, I., Manly, B., , and Lam, F., editors, *Proc. Pacific Statistical Congress*, pages 150–154. Elsevier.

Prokešová, M., Hahn, U., and Jensen, E. B. V. (2005). Statistics for locally scaled point processes. In Baddeley, A., Gregori, P., Mateu, J., Stoica, R., and Stoyan, D., editors, *Case Studies in Spatial Point Process Modelling*, Springer Lecture Notes in Statistics 185, Pages 99–123, Springer, New York.

Richardson, D. (1973). Random growth in a tessellation. *Proc. Camb. Phil. Soc.*

Schoenberg, F. P., Brillinger, D. R., and Guttorp, P. (2002). Point processes, spatial-temporal. In El-Shaarawi, A. and W.W., P., editors, *Encyclopedia of Environmetrics,*, pages 1573–1577. John Wiley & Sons, Ltd, Chichester.

Schurger, K. (1979). On the asymptotic geometrical behavior of a class of contact interaction processes with a monotone infection rate. *Zeitschrift für Wahrscheinlichkeitstheorie und verwandte Gebiete*, 48:35–48.

Stoyan, D., Kendall, S. W., and Mecke, J. (1995). *Stochastic Geometry and its Applications*. John Wiley & Sons, Chichester.

Stoyan, D. and Stoyan, H. (1998). Non homogeneous Gibbs process models for forestry – a case study. *Biometrical J.*, 40:521–531.

Van Lieshout, M. N. M. (2000). *Markov Point Processes and their Applications*. Imperial College Press, London.

CHAPTER 3

Using Transforms to Analyze Space-Time Processes

Montserrat Fuentes, Peter Guttorp, and Paul D. Sampson

Contents

3.1 Introduction

Transform tools, such as power spectra, wavelets, and empirical orthogonal functions, are useful tools for analyzing temporal, spatial, and space-time processes. In this chapter we develop some theory and illustrate its use in a variety of applications in ecology and air quality.

Harmonic (or frequency) analysis has long been one of the main tools in time series analysis. Application of Fourier techniques works best when the underlying process is stationary, and we develop and illustrate it here for stationary spatial and space-time processes. However, spatial stationarity is rather a severe assumption for air quality models, and we show how the theory can be generalized beyond that assumption. In some circumstances we can develop formal tests for stationarity.

Wavelets are another set of transform tools that recently have found much use in time series analysis. We illustrate here how wavelets can be used to estimate temporal trends, and to develop different models of nonstationary spatial processes.

In geophysics and meteorology, variants of principal components called empirical orthogonal functions (EOFs) have long been used to describe leading modes of variability in space-time processes. Here we use smoothed EOFs to model the spatio-temporal mean field of a random field, while yet a third type of spatially nonstationary model is used to describe the random part of the field.

Finally, we return to the frequency analysis of space-time processes, and describe a spectral representation and a parametric class of space-time covariances. We also show how to test for separability of a space-time covariance, another frequently made but rather severe assumption in the types of applications we discuss.

3.2 Spectral analysis of spatial processes

Spectral methods are a powerful tool for studying the spatial structure of random fields and generally offer significant computational benefits. This section offers a review of the Fourier transform and its properties, introduces the spectral representation of a stationary spatial process, and also presents Bochner's theorem to obtain a spectral representation for the covariance. We later describe some commonly used classes of spectral densities, and we introduce the periodogram, a nonparametric estimate of the spectrum, and we study its properties. We also present an approximation to the Gaussian likelihood using spectral methods.

3.2.1 Fourier analysis

Here we present a review of the Fourier transform and its properties and also discuss the aliasing phenomenon in the spectral domain. This aliasing effect is a result of the loss of information when we take a discrete set of observations on a continuous process.

3.2.1.1 Continuous Fourier transform

A Fourier analysis of a spatial process, also called a harmonic analysis, is a decomposition of the process into sinusoidal components (sine and cosine waves). The coefficients of these sinusoidal components are the Fourier transform of the process.

Suppose that g is a real or complex-valued function that is integrable over $\mathbb{R}^d$. Define

$$G(\omega) = \int_{\mathbb{R}^d} g(\mathbf{s}) \exp\{i\omega^t \mathbf{s}\} d\mathbf{s}, \tag{3.1}$$

for $\omega \in \mathbb{R}^d$. The function G in (3.1) is said to be the Fourier transform of g. Then, g has the representation

$$g(\mathbf{s}) = \frac{1}{(2\pi)^d} \int_{\mathbb{R}^d} G(\omega) \exp\{-i\omega^t \mathbf{s}\} d\omega \tag{3.2}$$

so that $|G(\omega)|$ represents the amplitude associated with the complex exponential with frequency ω. The right-hand side of (3.2) is called the Fourier integral representation of g. The functions g and G are said to be a Fourier transform pair.

The Fourier transform of $g(\mathbf{s})$ is often also defined as

$$G(\omega) = \int_{\mathbb{R}^d} g(\mathbf{s}) \exp\{i2\pi\omega^t \mathbf{s}\} d\mathbf{s} \tag{3.3}$$

and the inverse Fourier transform of $G(\omega)$ as

$$g(\mathbf{s}) = \int_{\mathbb{R}^d} G(\omega) \exp\{-i2\pi\omega^t \mathbf{s}\} d\omega. \tag{3.4}$$

It is often useful to think of functions and their transforms as occupying two domains. These domains are referred to as the upper and the lower domains in older texts, "as if functions circulated at ground level and their transforms in the underworld" (Bracewell, 1999). They are also referred to as the function and transform domains, but in most physics applications they are called the time (or space) and frequency domains, respectively. Operations performed in one domain have corresponding operations in the other. For example, the convolution operation in the time (space) domain becomes a multiplication operation in the frequency domain, that is, $f * g \rightarrow F \cdot G$. The reverse is also true, as we will see in the next section of this chapter: $F * G \rightarrow f \cdot g$. Such theorems allow one to move between domains so that operations can be performed where they are easiest or most advantageous.

3.2.1.2 Properties of the Fourier transform

3.2.1.2.1 Scaling property

If $\mathcal{F}f = F$ is a real, nonzero constant function, where $\mathcal{F}$ denotes the Fourier transform, then

$$
\begin{aligned}
\mathcal{F}\{f(ax)\} &= \int_{-\infty}^{\infty} f(ax) \exp(i\pi 2\omega x) dx \\
&= \frac{1}{|a|} \int_{-\infty}^{\infty} f(\beta) \exp(i\pi\beta 2\omega/a) d\beta \\
&= \frac{1}{|a|} F(\omega/a).
\end{aligned}
\tag{3.5}
$$

From this, the scaling property, it is evident that if the width of a function is decreased while its height is kept constant, then its Fourier transform becomes wider and shorter. If its width is increased, its transform becomes narrower and taller.

A similar frequency scaling property is given by

$$\mathcal{F}\left\{\frac{1}{|a|}f(x/a)\right\} = F(a\omega).$$

3.2.1.2.2 Shifting property

If $\mathcal{F}f = F$ and x_0 is a real constant, then

$$
\begin{aligned}
\mathcal{F}\{f(x - x_0)\} &= \int_{-\infty}^{\infty} f(x - x_0)\exp(i2\pi\omega x)dx \\
&= \int_{-\infty}^{\infty} f(\beta)\exp(i2\pi\omega(\beta + x_0))d\beta \\
&= \exp(i\pi 2x_0\omega)\int_{-\infty}^{\infty} f(\beta)exp(i\pi 2\omega\beta)d\beta \\
&= F(\omega)\exp(i2\pi x_0\omega).
\end{aligned}
\tag{3.6}
$$

This shifting property states that the Fourier transform of a shifted function is just the transform of the unshifted function multiplied by an exponential factor having a linear phase.

Likewise, the frequency shifting property states that if $F(\omega)$ is shifted by a constant ω_0, its inverse transform is multiplied by $\exp(-i\pi 2x\omega_0)$

$$\mathcal{F}\{f(x)\exp(-i2\pi x\omega_0)\} = F(\omega - \omega_0).$$

3.2.1.2.3 Convolution theorem

We now derive the previously mentioned convolution theorem. Suppose that $g = f * h$. Then, given that $\mathcal{F}g = G$, $\mathcal{F}f = F$, and $\mathcal{F}h = H$,

$$
\begin{aligned}
G(\omega) &= \mathcal{F}\{f * h(x)\} \\
&= \mathcal{F}\left\{\int_{-\infty}^{\infty} f(\beta)h(x - \beta)d\beta\right\} \\
&= \int_{-\infty}^{\infty}\left[\int_{-\infty}^{\infty} f(\beta)h(x - \beta)d\beta\right]\exp(i\pi 2\omega x)dx \\
&= \int_{-\infty}^{\infty} f(\beta)\left[\int_{-\infty}^{\infty} h(x - \beta)\exp(i\pi 2\omega x)dx\right]d\beta \\
&= H(\omega)\int_{-\infty}^{\infty} f(\beta)\exp(i\pi\beta 2\omega)d\beta \\
&= F(\omega)H(\omega).
\end{aligned}
\tag{3.7}
$$

This extremely powerful result demonstrates that the Fourier transform of a convolution is simply given by the product of the individual transforms; that is, the Fourier transform of $f * h$ is FH. Using a similar derivation, it can be shown that the Fourier transform of a product is given by the convolution of the individual transforms; that is, the Fourier transform of fh is $F * H$.

3.2.1.2.4 Parseval's theorem

Parseval's theorem states that the power of a signal represented by a function $h(x)$ is the same whether computed in signal space or frequency (transform) space; that is,

$$\int_{-\infty}^{\infty} h^2(x)dx = \int_{-\infty}^{\infty} |H(\omega)|^2 d\omega$$

(see Bracewell, 1999). The power spectrum, $P(\omega)$, is given by

$$P(\omega) = |H(\omega)|^2,$$

for $-\infty \le \omega \le +\infty$.

3.2.1.3 Aliasing

If we decompose a continuous process Z into a discrete superposition of harmonic oscillations, it is easy to see that such a decomposition cannot be uniquely restored from observations of Z in $\Delta \mathbb{Z}^2$ (this an infinite lattice with spacing Δ), where Δ is the distance between neighboring observations, and $\mathbb{Z}^2$ the integer lattice. The equal spacing in the space domain of the observations introduces an *aliasing* effect for the frequencies. Indeed,

$$\exp\{i\omega \mathbf{z}_1 \Delta\} = \exp\{i(\omega + \mathbf{z}_2 2\pi/\Delta)\mathbf{z}_1 \Delta\} = \exp\{i\omega \mathbf{z}_1 \Delta\} \exp\{i2\pi \mathbf{z}_2 \mathbf{z}_1\}$$

for any $\mathbf{z}_1$ and $\mathbf{z}_2$ in $\mathbb{Z}^2$. We simply cannot distinguish an oscillation with an angular frequency ω from all the oscillations with frequencies $\omega + 2\pi \mathbf{z}_2/\Delta$. The frequencies ω and $\omega' = \omega + 2\pi \mathbf{z}_2/\Delta$ are indistinguishable and hence are aliases of each other. The impossibility of distinguishing the harmonic components with frequencies differing by an integer multiple of $2\pi/\Delta$ by observations in the integer lattice with spacing Δ is called the *aliasing* effect.

Then, if observation of a continuous process Z is carried out only at uniformly spaced spatial locations Δ units apart, the spectrum of observations of the sample sequence $Z(\Delta \mathbf{z}_i)$, for $\mathbf{z}_i \in \mathbb{Z}^2$, is concentrated within the finite frequency band $-\pi/\Delta \le \omega < \pi/\Delta$. Every frequency not in that interval band has an alias in the band, termed its *principal alias*. The whole frequency spectrum is partitioned into bands of length $2\pi/\Delta$ by *fold points* $(2\mathbf{z}_i + 1)\pi/\Delta$, with $\mathbf{z}_i \in \mathbb{Z}^2$, and the power distribution within each of the bands distinct from the principal band $-\pi/\Delta \le \omega < \pi/\Delta$, is superimposed on the power distribution within the principal band. Thus, if we wish that the spectral characteristics of the process Z to be determined accurately enough from the observed sample, then the *Nyquist frequency* π/Δ must necessarily be so high that still higher frequencies ω make only a negligible contribution to the total

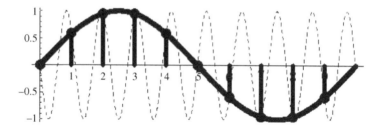

Figure 3.1 Example of aliasing due to undersampling in space.

power of the process. This means that we observe a dense sample of Z (small Δ). The Nyquist frequency is also called the *folding frequency*, because higher frequencies are effectively folded down into the band $-\pi/\Delta \leq \omega < \pi/\Delta$.

It should be noted that aliasing is a relatively simple phenomenon. In general, when one takes a discrete set of observations on a continuous function, information is lost. It is an advantage of the trigonometric functions that this loss of information is manifest in the easily understood form of aliasing. Figure 3.1 shows an example of aliasing. In the figure, the high-frequency sinusoid is indistinguishable from the lower frequency sinusoid due to aliasing. We say the higher frequency aliases to the lower frequency. Undersampling in the frequency domain gives rise to space-domain aliasing.

3.2.2 Spectral representation

In this section we introduce the spectral representation of a stationary spatial process using sine and cosine waves. We also present Bochner's theorem to obtain a spectral representation for the covariance.

3.2.2.1 The spectral representation theorem

Consider a weakly stationary process Z with mean 0 and covariance C. Before we can apply the ideas of Fourier series and Fourier integrals we must first ask: Can we represent a typical realization as a Fourier series? The answer to this question is clearly "No" because we have no reason to suppose that a realization of a general stationary process will be periodic in any way.

The next question is: Can we represent a typical realization as a Fourier integral? No. Maybe we cannot distribute the power over a continuous range of frequencies, but over a set of frequencies with discontinuities. This will lead to a Fourier type integral called Fourier-Stieltjes:

$$Z(\mathbf{s}) = \int e^{i\mathbf{s}^T \omega} dY(\omega),$$

for $\mathbf{s} \in \mathbb{R}^2$ and $\omega \in \mathbb{R}^2$, and where Y measures the average contributions from all components with frequencies less than or equal to ω.

3.2.2.1.1 The spectral representation theorem

To every mean zero stationary $Z(\mathbf{s})$ there can be assigned a process $Y(\omega)$ with orthogonal increments, such that we have for each fixed $\mathbf{s}$ the following stochastic integral that gives the spectral representation (e.g., Yaglom, 1987):

$$Z(\mathbf{s}) = \int_{\mathbb{R}^2} e^{i\mathbf{s}^T\omega} dY(\omega). \tag{3.8}$$

$Y(\omega)$ is defined up to an additive random variable. The Y process is called the spectral process associated with a stationary process Z. The random spectral process Y has the following properties:

$$E(Y(\omega)) = 0$$

(since mean of Z is 0), the process Y has orthogonal increments:

$$E[(Y(\omega_3) - Y(\omega_2))(Y(\omega_1) - Y(\omega_0))] = 0,$$

when (ω_3, ω_2) and (ω_1, ω_0) are disjoint intervals. If we define F as

$$E[|dY(\omega)|^2] = dF(\omega),$$

where $|dF(\omega)| < \infty$ for all ω. F is a positive measure, and

$$E[dY(\omega)dY(\omega')] = \delta(\omega - \omega')dF(\omega)d\omega'$$

where $\delta(\omega)$ is the Dirac δ-function.

The spectral representation theorem can be proved by various methods using Hilbert space theory or by means of trigonometric integrals. A good reference is Cramér and Leadbetter (1967).

3.2.2.1.2 Bochner's theorem

We derive the spectral representation of the autocovariance function C:

$$C(\mathbf{s}) = \int_{\mathbb{R}^2} e^{i\mathbf{s}^T\omega} dF(\omega).$$

Bochner's theorem says that a continuous function C is nonnegative definite if and only if it can be represented in the form above where F is real, never-decreasing, and bounded. Thus, the spatial structure of Z could be analyzed with a spectral approach or equivalently by estimating the autocovariance function; see Cramér and Leadbetter (1967) for more detail.

If we compare the spectral representation of $C(\mathbf{s})$ and $Z(\mathbf{s})$, e.g., Loéve, 1955,

$$C(\mathbf{s}) = \int_{\mathbb{R}^2} e^{i\mathbf{s}^T\omega} dF(\omega),$$

$$Z(\mathbf{s}) = \int_{\mathbb{R}^2} e^{i\mathbf{s}^T\omega} dY(\omega),$$

it will be seen that the elementary harmonic oscillations are, respectively, $e^{is^T\omega}dF(\omega)$ and $e^{is^T\omega}dY(\omega)$.

If we think of $Y(\omega)$ as representing the spatial development of some concrete physical systems, the spectral representation gives the decomposition of the total fluctuation in its elementary harmonic components. The spectral d.f. $F(\omega)$ determines the distribution of the total average power in the $Z(\mathbf{s})$ fluctuation over the range of angular frequency ω. The average power assigned to the frequency interval $A = [\omega_1, \omega_2]^2$ is $F(A)$, which for the whole infinite ω range becomes

$$E|Z(\mathbf{s})|^2 = C(0) = F(\mathbb{R}^2).$$

Thus, F determines the power spectrum of the Z process. We may think of this as a distribution of a spectral mass of total amount $C(0)$ over the ω axis. F only differs by a multiplicative constant from an ordinary d.f.

If F has a density with respect to Lebesgue measure, this density is the spectral density, $f = F'$, defined as the Fourier transform of the autocovariance function:

$$f(\omega) = \frac{1}{(2\pi)^2} \int_{\mathbb{R}^2} \exp(-i\omega^T \mathbf{x})C(\mathbf{x})d\mathbf{x}.$$

3.2.2.1.3 Spectral moments

The spectral moments

$$\lambda_k = \int \omega^k dF(\omega)$$

may or may not be finite. As a consequence of Bochner's theorem, the moment λ_{2k} is finite if and only if $C(x)$ has all partial derivatives of order $2k$ at the origin.

A process Z is m-times mean square differentiable if and only if $C^{(2m)}(0)$ exists and is finite and, if so, the autocovariance function of $Z^{(m)}$ is $(-1)^m C^{(2m)}$ (e.g., Stein, 1999, p. 21). Therefore, Z is m-times mean square differentiable if and only if the moment λ_{2m} is finite.

3.2.3 Some spectral densities

We describe in this subsection some commonly used classes of spectral densities. We consider a real process, so the spectral density is an even function. We also assume that the covariance is isotropic, so that the spectral density is a function of a single frequency (Matérn, 1960).

3.2.3.1 Triangular model

For a spatial process with a triangular isotropic covariance,

$$C(\mathbf{x}) = \sigma(a - |\mathbf{x}|)^+$$

for σ and a positive, where $(a)^+ = a$ if $a > 0$, otherwise $(a)^+ = 0$; and $|\mathbf{x}|$ denotes the Euclidean norm. The corresponding spectral density is

$$f(\omega) = \sigma\pi^{-1}\{1 - \cos(\alpha|\omega|)\}/|\omega|^2,$$

for $|\omega| > 0$, and $f(0) = \sigma\pi^{-1}\alpha^2/2$. The oscillating behavior of the spectral density would be probably quite unrealistic for many physical processes. There is usually no reason to assume the spectrum has much more mass near the frequency $(2n + 1)\pi$ than near $2n\pi$ for n large, which is the case for the spectral density $\{1 - \cos(\{\omega\})/|\omega|^2$. Some kriging predictors under this model have strange properties as a consequence of the oscillations of the spectral density at high frequencies.

3.2.3.2 Spherical model

One of the most commonly used models for isotropic covariance functions in geological and hydrological applications is the spherical

$$C(x) = \begin{cases} \sigma\left(1 - \frac{3}{2\rho}|x| + \frac{1}{2\rho^3}|x|^3\right) & |x| \le \rho \\ 0 & |x| > \rho \end{cases} \tag{3.9}$$

for positive constants σ and ρ. This function is not a valid covariance in higher dimensions than 3. The parameter ρ is called the range and is the distance at which correlations become exactly 0. This function is only once differentiable at $|x| = \rho$ and this can lead to problems when using likelihood methods for estimating the parameters of this model. In three dimensions, the corresponding isotropic spectral density has oscillations at high frequencies similar to the triangular covariance function in one dimension. Stein and Handcock (1989) show that when using the spherical model in three dimensions, certain prediction problems have rather pathological behavior.

3.2.3.3 Squared exponential model

The density of a spatial process with an isotropic squared exponential covariance

$$C(x) = \sigma e^{-\alpha|x|^2}$$

is

$$f(\omega) = \frac{1}{2}\sigma(\pi\alpha)^{-1/2}e^{-\omega^2/(4\alpha)}.$$

Note that both C and f are the same type of exponential functions when $\gamma = 2$. The parameter σ is the variance of the process and α^{-1} is a parameter that explains how fast the correlation decays.

3.2.3.4 Matérn-Whittle class

A class of practical variograms and autocovariance functions for a process Z can be obtained from the Matérn class of spectral densities

$$f(\omega) = \phi(\alpha^2 + |\omega|^2)^{\left(-\nu-\frac{d}{2}\right)} \tag{3.10}$$

with parameters $\nu > 0$, $\alpha > 0$, and $\phi > 0$ (the value d is the dimension of the spatial process Z). Here, the vector of covariance parameters is $\theta = (\phi, \nu, \alpha)$. The parameter α^{-1} can be interpreted as the autocorrelation range.

The parameter ν measures the degree of smoothness of the process Z, the higher the value of ν the smoother Z would be, in the sense that the degree of differentiability of Z would increase. The paramter ϕ is proportional to the ratio of the variance σ and the range (α^{-1}) to the $2\nu^{\text{th}}$ power, $\phi = \sigma\alpha^{2\nu}$.

The corresponding covariance for the Matérn class is

$$C_\theta(x) = \frac{\pi^{d/2}\phi}{2^{\nu-1}\Gamma(\nu+d/2)\alpha^{2\nu}}(\alpha|x|)^\nu \mathcal{K}_\nu(\alpha|x|), \qquad (3.11)$$

where $\mathcal{K}_\nu$ is a modified Bessel function of the third kind,

$$\mathcal{K}_\nu(x) = \frac{\pi}{2}\left(\frac{I_{-\nu}(x) - I_\nu(x)}{\sin(\pi\nu)}\right),$$

with I_ν the modified Bessel function of the first kind, that can be defined by a contour integral

$$I_\nu(x) = \oint e^{[x/2][t+1/t]}t^{-\nu-1}dt,$$

where the contour encloses the origin and is traversed in a counterclockwise direction (Arfken and Weber, 1995, p. 416). In the Matérn class, when $\nu = \frac{1}{2}$, we get the exponential covariance function

$$C_\theta(\mathbf{x}) = \pi\phi\alpha^{-1}\exp(-\alpha|x|).$$

When ν is of the form $m + \frac{1}{2}$ with m a nonnegative integer, the Matérn covariance function is of the form $e^{-\alpha|x|}$ times a polynomial in $|x|$ of degree m (Abramowitz and Stegun, 1965; Stein, 1999, p. 31).

Handcock and Wallis (1994) suggested the following parametrization of the Matérn covariance that does not depend on d:

$$C(x) = \frac{\sigma}{2^{\nu-1}\Gamma(\nu)\alpha^{2\nu}}(2\nu^{1/2}|x|/\rho)^\nu \mathcal{K}_\nu(2\nu^{1/2}|x|/\rho), \qquad (3.12)$$

but the corresponding spectral density does depend on d:

$$f(\omega) = \frac{\sigma g(\nu,\rho)}{(4\nu/\rho^2 + |\omega|^2)^{\nu+d/2}}$$

where

$$g(\nu,\rho) = \frac{\Gamma(\nu+d/2)(4\nu)^\nu}{\pi^{d/2}\rho^{2\nu}\Gamma(\nu)}$$

with $\sigma = \text{var}(Z(\mathbf{s}))$, the parameter ρ measures how the correlation decays with distance, and generally this parameter is called the *range*. The parameter α^{-1} has a very similar interpretation to ρ. But both parameters have different asymptotic properties under an infill asymptotic model. If we consider the limit as $\nu \to \infty$ we get the squared exponential covariance

$$K(x) = \sigma e^{-|x|^2/\rho^2}.$$

The smoothness of a random field, the parameter ν in the Matérn class, plays a critical role in interpolation problems. This parameter is difficult to estimate accurately from data. A number of the commonly used models for the covariance structure, including spherical, exponential, and squared exponential structures assume that the smoothness parameter is known *a priori*.

As an alternative to the Matérn covariance, sometimes the powered exponential model could be used:

$$C(x) = \sigma e^{-\alpha |x|^{\gamma}}$$

with $\alpha > 0$ and $\gamma \in (0, 2]$. The parameter γ (when $\gamma < 2$) plays the same role as 2ν in the Matérn; and for $\gamma = 2$, it corresponds to $\nu = \infty$. However, for values of $1 \leq \nu < \infty$, the powered exponential has no elements providing similar local behavior as the Matérn.

Figure 3.2 shows two Matérn covariances: a squared exponential covariance and an exponential covariance. The squared exponential is more flat at the origin; this indicates that the spatial process is very smooth. On the other hand, the exponential is almost linear at the origin, indicating that the corresponding spatial process is not very smooth; in fact, this process is not even once mean square differentiable.

Figure 3.3 shows two other Matérn covariances with $\nu = 1/2$ (exponential) and with $\nu = 3/2$, which corresponds to a process that is once differentiable. Figure 3.4 shows the corresponding spectral densities.

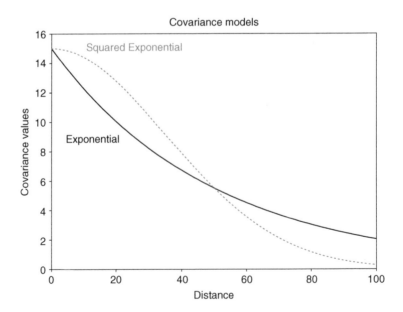

Figure 3.2 Covariance models: exponential and squared exponential.

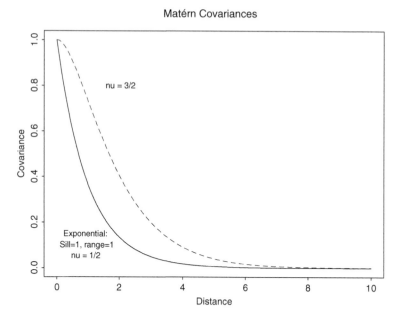

Figure 3.3 Covariances: Matérn class for $\nu = 1/2$ (exponential covariance) and $\nu = 3/2$.

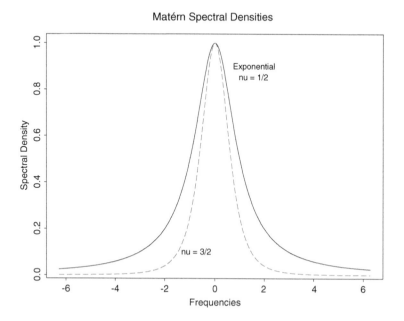

Figure 3.4 Spectral densities: Matérn class for $\nu = 1/2$ (exponential covariance) and $\nu = 3/2$.

3.2.4 Estimating the spectral density

The periodogram, a nonparametric estimate of the spectral density, is a powerful tool for studying the properties of stationary processes observed on a d-dimensional lattice. Use and properties of spatial periodograms for stationary processes have been investigated by Stein (1995, 1999), Guyon (1982,1992), Ripley (1981), Rosenblatt (1985), and Whittle (1954) among others. Pawitan and O'Sullivan (1994) proposed a nonparametric spectral density estimator using a penalized Whittle likelihood for a stationary time series. Guyon (1982) studied the asymptotic properties of various parameter estimation procedures for a general stationary process on a d-dimensional lattice, using spectral methods. Fuentes (2005b) uses spatial periodograms to test for the lack of separability of a spatial-temporal covariance function. Fuentes (2005a) also uses spatial periodograms to test for the lack of stationarity of a spatial covariance function.

This subsection is organized as follows. First, we introduce the periodogram, a nonparametric estimate of the spectral density. Then, using spectral tools we present an expression for the likelihood function that in practice is very easy to calculate. This version of the likelihood was proposed by Whittle (1954) and avoids computing the determinants and inverses of large matrices.

3.2.4.1 Periodogram

Consider a spatial stationary process Z with covariance parameter θ, which is assumed here to be known. We observe the process at N equally spaced locations in a regular grid D $(n_1 \times n_2)$, where $N = n_1 n_2$. The distance between neighboring observations is Δ. The periodogram is a nonparametric estimate of the spectral density, which is the Fourier transform of the covariance function. We define $I_N(\omega_0)$ to be the periodogram at a frequency ω_0,

$$I_N(\omega_0) = \Delta^2 (2\pi)^{-2}(n_1 n_2)^{-1} \left| \sum_{s_1=1}^{n_1} \sum_{s_2=1}^{n_2} Z(\Delta \mathbf{s}) \exp\{-i\Delta \mathbf{s}^T \omega\} \right|^2. \qquad (3.13)$$

If the spectral representation of Z is

$$Z(\mathbf{x}) = \int_{\mathbb{R}^2} e^{i\omega^T \mathbf{x}} dY(\omega),$$

we define $J(\omega)$, a discrete version of the spectral process $Y(\omega)$, which is the Fourier transform of Z (see, e.g., Priestley, 1981),

$$J(\omega) = \Delta(2\pi)^{-1}(n_1 n_2)^{-1/2} \sum_{x_1=1}^{n_1} \sum_{x_2=1}^{n_2} Z(\Delta \mathbf{x}) \exp\{-i\Delta \mathbf{x}^T \omega\}.$$

Using the spectral representation of Z and proceeding formally,

$$C(\mathbf{x}) = \int_{\mathbb{R}^2} \exp(i\omega^T \mathbf{x}) F(d\omega) \qquad (3.14)$$

where the function F is called the spectral measure or spectrum for Z. F is a positive finite measure, defined by

$$E\{d|Y(\omega)|^2\} = dF(\omega).\tag{3.15}$$

Thus, we get

$$I_N(\omega) = |J(\omega)|^2.\tag{3.16}$$

This expression for I_N is consistent with the definition of the spectral distribution F in (3.15), as a function of the spectral process Y. The periodogram (3.16) is simply the discrete Fourier transform of the sample covariance.

In practice, the periodogram estimate for ω is computed in the set of Fourier frequencies $(2\pi/\Delta)(\mathbf{f}/\mathbf{n})$, where $\mathbf{f}/\mathbf{n} = \left(\frac{f_1}{n_1}, \frac{f_2}{n_2}\right)$ and $\mathbf{f} \in J_N$, for

$$J_N = \{\lfloor-(n_1-1)/2\rfloor, \ldots, n_1 - \lfloor n_1/2\rfloor\} \times \{\lfloor-(n_2-1)/2\rfloor, \ldots, n_2 - \lfloor n_2/2\rfloor\},\tag{3.17}$$

where $\lfloor u\rfloor$ denotes the largest integer less than or equal to u.

3.2.4.2 Theoretical properties of the periodogram

The expected value of the periodogram at ω_0 is given by

$$E(I_N(\omega_0)) = (2\pi)^{-2}(n_1 n_2)^{-1} \int_{\Pi_\Delta^2} f_\Delta(\omega)W(\omega - \omega_0)d\omega,$$

where

$$\Pi_\Delta^2 = (-\pi/\Delta, \pi/\Delta)^2,$$

and

$$W(\omega) = \prod_{j=1}^2 \frac{\sin^2\left(\frac{n_j\omega_j}{2}\right)}{\sin^2\left(\frac{\omega_j}{2}\right)}$$

for $\omega = (\omega_1, \omega_2) = (2\pi/Delta)(\mathbf{f}/\mathbf{n})$ and $\mathbf{f} \in J_N\setminus\{0\}$, and $f_\Delta(\omega)$ is the spectral density of the process Z on the lattice with spacing Δ. The side lobes (subsidiary peaks) of the function W can lead to substantial bias in $I_N(\omega_0)$ as an estimator of $f_\Delta(\omega_0)$ because they allow the value of f_Δ at frequencies far from ω_0 to contribute to the expected value. Figure 3.5 shows a graph of W along the vertical axis. If the side lobes of W were substantially smaller, we could reduce this source of bias for the periodogram considerably. Tapering is a technique that effectively reduces the side lobes associated with the spectral window W. We form the product $h(\mathbf{x})Z_i(\mathbf{x})$ for each value of $\mathbf{x} = (x_1, x_2)$, where $\{h(\mathbf{x})\}$ is a suitable sequence of real-valued constants called a *data taper*, and then we compute the periodogram for the tapered data.

The periodogram values are approximately independent, and this facilitates the use of techniques such as nonlinear least squares (NLS) to fit a theoretical spectral model to the periodogram values.

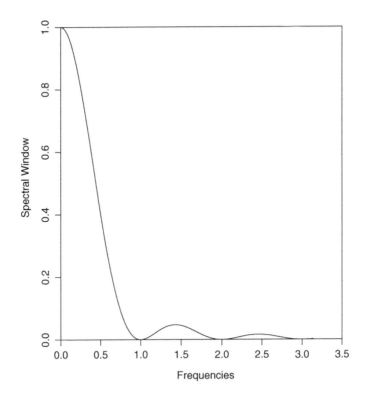

Figure 3.5 Spectral window along the vertical axis. The horizontal axis shows the frequencies, while the vertical axis shows the spectral window along the vertical axis for the periodogram (without tapering), for $n_2 = 500$.

3.2.4.2.1 Asymptotic properties of the periodogram

Theorem 1 (Brillinger, 1981): Consider a Gaussian stationary process Z with spectral density $f(\omega)$ on a lattice D. We assume Z is observed at N equally spaced locations in D $(n_1 \times n_2)$, where $N = n_1 n_2$, and the spacing between observations is Δ. We define the periodogram function, $I_N(\omega)$, as in (3.34).

Assume $n_1 \to \infty$, $n_2 \to \infty$, $n_1/n_2 \to \lambda$, for a constant $\lambda > 0$. Then, we get:

1. The expected value of the periodogram, $I_N(\omega)$, is asymptotically $f_\Delta(\omega)$.
2. The asymptotic variance of $I_N(\omega)$ is $f_\Delta^2(\omega)$.
3. The periodogram values $I_N(\omega)$, and $I_N(\omega')$ for $\omega \neq \omega'$, are asymptotically independent.

By Part (1) the periodogram I_N using *increasing domain asymptotics* is asymptotically an unbiased estimate of the spectral density f_Δ on the lattice. Note that if f is the continuous process Z and f_Δ the spectral density on the lattice, then using *increasing-domain* asymptotics, I_N is not asymptotically an unbiased estimate of f but of f_Δ, the spectral density of the sampled sequence $Z(\Delta\mathbf{x})$. By Theorem 1 part (2), the variance of the periodogram at ω is asymptotically $f_\Delta^2(\omega)$. The traditional approach to this inconsistency problem is to smooth the periodogram across frequencies.

By Theorem 1 part (3), the periodogram values are approximately independent. This property allow us easily to fit in the spectral domain a parametric model to the periodogram values. However, in the space domain, the correlation among the empirical covariance or variogram values thwarts the use of least squares.

3.2.4.2.2 Asymptotic distribution of the periodogram

If the process Z is stationary, such that the absolute value of the joint cumulants of order k are integrable (for all k), then the periodogram has asymptotically a distribution that is a multiple of χ_2^2. More specifically, the periodogram $I_N(\omega_j)$ where ω_j is a Fourier frequency has asymptotically a $f(\omega_j)\chi_2^2/2$ distribution.

3.2.4.3 Examples of spectral analysis of 2-d processes

3.2.4.3.1 A simulated example

To get a feel for how bivariate spectra look, we borrow an example from Renshaw and Ford (1983). The process is in essence a weighted average of cosines with random phase. Figure 3.6 shows the random field, which is generated by

$$Z(s) = \sum_{i=0}^{16} \sum_{j=-16}^{15} g_{ij} \cos\left(2\pi\left(\frac{is_1}{m} + \frac{js_2}{n}\right) + U_{ij}\right)$$

where

$$g_{ij} = \exp(-|i + 6 - j\tan(20°)|).$$

The resulting spectrum, Figure 3.7, shows a single ridge in the $20°$ direction, going from $(0, -6)$ to $(16, 0)$.

3.2.4.3.2 A forestry example

The main topic of the paper by Renshaw and Ford (1983) is the analysis of canopy heights in a managed forest of Scots pine at Thetford Forest, U.K. The forest was about 40 years old when the measurements were taken, and had about 1000 tress per hectare. The horizontal extent of individual tree crowns was measured along parallel transects at 1-m intervals, and the crown height, location, and perimeter crown height of each tree were determined. Intermediate crown heights were estimated by geometric projection. We focus on a 32×32 submatrix of canopy heights. The spectrum is given in Figure 3.8.

The spectrum shows a low-frequency ridge in the positive quadrant from $(1,0)$ to $(8,4)$, with large elements at $(4,1)$ and $(8,2)$, corresponding to an angle of $76°$. Separated from this ridge by a single row is a ridge of high-frequency

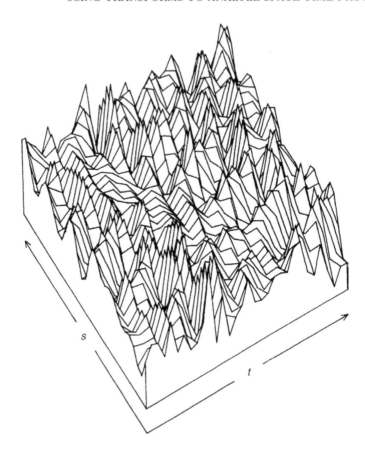

Figure 3.6 A simulated process with a strong directional component. (From Renshaw and Ford, 1983.)

elements with peak value at (12,4), or 72°. This strong directional feature was not obvious when observing the canopy from above. There is also a ridge in the negative quadrant from (4,−4) to (15,1). The high values are clustered around 90° and constitute a basic row effect.

When looking at the entire data set of 36 by 120 m, there is clear evidence of a changing spectral structure by looking at partially overlapping submatrices. The high-frequency components remain relatively stable, but the low-frequency features shift between dominating for negative values of ω_2 to positive values as one moves along the forest. This is probably related to a thinning that was less severe around a measurement mast used for microclimate measurements in the middle of the study area. Where there has been least thinning, the wave-like aggregation of tree crowns is strongest. Thus, for the entire data set, a model allowing for nonstationary spectral analysis is needed.

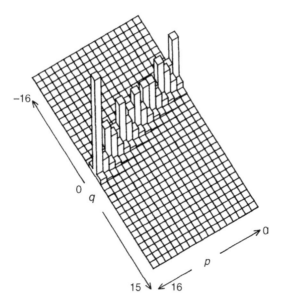

Figure 3.7 The periodogram for the process in Figure 3.6. (From Renshaw and Ford, 1983.)

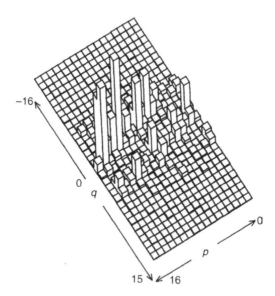

Figure 3.8 The spectrum of a 32×32 submatrix of the Thetford forest canopy heights. The notation (p, q) corresponds to our (ω_1, ω_2). (From Renshaw and Ford, 1983.)

3.2.4.4 Nonlinear WLS estimation in the spectral domain

Consider modeling the spatial structure of Z by fitting a spectral density f to the periodogram values. We could use a weighted nonlinear least squares (WNLS) procedure, which gives more weight to higher frequency values because high frequencies are important for interpolation. An approximate expression for the spectral density of the Matérn class for high-frequency values is obtained from (3.10) by letting $|\omega|$ go to ∞:

$$f(\omega) = \phi(|\omega|^2)^{\left(-\nu - \frac{d}{2}\right)}. \tag{3.18}$$

Thus, the degree of smoothness, ν, and ϕ are the critical parameters (and not the range α^{-1}).

Thus, we propose using the high-frequency parametric model for f as in (3.18) and the weights $f(\omega)^{-1}$ to give higher weight to higher frequencies. This is reasonable because for large N the approximate standard deviation of the periodogram I_N is $f(\omega)$.

The proposed weights $f(\omega)^{-1}$ also stabilize the variance of the periodogram values. This is similar to the weighted least squares method used in the space domain to fit a variogram model (Cressie, 1985). We recommend to use weighted least squares in the spectral domain rather than in the space domain, because periodogram values are approximately independent while variogram values are not.

An alternative approach would be to fit on the log scale a linear model

$$\log(f(\omega)) = \beta_0 + \beta_1 X, \tag{3.19}$$

where $X = \log(\omega)$, $\beta_{0i} = \log(\phi)$, and $\beta_1 = 2\left(-\nu - \frac{d}{2}\right)$.

3.2.4.5 Likelihood estimation in the spectral domain

For large data sets, calculating the determinants that we have in the likelihood function can be often infeasible. Spectral methods could be used to approximate the likelihood and obtain the maximum likelihood estimates (MLE) of the covariance parameters: $\theta = (\theta_1, \ldots, \theta_r)$.

Spectral methods to approximate the spatial likelihood have been used by Whittle (1954), Guyon (1982), Dahlhaus and Künsch, (1987), and Stein (1995, 1999), among others. These spectral methods are based on Whittle's (1954) approximation to the Gaussian negative log likelihood:

$$\frac{N}{(2\pi)^2} \int_{-\pi}^{\pi} \int_{-\pi}^{\pi} \log f(\omega) + I_N(\omega) f(\omega)^{-1} d\omega, \tag{3.20}$$

where the integral is approximated with a sum that is evaluated at the discrete Fourier frequencies, I_N is the periodogram, and f is the spectral density of the lattice process. The approximated likelihood can be calculated very efficiently using the fast Fourier transform. This approximation requires only $O(N log_2 N)$ operations. Simulation studies conducted by the authors seem to indicate that N needs to be at least 100 to get good estimated MLEs using Whittle's approximation.

The asymptotic covariance matrix of the MLEs of $\theta_1, \ldots, \theta_r$ is

$$\left\{ \frac{2}{N} \left[\frac{1}{4\pi^2} \int_{[-\pi,\pi]} \int_{[-\pi,\pi]} \frac{\partial \log f(\omega)}{\partial \theta_j} \frac{\partial \log f(\omega)}{\partial \theta_k} d\omega \right]^{-1} \right\}_{jk}. \tag{3.21}$$

This is much easier to compute than the inverse of the Fisher information matrix.

Guyon (1982) proved that when the periodogram is used to approximate the spectral density in the Whittle likelihood function, the periodogram bias contributes a nonnegligible component to the mean squared error (mse) of the parameter estimates for two-dimensional processes; and for three dimensions this bias dominates the mse. Thus, the MLEs of the covariance function based on the Whittle likelihood are only efficient in one dimension, but not in two and higher-dimensional problems. However, they are consistent. Guyon demonstrated that this problem can be solved using a different version of the periodogram, an "unbiased peridogram," which is the discrete Fourier transform of an unbiased version of the sample covariance. Dahlhaus and Künsch (1987) demonstrated that tapering also solves this problem.

3.2.5 Model for nonstationary processes

3.2.5.1 Convolution of locally stationary processes

In this section we propose a class of nonstationary covariances, based on a convolution of local stationary covariance functions (Fuentes, 2001, 2002; Fuentes and Smith, 2001). This model has the advantage that it is simultaneously defined everywhere, unlike moving window approaches (Haas, 1998), but it retains the attractive property that, locally in small regions, it behaves like a stationary spatial process.

We model a nonstationary covariance function $C(\mathbf{x_1}, \mathbf{x_2}; \theta)$ corresponding to a Gaussian spatial process Z observed on a region D, as a convolution of stationary covariance functions $C_{\theta(\mathbf{s})}(\mathbf{x_1} - \mathbf{x_2})$ corresponding to underlying stationary processes $Z_{\theta(\mathbf{s})}(x)$, $x \in D$; each process $Z_{\theta(\mathbf{s})}$ explains the *local* stationary spatial structure in a neighborhood of $\mathbf{s}$ and has covariance parameters $\theta(\mathbf{s})$ (Fuentes, 2001, 2002):

$$C(\mathbf{x_1}, \mathbf{x_2}; \theta) = \int_D K(\mathbf{x_1} - \mathbf{s}) K(\mathbf{x_2} - \mathbf{s}) C_{\theta(\mathbf{s})}(\mathbf{x_1} - \mathbf{x_2}) d\mathbf{s}, \tag{3.22}$$

where K is a kernel function. The function in (3.22) is a positive-definite function because it is an integral of valid covariance functions. The covariance function $C_{\theta(\mathbf{s})}$ is stationary with parameter $\theta(\mathbf{s})$, and we assume that $\theta(\mathbf{s})$ is a continuous function on $\mathbf{s}$. In the proposed nonstationary model (3.22), if K is a sharply peaked kernel function and $\theta(\mathbf{s})$ varies slowly with $\mathbf{s}$, this has the property that for $\mathbf{x}$ near $\mathbf{s}$, the covariance "looks like" a stationary covariance with parameter $\theta(\mathbf{s})$. On the other hand, because $\theta(\mathbf{s})$ can vary substantially over the whole space, it also allows significant nonstationarity. Another attraction of model (3.22) is that it allows all model parameters θ to vary over D.

The covariance $C_{\theta(\mathbf{s})}$ could be a Matérn isotropic covariance function of the form

$$C_{\theta(\mathbf{s})}(\mathbf{x}) = \frac{\pi^{d/2}\phi_s}{2^{\nu_s-1}\Gamma(\nu_s+d/2)\alpha_s^{2\nu_s}}(\alpha_s|x|)^{\nu_s}\mathcal{K}_{\nu_s}(\alpha_s|x|), \qquad (3.23)$$

where $\mathcal{K}_{\nu_s}$ is a modified Bessel function, and d is the dimension of $\mathbf{s}$, $\theta(\mathbf{s}) = (\nu_s, \alpha_s, \phi_s)$. The parameter α_s^{-1} can be interpreted as the autocorrelation range, ϕ_s is a scale parameter, and the parameter ν_s measures the degree of smoothness of the underlying process $Z_{\theta(\mathbf{s})}$.

In (3.22) every entry requires an integration. Because each such integration is actually an expectation with respect to a uniform distribution, we could use Monte Carlo integration or other numerical schemes to approximate the integral (3.22). We propose to draw an independent set of locations $\mathbf{s}_i$, $i = 1, 2, ..., k$, on D. Hence, we replace $C(\mathbf{x_1}, \mathbf{x_2}; \theta)$ with

$$\hat{C}(\mathbf{x_1}, \mathbf{x_2}; \theta) = k^{-1}\sum_{i=1}^{k} K(\mathbf{x_1}-\mathbf{s}_i)K(\mathbf{x_2}-\mathbf{s}_i)C_{\theta(\mathbf{s}_i)}(\mathbf{x_1}-\mathbf{x_2}). \qquad (3.24)$$

In this notation, the "hat" denotes an approximation that can be made arbitrarily accurate and has nothing to do with the data Z. The kernel function $K(x-\mathbf{s}_i)$ centered at $\mathbf{s}_i$ could be positive for all $x \in D$, or could have compact support. In the latter case, $K(x - \mathbf{s}_i)$ would be only positive when x is in a subregion S_i centerd at $\mathbf{s}_i$, and this would simplify the calculations.

The size of the sample, k, is selected using the following iterative algorithm. We first start with a systematic sample of size k, where k is small, and we increase k by adding a new sample point at a time. At each step of the iterative approach we draw a new sample point between two neighboring points in the current sample sequence. Thus, in each iteration we decrease by half the distance between two neighboring draws. We iterate this process until an Akaike information criterion (AIC; Akaike 1974) or Bayesian information criterion (BIC; Schwarz, 1978) suggests no significant improvement in the estimation of the nonstationary covariance of Z by increasing k, equivalent to decreasing the distance between draws in the sample sequence.

Throughout the remainder of this subsection we simplify the notation by writing Z_i to denote $Z_{\theta(\mathbf{s}_i)}$, the process with covariance $C_{\theta(\mathbf{s}_i)}$, and $K_i(x)$ to represent $K(x - \mathbf{s}_i)$, the kernel or weight function centerd at $\mathbf{s}_i$.

3.2.5.2 The spectrum for the convolution model

Consider $S_1, \ldots, S_k$, k subregions of stationarity on D. The nonstationary process Z is modeled here as a mixture of weakly stationary processes Z_j, $j = 1, \ldots, k$, with $\text{cov}\{Z_j(x), Z_i(\mathbf{y})\} = 0$ for $j \neq i$. Each underlying stationary process, Z_j explains the spatial structure of Z in a subregion of stationarity S_j. This is a discrete version of the nonstationary model previously presented. We have

$$Z(x) = \sum_{j=1}^{k} Z_j(x)K_j(x),$$ (3.25)

and we choose k using the BIC (or AIC) approach discussed in the previous subsection.

Each stationary process Z_j has the representation

$$Z_j(x) = \int_{\mathbb{R}^2} \exp(ix^T\omega)dY_j(\omega),$$ (3.26)

where the Y_j are random functions with uncorrelated increments.

Thus, the spectral representation of Z is $Z(x) = \int_{\mathbb{R}^2} \exp(ix^T\omega)dY(\omega)$, where the spectral process Y does not have uncorrelated increments, and we have

$$Y(\omega) = \sum_{i=1}^{k} \mathcal{F}K_i * Y_i(\omega);$$ (3.27)

$\mathcal{F}K_i$ is the Fourier transform of K_i, and $*$ denotes the following convolution function:

$$\mathcal{F}K_i * Y_i(\omega) = \int_{\mathbb{R}^2} \mathcal{F}K_i(\mathbf{h})Y_i(\omega - \mathbf{h})d\mathbf{h}.$$

The covariance function of Z can be defined in terms of the covariance function of the orthogonal stationary processes Z_i:

$$\mathrm{cov}\{Z(x_1), Z(x_2)\} = \sum_{i=1}^{k} K_i(x_1)K_i(x_2)\mathrm{cov}\{Z_i(x_1), Z_i(x_2)\}.$$ (3.28)

This is a valid nonstationary covariance function. Then, the corresponding nonstationary spectral density is

$$f(\omega_1, \omega_2) = \sum_{i=1}^{k} f_i * \{\mathcal{F}K_i(\omega_1)\mathcal{F}K_i(\omega_2)\},$$ (3.29)

where f_i is the spectral density of the process Z_i, and

$$f_i * \{\mathcal{F}K_i(\omega_1)\mathcal{F}K_i(\omega_2)\} = \int_{\mathbb{R}^2} f_i(\mathbf{h})\mathcal{F}K_i(\omega_1 - \mathbf{h})\mathcal{F}K_i(\omega_2 - \mathbf{h})d\mathbf{h}.$$

For stationary processes the spectrum is a function of only one argument, that is, just one frequency ω; but because Z is nonstationary, the spectral density is a function of two spectral coordinates (ω_1 and ω_2).

3.2.5.3 Nonparametric spectral estimation

We present here an asymptotically unbiased nonparametric estimator, $\tilde{I}_N$, of the spectral density f of a nonstationary process Z. We model Z as in (3.25). Thus, a natural way of defining $\tilde{I}_N$ is as a convolution of the periodograms $I_{i,N}$ of the stationary processes Z_i with domain D:

$$\tilde{I}_N(\omega_1, \omega_2) = \sum_{i=1}^{k} I_{i,N} * \{\mathcal{F}K_i(\omega_1)\mathcal{F}K_i(\omega_2)\}, \tag{3.30}$$

where $*$ denotes the convolution

$$I_{i,N} * \{\mathcal{F}K_i(\omega_1)\mathcal{F}K_i(\omega_2)\} = \sum_{\omega \in J_N} I_{i,N}(\omega)\mathcal{F}K_i(\omega_1 - \omega)\mathcal{F}K_i(\omega_2 - \omega),$$

with J_N the set of the Fourier frequencies (3.17). The weights K_i have compact support (they are only positive in the corresponding subregion S_i of stationarity) and they help to identify the processes Z_i that are being used. By the definition of f in (3.29) as a function of the spectral densities f_i, $i = 1, \ldots, k$, and the fact that the periodograms $I_{i,N}$ are asymptotically unbiased estimators of f_i, we obtain that $\tilde{I}$ is asymptotically unbiased. The asymptotic variance of $\tilde{I}_N(\omega_1, \omega_2)$ can be easily obtained because the processes Z_i are orthogonal. Thus, when $n_i \to \infty$, for $i = 1, 2$, $\Delta \to 0$ and $\Delta n_1 \to \infty$, $\Delta n_2 \to \infty$, the variance of $\tilde{I}_N(\omega_1, \omega_2)$ becomes

$$\sum_{i=1}^{k} f_i^2 * \{\mathcal{F}K_i^2(\omega_1)\mathcal{F}K_i^2(\omega_2)\}.$$

Furthermore, because $\tilde{I}_N$ is a convolution of independent stationary periodograms, we obtain $\text{cov}\{\tilde{I}_N(\omega_1, \omega_2), \tilde{I}_N(\omega_1', \omega_2')\} = 0$ asymptotically.

In practice, we compute $I_{i,N}$ as the periodogram of the observed values of Z in the subregion of stationarity S_i.

The rate of convergence for this nonparametric estimator has been studied by Fuentes (2002).

3.2.5.4 Parametric spectral estimation

Suppose again that Z takes the form (3.25), so that we use the expression in (3.29) for f. The spectral density f is then modeled as a function of the spectral densities f_i, $i = 1, \ldots, k$.

A parametric estimator $\hat{f}$ of the spectral density is easily obtained from parametric estimators of the spectral densities f_i, $i = 1, \ldots, k$:

$$\hat{f}(\omega_1, \omega_2) = \sum_{i=1}^{k} \hat{f}_i * \{\mathcal{F}K_i(\omega_1)\mathcal{F}K_i(\omega_2)\}. \tag{3.31}$$

The properties of this estimator have not been studied, but they depend on the parametric estimates chosen for each f_i. The f_i could be estimated with their Bayesian posterior distribution, or using a likelihood approach.

We study now parametric models for the f_i. A class of practical variograms and autocovariance functions for the spatial stationary processes Z_i can be obtained from the Matérn class of spectral densities

$$f_i(\omega) = \phi_i(\alpha_i^2 + \|\omega\|^2)^{(-\nu_i - 1)} \tag{3.32}$$

with parameters $\nu_i > 0$, $\alpha_i > 0$, and $\phi_i > 0$. Here, the vector of covariance parameters is $\theta_i = (\phi_i, \nu_i, \alpha_i)$.

3.2.5.5 An example in air quality

In the air pollution example presented here (see Fuentes, 2002), the main objective is to understand and quantify the weekly spatial structure of air pollutants using the output of a regional scale air quality model, known as Models-3. Models-3 estimates hourly concentrations and fluxes of different pollutants. We study here nitric acid. The spatial domain D (Figure 3.9) is a regular 81×87 grid, where the dimensions of each cell on the grid are 36 km $\times$ 36 km. The 81×87 lattice for Models-3 is a two-dimensional grid that takes account of the earth's curvature. Models-3 provides the estimated concentration for the

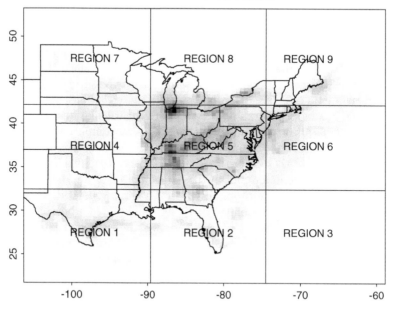

Figure 3.9 Output of EPA Models-3 showing the estimated concentrations of nitric acid (ppb) for the week starting July 11, 1995. The resolution is 36 km.

middle point of each cell. In this example we analyze the spatial structure of the hourly averaged nitric acid concentrations for the week starting July 11, 1995. We fit model (3.25), taking K to be the Epanechnikov kernel $K(u) = \frac{2}{\pi}(1 - ||u||^2/h^2)$ with h an arbitrary bandwidth, and replacing the integral over D by a sum over a grid of cells covering the observation region.

In practice, the choice of h is crucial. The bandwidth should be small to preserve the general "shape" of the data (Clark, 1977). In a regression setting, reducing the size of the bandwidth reduces the bias but increases the variance. In our spatial setting, because the variance of the process (the sill parameter) might change with location, due to the lack of stationarity, we do not gain much by increasing h. Because we still need to model and estimate the local and different spatial structure over our domain of interest, too much smoothing (large h) would not allow us to capture these local spatial patterns. The shape of the process is represented by the parameter θ, which accounts for the lack of stationarity of Z. Thus, we need to choose h as small as possible to preserve this general shape. However, we also need to ensure that for all $x \in D$ there is at least one $\mathbf{s}_i$, with $K(x - \mathbf{s}_i) > 0$, where the $\{\mathbf{s}_i\}$ are the k draws on D to calculate the covariance (3.24). In this application we choose the smallest value of h that satisfies this condition. When the distance between neighboring points of the sample sequence $\mathbf{s}_1, \ldots, \mathbf{s}_k$ varies, we could also allow the bandwidth to change with location. If we have k draws from a systematic sample with a distance l between sampling points, then the recommended value for h is $l/\sqrt{2}$. Note that the value of h depends on k.

In this example $k = 9$, which is the optimal value for k based on the AIC criterion. The sample points $\mathbf{s}_1, \ldots, \mathbf{s}_9$ are a systematic sample over our spatial domain. The distance l between the sampling points is 972 km. The value of h in this application is $h = l/\sqrt{2} = 687$ km. We used a likelihood approach to estimate the parameters of the nonstationary covariance matrix, which is a mixture of nine stationary Matérn models of the form (3.61). Because the kernel function K has compact support, the covariance matrix of Z is approximately a block matrix, which simplifies the calculations; otherwise the evaluation of the likelihood function requires us to compute the inverse and determinant of a 7209×7209 matrix.

Figure 3.10 shows the Matérn fitted models (using the likelihood function of Z in the spectral domain and in the subregions S_i's) for the spectral densities, f_i, of the stationary processes Z_i, for $i = 1, \ldots, 9$. The nonstationary spectral density $f(\omega_1, \omega_2)$, defined in (3.29), is obtained as a convolution of the densities f_i, $i = 1, \ldots, 9$. Table 3.1 shows the estimated parameters for the spectral densities f_i and the corresponding standard errors.

The smoothing parameter represents the rate of decay of the spectral density at high frequencies; this is an indication of how smooth the corresponding process is. The smoothing parameter is approximately 0.5, corresponding to the exponential model, for the processes Z_1, Z_4, Z_7 and Z_8; these processes explain the spatial structure of the nitric acid concentrations on the eastern part of our domain. We observe a relatively faster rate of decay at high frequencies for the processes Z_5, Z_6, Z_9, Z_2, and Z_3, with a smoothing parameter

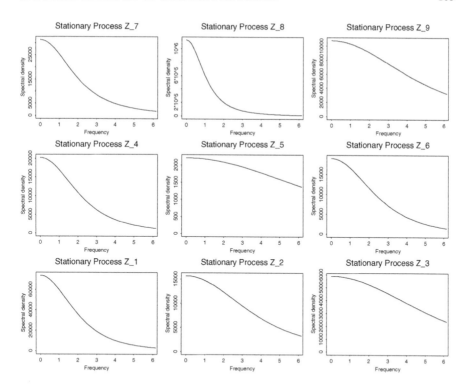

Figure 3.10 The Matérn spectral densities, obtained using Whittle's approximation to the likelihood, for the nine equally-dimensioned regions shown in Figure 3.9. Region 1 represents the lower left subarea (Texas) of Figure 3.9. (From Fuentes, 2002.)

Table 3.1 Estimated parameters for the spectral densities of the processes Z_i. The values in parentheses are the standard errors of the estimated parameters. The parameters have been estimated using a likelihood approach.

Process	Sill	Range	Smoothness
Z_1	2.9 (1.1)	566 (282)	0.74 (0.03)
Z_2	2.1 (0.5)	315 (131)	0.98 (0.10)
Z_3	1.8 (0.1)	231 (18)	1.01 (0.05)
Z_4	1.1 (0.3)	480 (193)	0.63 (0.03)
Z_5	1.22 (0.07)	150 (20)	1.19 (0.22)
Z_6	1.17 (0.01)	476 (17)	0.91 (0.01)
Z_7	1.4 (0.3)	500 (139)	0.46 (0.03)
Z_8	15 (3)	975 (292)	0.67 (0.01)
Z_9	3 (0.2)	252 (45)	0.84 (0.08)

Note: From Fuentes, 2002.

of approximately 1, corresponding to the Whittle model. These processes explain the spatial structure of the nitric acid concentrations on the western part of our domain, mainly over water; the nitric acid seems to be a smoother process over water than over the land surface. The nitric acid is a secondary pollutant, as the result of photochemical reactions in the atmosphere rather than emitted directly from sources on the surface. It therefore usually remains in the atmosphere for long periods of time and travels long distances across water.

When the range parameter is large (e.g., for process Z_8), there is a faster decay of the spectral density at short frequencies. We can appreciate this phenomenon by comparing the spectral density of Z_6, large range, to the spectral density of Z_5, small range. In general we observe larger ranges of autocorrelation on the western part of the grid. Furthermore, on the eastern part we should not expect large ranges because of the discontinuity of the nitric acid concentration that results from transition from land to ocean.

The variance of the process, also called the sill parameter, is the integral of the spectral density function, $\int_{\mathbb{R}^2} f(\omega)d\omega$. In this example, the sill is relatively large for Z_8. There is higher spatial variability, large sill, mainly on the Great Lakes area, process Z_8, because the area is downwind from sources of pollution, primarily Chicago. One of the main objectives of this analysis was to understand the changes in the variance of the process, by allowing the sill to vary with location. However, one could have forced all underlying processes to have the same variance. That way, when comparing their spectral density functions, their smoothness properties would be on the same scale.

3.2.6 Testing for stationarity

We present now a formal test for stationarity of a spatial process. This test is a generalization of the test for stationarity of time series presented by Priestley and Rao (1969) to spatial processes. We first need to introduce an estimate of the spatial spectral density for a nonstationary process.

3.2.6.1 Nonparametric estimation of a spatial spectrum

Assume we observe the process Z at N equally spaced locations in a regular grid D $(n_1 \times n_2)$, where $N = n_1 n_2$, and the spacing between observations is Δ. In this subsection we propose a nonparametric estimate of the spectral density (in a neighborhood of x), f_x. We allow the spectrum to vary from part to part of our domain, by having a spectral density that is a function of location. We use the following representation of a nonstationary process Z (Fuentes, 2005a), and allow the spectrum to be space dependent,

$$Z(\mathbf{x}) = \int_{\mathbb{R}^2} \exp(i\mathbf{x}^T \omega)\phi_x(\omega)dY(\omega), \qquad (3.33)$$

where Y is an orthogonal process. The representation can be interpreted as representation of the process Z in the form of a superposition of sinusoidal oscillations with different frequencies ω and random amplitudes $\phi_x(\omega)dY(\omega)$ varying over space. According to this interpretation, the variance of the process, $\int_{\mathbb{R}^2} |\phi_x(\omega)|^2 d\mu(\omega)$, describes the distribution of the "total power" of the process $Z(x)$ at location x over the frequencies ω; hence the contribution from the frequency ω is $|\phi_x(\omega)|^2 d\mu(\omega)$. Therefore, the function $F_x(\omega)$ defined by the relation $dF_x(\omega) = |\phi_x(\omega)|^2 d\mu(\omega)$, will be called the spatial spectral distribution function of the process Z, and $f_x(\omega) = |\phi_x(\omega)|^2 h(\omega)$ is the spatial spectral density of Z, where h is the density associated with the measure μ.

In this subsection we propose a nonparametric estimate of the spectral density (in a neighborhood of x), f_x. This estimate is simply a spatial periodogram with a filter function to give more weight to neighboring values of x.

We introduce now some notation. Let $dH_\omega(\theta)$ denote the Fourier transform of $\phi_x(\omega)$ as a function of x. We define the characteristic width B_Z of the process Z,

$$B_Z = [\sup_\omega \{B_H(\omega)\}]^{-1},$$

where

$$B_H(\omega) = \int_{\mathbb{R}^2} |\theta| |dH_\omega(\theta)|.$$

The characteristic width of Z can be interpreted roughly as the maximum area over which the process can be treated as approximately stationary.

We first define $J_x(\omega_0)$,

$$J_x(\omega_0) = \Delta \sum_{u_1=0}^{n_1} \sum_{u_2=0}^{n_2} g(x - \Delta \mathbf{u}) Z(\Delta \mathbf{u}) \exp\{-i\Delta \mathbf{u}^T \omega_0\}, \tag{3.34}$$

where $\mathbf{u} = (u_1, u_2)$ and $\{g(\mathbf{s})\}$ is a filter satisfying the following conditions **B.1** and **B.2**. Thus, the function J_x is simply the discrete Fourier transform of the process Z considered in a neighborhood of x.

B.1 $\{g(\mathbf{s})\}$ is a square integrable and normalized filter, so that

$$(2\pi)^2 \int_{\mathbb{R}^2} |g(\mathbf{s})|^2 d\mathbf{s} = \int_{\mathbb{R}^2} |\Gamma(\omega)|^2 d\omega = 1.$$

Here

$$\Gamma(\omega) = \int_{\mathbb{R}^2} g(\mathbf{s}) \exp\{-i\mathbf{s}^T \omega\} d\mathbf{s}$$

denotes the Fourier transform of $\{g(\mathbf{s})\}$.

B.2 $\{g(\mathbf{s})\}$ has finite "width" B_g defined by

$$B_g = \int_{\mathbb{R}^2} |\mathbf{s}| |g(\mathbf{s})| d\mathbf{s}$$

where B_g is smaller than B_Z.

We refer to $|J_x(\omega)|^2$ as the spatial periodogram at a location x for a frequency ω.

The spectral estimate $|J_x(\omega)|^2$ is an approximately unbiased estimate of $f_x(\omega)$ (see Theorem 1 in Fuentes, 2005a), but as its variance may be shown to be independent of N it will not be a very useful estimate in practice. We therefore estimate $f_x(\omega)$ by "smoothing" the values of $|J_x(\omega)|^2$ over neighboring values of x. More precisely, let W_ρ be a weight function or "window," depending on the parameter ρ, which has integral 1 and the square of W is integrable.

We write

$$w_\rho(\lambda) = \int_{-\infty}^{+\infty} \int_{-\infty}^{+\infty} \exp\{is^T\lambda\}W_\rho(s)ds.$$

We assume that there exists a constant C such that

$$\lim_{\rho \to \infty} \left\{ \rho^2 \int_{\mathbb{R}^2} |w_\rho(\lambda)|^2 d\lambda \right\} = C.$$

The parameter ρ determines the bandwidth of $\{W_\rho\}$, and it is chosen such that it is larger than the width B_g.

Then we estimate $f_x(\omega_0)$ by

$$\hat{f}_x(\omega_0) = \int_{-\infty}^{\infty} \int_{-\infty}^{\infty} W_\rho(x - s) |J_s(\omega_0)|^2 \, ds. \tag{3.35}$$

We evaluate (3.35) by Monte Carlo integration.

In the application presented in this subsection we consider $\{g(s)\}$ for $s = (s_1, s_2)$ to be a multiplicative filter, that is, the tensor product of two one-dimensional filters, $g(s) = g_1(s_1)g_1(s_2)$, where g_1 is of the form

$$g_1(s) = \begin{cases} 1/\{2\sqrt{h\pi}\} & |s| \leq h \\ 0 & |s| > h \end{cases} \tag{3.36}$$

corresponding to the Barlett window. Then, $\Gamma(\omega) = \Gamma_1(\omega_1)\Gamma_1(\omega_2)$ for $\omega = (\omega_1, \omega_2)$. We also choose W_ρ to be of the form $W_\rho(s) = W_{1,\rho}(s_1)W_{1,\rho}(s_2)$, where

$$W_{1,\rho}(s) = \begin{cases} 1/\rho & -1/2\rho \leq s \leq 1/2\rho \\ 0 & \text{otherwise,} \end{cases} \tag{3.37}$$

corresponding to the Daniell window.

The asymptotic properties of $\hat{f}_x(\omega_0)$ using a *shrinking asymptotics* model were studied by Fuentes (2005a).

3.2.6.2 Testing for stationarity

We calculate our estimate of the spatial spectral density $f_{s_i}(\omega)$ at m nodes $s_1, \ldots, s_m$ that constitute a systematic sample on D. We write

$$U(s_i, \omega) = \log \hat{f}_{s_i}(\omega_j) = \log f_{s_i}(\omega) + \epsilon(s_i, \omega).$$

We obtain that asymptotically $E(\epsilon(\mathbf{s}_i, \omega)) = 0$ and $\text{var}\{\epsilon(\mathbf{s}_i, \omega)\} = \sigma^2$ where

$$\sigma^2 = (C/\rho^2) \left\{ \int_{-\infty}^{+\infty} \int_{-\infty}^{+\infty} |\Gamma(\theta)|^4 d\theta \right\}, \tag{3.38}$$

for $\omega \notin \partial \Pi_\Delta^2$, where $\Pi_\Delta = [-\pi/\Delta, \pi/\Delta]$ and $\partial \Pi_\Delta^2$ denotes the boundary of the region Π_Δ^2. The variance σ^2 is clearly independent of x and ω.

Now we evaluate the estimated spatial spectra, $\hat{f}_{\mathbf{s}_i}(\omega)$, at the m nodes $\mathbf{s}_1, \ldots, \mathbf{s}_m$ and a set of frequencies $\omega_1, \omega_2, \ldots, \omega_n$ that cover the range of locations and frequencies of interest. Assuming the $\mathbf{s}_i$ and ω_j are spaced "sufficiently wide apart," then the $\epsilon(\mathbf{s}_i, \omega_j)$ will be approximately uncorrelated. This result is based on asymptotic properties of $\hat{f}_{\mathbf{s}_i}(\omega_j)$. The spatial periodogram values $\hat{f}_x(\omega)$ and $\hat{f}_y(\omega')$ are asymptotically uncorrelated if either $\|\omega \pm \omega'\| \gg$ bandwidth of $|\Gamma(\theta)|^2$ or $\|x - \mathbf{y}\| \gg$ bandwidth of the function $\{W_\rho(\mathbf{u})\}$. In the application presented in this subsection we used the distance between the "half-power" points on the main lobe of $|\Gamma(\omega)|^2$ to approximate the bandwidth.

The logarithmic transformation brings the distribution of a smoothed spatial periodogram closer to normality (Jenkins, 1961). Thus, we can treat the $\epsilon(x_i, \omega_j)$ as independent $N(0, \sigma^2)$. We write

$$U_{ij} = U(\mathbf{s}_i, \omega_j), \ \log f_{\mathbf{s}_i}(\omega_j) = f_{ij}, \ \text{and } \epsilon_{ij} = \epsilon(\mathbf{s}_i, \omega_j).$$

Then we have the model

$$U_{ij} = f_{ij} + \epsilon_{ij}. \tag{3.39}$$

Equation (3.39) becomes the usual "two-factor analysis of variance" model, and could be rewritten in the more conventional form:

$$H_1 : U_{ij} = \mu + \alpha_i + \beta_j + \gamma_{ij} + \epsilon_{ij}$$

for $i = 1, \ldots, m$ and $j = 1, \ldots, n$. Then we test for stationarity of Z using the standard techniques to test the model

$$H_0 : U_{ij} = \mu + \beta_j + \epsilon_{ij}$$

against the more general model H_1. Because we know the value of $\sigma^2 = \text{var}\{\epsilon_{ij}\}$, we can test for the presence of the interaction term, γ_{ij}, with one realization of the process. If the model H_0 is rejected, then there is a significant difference between the parameters α_i, for $i = 1, \ldots, m$, which is evidence of lack of stationarity for Z at the m nodes. Thus, the complex and challenging problem of testing for nonstationarity is reduced to a simple two-factor analysis of variance.

The parameters $\{\alpha_i\}$, $\{\beta_j\}$ represent the main effects of the space and frequency factors, and $\{\gamma_{ij}\}$ represents the interaction between these two factors. A test for the presence of interaction is equivalent to testing if Z is a uniformly modulated process; this means $\log f_x(\omega)$ is additive in terms of space and frequency, then $f_x(\omega)$ is multiplicative, that is, $f_x(\omega) = c^2(x)f(\omega)$, so the process

Z is of the form: $Z(x) = c(x)Z_0(x)$, where Z_0 is stationary with spectral function f and c is a function of space. If the interaction is not significant, we conclude that Z is a uniformly modulated process. If the interaction is significant, we conclude that Z is nonstationary, and nonuniformly modulated. We can study if the nonstationarity of Z is restricted only to certain frequency components by selecting those frequencies (e.g., $\{\omega_{j_1}, \ldots, \omega_{j_k}\}$) and testing for stationarity at these frequencies.

If Z is an isotropic process, then $f_x(\omega)$ depends on its vector argument ω only through its length $\|\omega\|$. Then, we could test for isotropy by selecting a set of frequencies with the same absolute values, say $\{\omega_{j_1}, \omega_{j_2}\}$ where $\omega_{j_1} \neq \omega_{j_2}$ but $\|\omega_{j_1}\| = \|\omega_{j_2}\|$, and examine whether the "main-effect" effect β is significant.

We could test for "complete randomnes" (i.e., constant spectra for the spectral density on the lattice) by testing the "main-effect" β, either at all locations on the lattice when the interaction term is not significant, or at a particular subset of locations. All these comparisons are based on χ^2 rather than F-tests because σ^2 is known.

3.2.6.3 An example in air quality

For the purpose of illustrating the techniques presented in this subsection, we take a systematic sample of locations in D, the sample points $x_1, \ldots, x_9$ are the centroids of the nine equally-dimensioned regions $S_1, \ldots, S_9$, in our domain of interest (Figure 3.9); see Fuentes (2005a). The tables in this subsection are from Fuentes (2005a).

Now we implement our test for stationarity. We select values of locations x and frequencies ω that are sufficiently apart. The estimates, $\hat{f}_x(\omega)$ (Table 3.2) were obtained using expression (3.35), in which $W_\rho(\mathbf{u})$ is given by (3.37) with $\rho = 20$ units (1 unit $= 36$ km), and $g(\mathbf{u})$ is of the form (3.36) with $h = 3$. The window $|\Gamma(\omega)|^2$ has a bandwidth of approximately $\pi/h = \pi/3$.

Table 3.2   Values of $\hat{f}_x(\omega)$

$x \backslash \omega$	ω_1	ω_2	ω_3	ω_4	ω_5	ω_6	ω_7	ω_8	ω_9
x_1	4315.92	31.49	5.35	28.01	9.18	0.20	3.89	5.12	1.02
x_2	3376.27	35.46	3.30	35.81	8.73	4.67	4.58	6.29	4.31
x_3	2670.07	38.01	6.54	40.99	12.38	10.68	7.36	2.86	2.14
x_4	1617.05	13.28	3.52	14.90	2.98	2.08	1.24	4.91	3.20
x_5	1256.20	38.80	5.69	30.58	13.56	3.05	9.66	3.30	2.64
x_6	1765.69	14.20	0.52	13.93	7.29	1.45	1.16	2.36	4.79
x_7	2016.57	12.97	2.55	17.88	0.69	3.27	0.37	0.13	2.21
x_8	13597.65	70.37	10.67	75.72	23.63	8.20	12.55	6.89	6.36
x_9	4618.01	63.28	12.09	56.93	21.94	7.71	10.32	1.23	1.21

The distance between the "half-power" points on the main lobe of $|\Gamma(\omega)|^2$ was used as an approximation of the bandwidth. The window $\{W_\rho(x)\}$ has a bandwidth of $\rho = 20$. Thus, to obtain approximately uncorrelated estimates, the points ω_j and x_i should be chosen so that the spacings between the ω_j are at least $\pi/3$ and the spacings between the x_i are at least 20 units, the sample points $x_1, \ldots, x_9$ are the centroids of the nine equally dimensioned regions $S_1, \ldots, S_9$, covering the domain shown in Figure 3.9. The ω_j were chosen as follows: $\omega_j = (\omega_{j_1}, \omega_{j_2}) = (\pi j/20, \pi j_2/20)$ with $j_1 = 1\,(7)\,15$, $j_2 = 1\,(7)\,15$, corresponding to a uniform spacing of $7\pi/20$ (which just exceeds $\pi/3$). The values of $\hat{f}_x(\omega)$ are shown in Table 3.2, where $\omega_1 = (\pi/20, \pi/20)$, $\omega_2 = (8\pi/20, \pi/20)$, $\omega_3 = (15\pi/20, \pi/20)$, $\omega_4 = (\pi/20, 8\pi/20)$, $\omega_5 = (8\pi/20, 8\pi/20)$, $\omega_6 = (15\pi/20, 8\pi/20)$, $\omega_7 = (\pi/20, 15\pi/20)$, $\omega_8 = (8\pi/20, 15\pi/20)$, and $\omega_9 = (15\pi/20, 15\pi/20)$.

We need to calculate σ^2 (see Equation (3.38)) to perform the test of stationarity for Z. In this application, $\sigma^2 = 16h^2/(9\rho^2) = 0.04$.

The interaction is significant (see Table 3.3) (χ^2 is very large compared to $\chi^2_{64}(0.05) = 83.67$), confirming that we do not have a uniformly modulated model, and both the "between spatial points" and "between frequencies" sums of squares are highly significant (χ^2 is extremely large compared to $\chi^2_8(0.05) = 15.51$), confirming that the process is nonstationary and that the spectra are nonuniform.

If isotropy is a reasonable assumption, then columns 3 and 7 in Table 3.2 should have similar values. We present now an approach to test if there is any significant difference between columns 3 and 7 in Table 3.2.

The next table (Table 3.4) presents an analysis of variance to study the significance of the difference between columns 3 and 7 in Table 3.2. The spatial points are the same as in Table 3.2, $x_1, \ldots, x_9$, and the frequency values used here are ω_3 and ω_7, both having the same absolute value.

The "between frequencies" effect is not significant (see Table 3.4) (χ^2 is smaller than $\chi^2_1\,(0.05) = 3.84$), suggesting that there is no evidence of anisotropy. This is not surprising because for air pollution the lack of anisotropy is usually detected at higher spatial resolutions (here the resolution of the

Table 3.3 Analysis of variance

Item	Degrees of freedom	Sum of squares	$\chi^2 = $ (sum of squares/σ^2)
between spatial points	8	26.55	663.75
between frequencies	8	366.84	9171
interaction + residual	64	30.54	763.5
total	80	423.93	10598.25

Source: Reprinted from *Journal of Multivariate Analysis*, 96, Fuentes, M., A formal test for nonstationarity of spatial stochastic processes, 30–54, Copyright 2005, with permission from Elsevier.

Table 3.4 Analysis of variance

Item	Degrees of freedom	Sum of squares	$\chi^2 = $ (sum of squares/σ^2)
between spatial points	8	15.96	399.18
between frequencies	1	0.12	3.17
interaction + residual	8	2.87	71.99
total	17	18.95	473.75

Source: Reprinted from *Journal of Multivariate Analysis*, 96, Fuentes, M., A formal test for nonstationarity of spatial stochastic processes, 30–54, Copyright 2005, with permission from Elsevier.

models is 1296 km^2). However, the "between spatial locations" sums of squares is highly significant (χ^2 is extremely large compared to $\chi_8^2(0.05) = 15.51$), confirming that the process is nonstationary.

We could test for stationarity within the subregions $S_1, \ldots, S_9$ by drawing a larger systematic sample in D with more than one sample point within each subregion. Further testing suggests that $\theta(\mathbf{x})$ does not change significantly within the subregions $S_1, \ldots, S_9$.

3.2.6.3.1 Power of the test

Several simulation studies were conducted to estimate the power of our test to detect deviations from stationarity when the underlying covariance was not stationary. We simulated 400 versions of a spatial Gaussian process on the same grid as in the air quality application shown here, with a nonstationary covariance. The covariance function used was a weighted average of two exponential covariances: $C(x, y) = w_1(x)w_1(y)C_1(x - y) + w_2(x)w_2(y)C_2(x - y)$, where s_1 and s_2 are the two centers of gravity of two subregions of the same size that cover the entire domain, and $K_i(x)$ is the inverse square distance between $\mathbf{x}$ and s_i. The covariance functions C_1 and C_2 had a nugget of .01, a range of 2 and 6, respectively, and a partial sill of 2 and 4. We used the same pairs and frequencies as in our application. Out of the 400 simulated fields, 346 (87%) rejected the null hypothesis of stationarity (for a level $\alpha = .05$ test). We repeated this simulation 50 times; we consistently got similar results, an average of .85 power. The estimated probability of rejecting the null hypothesis of stationarity when the underlying process was stationary (an exponential with nugget of .01, range of 2, and partial sill of 2) was .078 (Type I error).

The approach proposed here is designed to test the stationarity of the covariance function when the mean of the process has been removed. If the mean of the process is a function of space, the test will be sufficiently powerful to detect the nonstationarity of the underlying process, even if we do not remove the mean surface. We conducted a simulation study as above but with a nonstationary mean function that was a third-order polynomial in space; the power was 1 to detect lack of stationarity when the underlying covariance was stationary, but it also rejected 64% time stationarity (due to the non-stationary mean) when the covariance was nonstationary. However, the test did not appear to be sensitive to a spatial temporal trend when this trend was separable. We conducted the same simulation study with a linear trend and a separable covariance; the power was still .9 and we rejected separability 9% of the time ($\alpha = 0.05$) when the covariance was separable, despite the space-time linear trend surface.

3.2.6.3.2 Selection of bandwidth

Before the spectral estimate $\hat{f}(\omega_0)$ can be evaluated, we have to choose the form of the filter $\{g(\mathbf{s})\}$ and the form of the weight function $\{W_\rho(\mathbf{s})\}$. Generally, $\{g(\mathbf{s})\}$ will be chosen from the standard collection of "windows" and will involve a parameter h, so that by adjusting the value of h we can vary the values of B_g. Similarly, we may choose $\{W_\rho(\mathbf{s})\}$ from the same collection of windows with a parameter ρ. Suppose now that we have chosen the mathematical forms of $\{g(\mathbf{s})\}$ and $\{W_\rho(\mathbf{s})\}$. The problem arises as to how we should choose the parameters h and ρ so that the estimate $\hat{f}(\omega_0)$ possesses certain required properties. The asymptotic results obtained by Fuentes (2005a) provide an approximation for the mean squared error (mse) of $\hat{f}(\omega_0)$. We suggest using a plug-in approach and replace $f(\omega_0)$ with $\hat{f}(\omega_0)$ in the expression for the mse, to obtain the values of h and ρ that minimize the relative mse of $\hat{f}(\omega_0)$. An alternative method to choose h and ρ is using a Bayesian approach, and putting priors on the hyperparameters h and ρ.

The power of the test did not appear to be sensitive to the chosen bandwidth parameters. However, further study remains to be done.

3.3 Wavelet analysis

Wavelet analysis, as opposed to frequency analysis, attempts to decompose the variability of a process both in time and frequency. It is generally an orthogonal expansion. In this section we first describe the continuous and discrete wavelet transforms, following Percival and Walden (2000).

There are some advantages of wavelet analysis over Fourier analysis. First, wavelet analysis really aims at handling nonstationary processes. The frequency decomposition (in wavelet theory the corresponding term is "scale decomposition") is allowed to change over time. The price to pay is that not all possible scales can be studied. Second, wavelet theory can handle long-term memory processes, where Fourier theory breaks down. In particular, wavelet coefficients for different scales are approximately uncorrelated even for long-term memory processes (Craigmile and Percival, 2005).

While wavelets can be applied to spatial fields, we focus first for simplicity on a description for a temporal process. Then we describe how one can use the discrete wavelet transform to estimate temporal trends, and derive confidence bands and formal tests for nonstationary time series with long-term dependence structure. Finally we look at a wavelet transform in the plane and show how it can be used to create a nonstationary spatial covariance function.

3.3.1 The continuous wavelet transform

Consider a real-valued time series $x(t)$. For a scale λ and a time t, look at the average

$$A(\lambda, t) = \frac{1}{\lambda} \int_{t-\lambda/2}^{t+\lambda/2} x(u)du.$$

Essentially, wavelet analysis looks at how much such averages change over time. Let

$$D(\lambda, t) = A(\lambda, t + \tfrac{\lambda}{2}) - A(\lambda, t - \tfrac{\lambda}{2})$$

$$= \frac{1}{\lambda} \int_{t}^{t+\lambda} x(u)du - \frac{1}{\lambda} \int_{t-\lambda}^{t} x(u)du.$$

The difference D at scale $\lambda = 1$ and time $t = 0$ can be written

$$D(1,0) = \sqrt{2} \int_{-\infty}^{\infty} \psi^{(H)}(u)x(u)du$$

where

$$\psi^{(H)}(u) = \begin{cases} -\frac{1}{\sqrt{2}}, & -1 < u \leq 0 \\ \frac{1}{\sqrt{2}}, & 0 < u \leq 1 \\ 0, & otherwise \end{cases}$$

is the Haar wavelet. We get a family of Haar wavelets by translation

$$\psi_{1,t}^{(H)}(u) = \psi^{(H)}(u - t)$$

and scaling

$$\psi_{\lambda,t}^{(H)}(u) = \frac{1}{\sqrt{\lambda}} \psi^{(H)}\left(\frac{u - t}{\lambda}\right).$$

Then the continuous Haar wavelet transform of the process $x(t)$ is

$$\widehat{W}(\lambda, t) = \int_{-\infty}^{\infty} \psi_{\lambda,t}^{(H)}(u)x(u)du \propto D(\lambda, t).$$

Figure 3.11 shows the signal $x(t)$, the wavelet $\psi^{(H)}(t)$, and the integrand in the wavelet transform (i.e., the product of the signal and the wavelet). A general (mother) wavelet is a function that integrates to zero and square integrates to one. A family of wavelets is generated by translation and scaling; $\psi_{a,b}(x) = \frac{1}{\sqrt{a}}\psi((x - b)/a)$. There is an additional technical constraint, namely

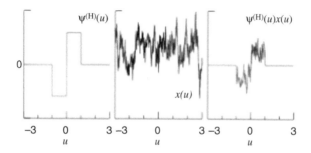

Figure 3.11 The Haar wavelet (left), a time series (middle), and the product of the two (right). The latter is what is integrated to calculate the wavelet transform at the scale of this wavelet.

that

$$C_\psi = \int |u|^{-1}|\mathcal{F}(\psi)(u)|^2 du < \infty.$$

A few different classes of wavelets are given in a later subsection. The general continuous wavelet transform for a wavelet ψ is

$$\hat{W}(\lambda, t) = \int\limits_{-\infty}^{\infty} \psi_{\lambda,t}(u)x(u)du.$$

The continuous wavelet transform is equivalent to the original process x. We get the process back from the transform by the formula

$$x(t) = \frac{1}{C_\psi} \int\limits_{0}^{\infty} \left[\int\limits_{-\infty}^{\infty} \hat{W}(\lambda, u)\psi_{\lambda,t}(u)du \right] \frac{d\lambda}{\lambda^2},$$

where C_ψ is the finite constant given above.

Like the Fourier transform, the wavelet transform decomposes energy, in the sense (for a mean zero process x) that

$$\int\limits_{-\infty}^{\infty} x^2(t)dt = \int\limits_{0}^{\infty} \int\limits_{-\infty}^{\infty} \frac{\hat{W}^2(\lambda, t)}{C_\psi \lambda^2} dt d\lambda.$$

Here we think of the left-hand side as the total energy, while the integrand on the right-hand side corresponds to the energy at time t and frequency λ.

3.3.2 The discrete wavelet transform

Usually we do not get to observe our process $x(t)$ continuously. Rather, we observe it at discrete times (for spatial processes, at a grid). Hence we can consider dyadic scales $\lambda = \tau_j = 2^{-j}$ and times restricted to the integers.

The discrete wavelet transform then has coefficients $W_{j,k} \propto \hat{W}(\tau_j, k)$. We write $\mathbf{W} = (W_{j,n})$ for the wavelet coefficients, and $\mathcal{W}$ for the wavelet transform, so

$$\mathbf{W} = \mathcal{W}\mathbf{Y}.$$

In many cases we want $(\tau_{j,n})_{n=1}^{\infty}$ to be an orthogonal basis for each j, so that $\mathcal{W}^{-1} = \mathcal{W}^T$. This puts some further restrictions on the choice of wavelets, as we will see in the next subsection.

Following Tukey, we can think of $r_n = \sum_{j=1}^{J} W_{j,n}$ as the *rough* of the series, while $s_n = x_n - r_n$ is the *smooth*. A *multiwavelet analysis* shows the *details* or the wavelet coefficients for each scale $j < \log_2(N)$, where N is the number of observations, as well as the smooth. Each of the details describes the variability of the time series at the corresponding scale.

Example

Cole et al. (2000) studied a 194-year record of the $\delta^{(18)}O$ oxygen isotope measured from a 4-m-high coral colony growing at a depth of 6-m (at low tide) in Malindi, Kenya. The purpose of the study was to measure changes in sea surface temperature over time.

A decrease in the oxygen value corresponds to an increase in the sea surface temperature (SST) (roughly a change of -0.24 ppm concentration corresponds to an SST increase of $1°C$). The issue is whether there is a significant decadal trend after we adjust for the variability in the process. Figure 3.12 shows a multiscale analysis of the negative of these data, computed using the pyramid scheme (Mallat, 1989) with the LA8 wavelet (see next subsection). We see a tendency toward an increase in the smooth of the series toward the end of the period, corresponding to a possible increase in SST.

3.3.3 Some wavelets

The requirement that a wavelet needs to generate an orthogonal basis affects what functions one can use. There is of course a variety of basis functions. For computational reasons it is desirable that the wavelet has compact support. It turns out that the orthogonality requirement and the compact support together force the wavelet to be asymmetric (if it is not the Haar wavelet). In this subsection we first present a wavelet family without compact support, and then a family with compact support.

3.3.3.1 Mexican hat

The Mexican hat is proportional to the second derivative of the Gaussian density function, that is, $\psi(x) \propto (1 - x^2) \exp(-x^2/2)$. Its name comes from the effect of rotating this function around the origin, thus creating a two-dimensional wavelet that has been popular in image analysis. Figure 3.13 shows this wavelet.

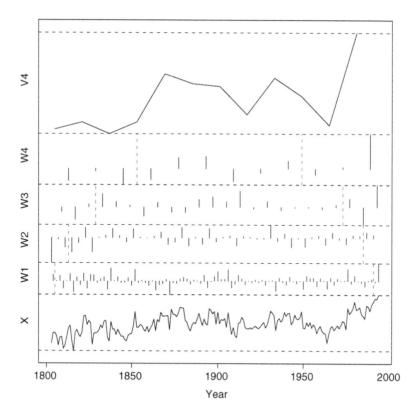

Figure 3.12 Multiscale analysis of the Malindi time series of oxygen isotope. The smooth is on top, followed successively by the wavelet coefficients at scale 16, 8, 4, and 2 years. The bottom graph shows the time series.

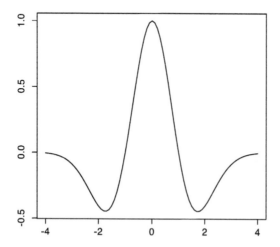

Figure 3.13 The Mexican hat wavelet, arbitrarily scaled.

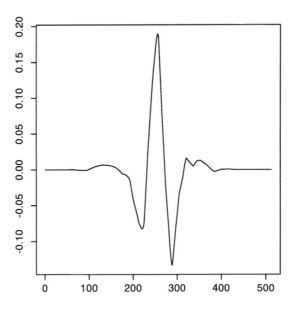

Figure 3.14 The LA(8) wavelet, arbitrarily scaled.

3.3.3.2 Daubechies' least asymmetric compactly supported wavelet

The least asymmetric compactly supported wavelet can be written as a low-pass filter supported on $2L$ points. The calculation of the wavelet is done according to a recipe in Daubechies (1992). Figure 3.14 shows the wavelet corresponding to $L = 8$, commonly called the LA(8)-wavelet. It is the default wavelet in software packages such as waveslim in R (R Development Core Team (2004)).

3.3.4 A nonparametric trend estimator and its properties

The smooth part of the wavelet transform (called the scaling coefficients r_n) is in essence a nonparametric trend estimator, where the trend corresponds to variability at a larger scale than what is captured in the details. In this subsection we outline how Craigmile et al. (2004) have developed tests for polynomial trends and simultaneous confidence bands for a trend estimator which is related to the scaling part of the wavelet transform. The statistical properties are valid even when the underlying time series exhibits long-term memory. We consider the usual model $X_t = T_t + Y_t$ where $\mathbf{T} = (\mathbf{T_t})$ is the trend, assumed non-stochastic, and $\mathbf{Y} = (\mathbf{Y_t})$ is the error process.

3.3.4.1 Fractional difference model for long-term memory

Many atmospheric and climatological time series exhibit *long-term memory*, that is, the autocovariance function decays very slowly. This creates difficulties for Fourier analysis, in that the convenient lack of correlation between periodogram values at different Fourier frequencies no longer holds, but turns out to be handled rather well by wavelet analysis.

We will focus here on a simple class of Gaussian long-term memory models, namely the *fractional difference* class. It is a mean zero Gaussian time series, with Fourier density

$$f(\omega) = \sigma^2 \left|2\sin(\pi\omega)\right|^{-2\delta}. \tag{3.40}$$

One can think of this as an ARIMA$(0,\delta,0)$-process, where δ is not necessarily an integer. The parameter can be estimated from the time series (see, e.g., Craigmile et al., 2004). General description and theory of long-term memory processes can be found in Beran's book (1994)). Using the representation (3.40), and taking logarithms on both sides, we see that for small values of ω we have

$$\log f(\omega) \approx c - 2\delta \log(\pi\omega)$$

so if we plot $\log f$ against $\log \omega$ we should get a straight line for small values of ω. Figure 3.15 shows this plot for the Malindi data.

3.3.4.2 The trend estimator

Using one of Daubachies' wavelets with compact support and letting the time series wrap around at the end (this is only needed if the sample size is not a power of two), we can divide the wavelet coefficients into those that are affected by boundary values (those that were wrapped) and those that are not. In fact, we can write

$$\mathbf{W} = \mathbf{W_s} + \mathbf{W_b} + \mathbf{W_{nb}}$$

where $\mathbf{W_s}$ has the scaling coefficients and zeros elsewhere, $\mathbf{W_b}$ has the wavelet coefficients affected by the boundary, and zeros elsewhere, while $\mathbf{W_{nb}}$ are the

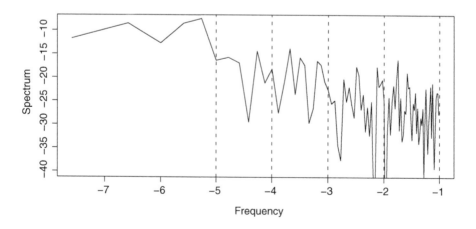

Figure 3.15 Spectral density function for the Malindi oxygen isotope series on a log-log scale.

non-boundary wavelet coefficients. Noting that $\mathbf{X} = \mathcal{W}^T \mathbf{W}$, we can write

$$\mathbf{X} = \mathcal{W}^T(\mathbf{W_s} + \mathbf{W_b}) + \mathcal{W}^T \mathbf{W_{nb}} = \tilde{\mathbf{T}} + \tilde{\mathbf{Y}},$$

where $\tilde{\mathbf{T}}$ is an estimate of trend and $\tilde{\mathbf{Y}}$ is a tapered estimate of error. Details of this calculation are in Craigmile et al. (2004). If the true trend is polynomial of degree K, and we use a wavelet filter of length $L \geq 2(K+1)$, then nonboundary wavelet coefficients will not contain the trend. We can then use $\tilde{\mathbf{Y}}$ to estimate parameters of the error process, and these will be uncorrelated with the trend estimate.

3.3.4.3 Testing for polynomial trend

To test the null hypothesis of no trend $(\mathbf{T} = \mathbf{0})$, we use, for a FD error process with parameter δ, the test statistic

$$P(\delta) = \frac{\|\tilde{\mathbf{T}}\|^2}{\|\tilde{\mathbf{Y}}\|^2} = \frac{\|\mathbf{AW}\|^2}{\|(\mathbf{I} - \mathbf{A})\mathbf{W}\|^2} = \frac{\|\mathbf{X}\|^2}{\|(\mathbf{I} - \mathbf{A})\mathbf{W}\|^2} - 1$$

where $\mathbf{A}$ contains the indicators of boundary and scaling coefficients. For a given value of δ we can simulate the distribution of $P(\delta)$. To take into account the variability in the parameter estimates, we repeatedly simulate FD processes from random samples of the limit law of the parameter estimates, and then compute the test statistic for each simulated path. In the case of the Malindi data, the estimated parameters are $\hat{\sigma} = 0.0667$ and $\hat{\delta} = 0.359$, with a confidence band for δ of $(0.143, 0.597)$, indicating that the series does indeed have long-term memory. We cannot rule out nonstationarity. Performing the simulation as outlined above, the resulting P-value is below 0.000, indicating strong evidence of a non-constant mean value (in order to look at non-zero mean, one can look at anomalies; i.e., first take out the overall mean of the series).

3.3.4.4 Computing simultaneous confidence band for the trend estimator

The rejection of a constant mean value only gives part of the story. Ideally we would like to derive a simultaneous confidence band for the trend. Let $\mathbf{v} = (\mathrm{var}\tilde{\mathbf{T}}_0, ..., \mathrm{var}\tilde{\mathbf{T}}_{\mathbf{N}-1})$ and write $\mathbf{U} = \tilde{\mathbf{T}} - \mathbf{T}$. We want to find λ so that

$$1 - \alpha = P(\tilde{\mathbf{T}} - \lambda\mathbf{v} \leq \mathbf{T} \leq \tilde{\mathbf{T}} + \lambda\mathbf{v}) = 1 - 2\mathbf{P}(\mathbf{U} > \lambda\mathbf{v}).$$

Given the process parameters, $\mathbf{U}$ has a known multivariate normal distribution, and conditionally we can do a bootstrap. To get a confidence band that takes into account the estimation of process parameters, we do again a parametric bootstrap from the limiting distribution. The result for the Malindi data is seen in Figure 3.16, where we see that the temperatures are high in the later years, but also somewhat low in the earlier years of the study.

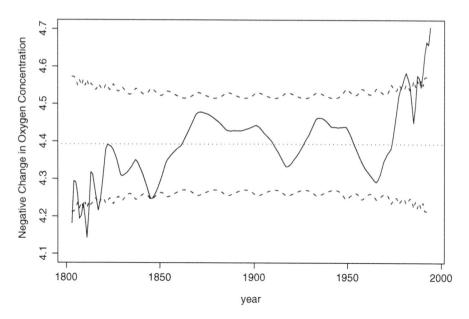

Figure 3.16 95% confidence limits for the trend of the Malindi oxygen isotope series.

3.3.4.5 Some spatial wavelets

To create spatial wavelets we can start with two wavelet funcions, the "mother wavelet" ψ and the "father wavelet" ϕ. We then create the following functions:

$$S(x_1, x_2) = \phi(x_1)\phi(x_2)$$
$$H(x_1, x_2) = \psi(x_1)\phi(x_2)$$
$$V(x_1, x_2) = \phi(x_1)\psi(x_2)$$
$$D(x_1, x_2) = \psi(x_1)\psi(x_2)$$

where S stands for smooth, H for horizontal, V for vertical, and D for diagonal. Figure 3.17 shows the four functions. The horizontal, vertical, and diagonal are called *detail functions*.

The key idea behind the wavelet transform is recursion. At each step an image of size (say, $n_1 \times n_2$) is decomposed through four finite-length separable, linear filters into four equal submatrices of smooth, horizontal, vertical, and diagonal terms. The three matrices for the H, V, and D components at this level of resolution are saved. The submatrix of $(n_1/2) \times (n_2/2)$ smoothed coefficients now becomes the image for the next step and the filtering is repeated. This process continues until one reaches a smoothed image of a particular size.

Technically we create a set of basis functions by starting with a coarsest level of resolution, say J. A basis of 32 functions is displayed in Figure 3.18. The first J basis functions are similar to the father wavelet translated to J equally spaced locations. These are given in plot (a) of Figure 3.18 for

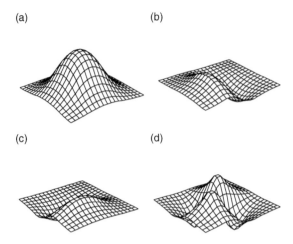

Figure 3.17 The four wavelet functions for spatial analysis: (a) the smooth S; (b) the horizontal detail H; (c) the vertical detail V; and (d) the diagonal detail D. (From Nychka et al., 2002.)

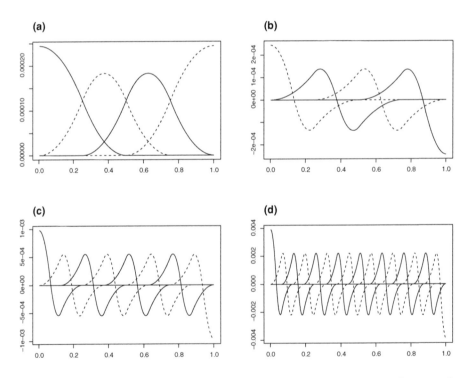

Figure 3.18 Family of 32 basis functions based on an approximate translation and scaling of father (a) and mother (b)–(d) wavelets. See text for details. (From Nychka et al., 2002.)

$J = 4$. The father wavelet appears only in this first J set and all subsequent basis functions are similar in form to the mother wavelet. The next J basis functions are the mother wavelets translated in the same manner and are in plot (b). The next generation of basis functions has twice the resolution and twice as many members (8) and is similar to a scaling and translation of the mother wavelet. Plot (c) of Figure 3.18 shows this generation. This cascade continues with the number of members in each subsequent generation and the resolution increasing by a factor of two. Plot (d) completes the basis of size 32.

3.3.5 A nonstationary covariance structure

The Karhunen-Loève expansion of a Gaussian process $Z(s)$ is given by

$$Z(s) = \sum_{i=1}^{\infty} \sqrt{\lambda_i} A_i \psi(s)$$

where

$$Cov(Z(s_1), Z(s_2)) = C(s_1, s_2)$$
$$= \sum_{i=1}^{\infty} \lambda_i \psi_i(s_1) \psi_i(s_2)$$

and the A_i are iid N(0,1) random variables. Instead of using the eigenfunctions of the covariance Nychka et al. (2002) suggested using the wavelet basis described in the previous subsection. This, of course, would lead to dependent coefficients A_i. Specifically, we can write the covariance

$$\mathbf{\Sigma} = \mathbf{\Psi}\mathbf{D}\mathbf{\Psi^T}; \mathbf{D} = \mathbf{\Psi^{-1}}\mathbf{\Sigma}(\mathbf{\Psi^T})^{-1}. \tag{3.41}$$

Letting the vector $\mathbf{\Psi}$ consist of the wavelet functions evaluated on a grid and stacked into a column vector, the matrix $\mathbf{D}$, which in the eigenfunction expansion would be diagonal, is no longer so. However, one may be able to approximate the covariance by an expansion of type (3.41) with a nearly diagonal $\mathbf{D}$. Something nearly diagonal will be needed, because even for a small grid $\mathbf{\Sigma}$ is huge.

Assume now that a random field Z is observed at m points on a regular grid, and that we have K independent replications. Typically m would be much larger than K. Write the data $\mathbf{Z}$ as an $m \times K$ matrix, and remove the spatial mean for each time point. Then we can estimate $\mathbf{\Sigma}$ by the sample covariance, and consequently we have $\hat{\mathbf{D}} = (1/K)(\mathbf{\Psi^{-1}Z})(\mathbf{\Psi^{-1}Z})^\mathbf{T}$. It is convenient to work with the square root $\mathbf{H}$ of $\mathbf{D}$, that is, a matrix such that $\mathbf{H^2} = \mathbf{D}$. Using the singular value decomposition of $(\mathbf{\Psi^{-1}Z})$ we can then write $\hat{\mathbf{H}} = \mathbf{V\Lambda^{1/2}V^T}$, where Λ contains the singular values and $\mathbf{V}$ is the left singular vector. Numerical studies indicate that if $\hat{\mathbf{H}}$ has a small number of nonzero values, one can achieve a good approximation across a family of covariance values (Nychka et al., 2002). Once the nonzero elements are determined, one can further decimate them by smoothing over spatially adjacent values.

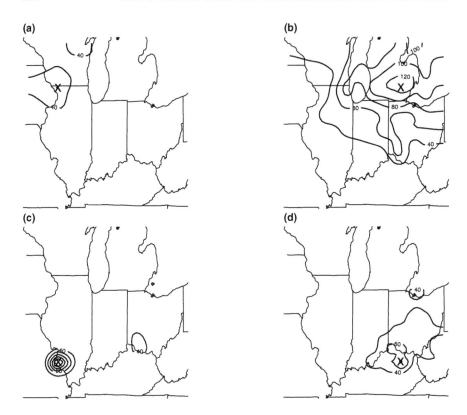

Figure 3.19 Estimated covariance surface at four sample locations for the ROM output. The image plots indicate the estimated covariance between points in the domain and the point location denoted by an x. Contour levels are at (40, 60, 80, 100, 120). (From Nychka et al., 2002.)

Example

To illustrate this process, Nychka et al. (2002) apply it to the output of a model for tropospheric ozone, the regional oxidant model (ROM), studied on a 48 × 48 grid over Illinois and Ohio in the summer of 1987. Each grid square is 16 km × 16 km, and because the correlation of ozone decays at about 300 km, we start with a 3 × 3 grid father wavelet at the coarsest resolution. In this application the leading 12 × 12 block of $\hat{\mathbf{H}}$ was decimated by 90%, and the diagonal elements were retained for the remaining levels. The resulting covariance structure is shown in Figure 3.19, and is highly nonstationary. This shows isocovariance curves for four different locations (each marked by an x). In particular, we notice that the sites (a) and (b) have covariances that are longer ranged to the west, presumably related to the dominant meteorological structure of weather systems moving from west to east.

3.4 Analyzing space-time processes

3.4.1 Empirical orthogonal function analysis

In geophysics and meteorology, variants of principal components called empirical orthogonal functions (EOFs) have long been used to describe leading modes of variability in space-time processes. Here we use smoothed EOFs to model the spatio-temporal mean of a random field viewed as spatially varying systematic temporal trends. In contrast to the following section (3.4.2), which also concerns space-time modeling, this section does not require the data to derive from a regular grid for spectral modeling; on the other hand, it is not concerned with temporal correlation structure or, more generally, the modeling of nonseparable spatio-temporal correlation structure.

It is common in the space-time modeling literature to decompose observations into the sum of a systematic trend component and residuals

$$Z(x,t) = \mu(x,t) + \varepsilon(x,t). \tag{3.42}$$

We saw such a decomposition for the purely temporal case in Section 2.4 where the trend was denoted T_t and modeled nonparametrically using wavelets. By contrast, the temporal trend for hourly averaged ozone fluxes in the application of Section 3.2.5 is a simple trigonometric model, which was fitted and removed at the regular grid of locations for the spatio-temporal air quality model predictions. The trend in that application was believed to be constant in space, although that assumption is not generally necessary. The trigonometric model is one of the most common examples of the decomposition of temporal trend in terms of a series of orthogonal basis functions. Where details of the trend structure vary spatially, we write such a decomposition more generally, with spatially varying coefficients, as

$$\mu(x,t) = \beta_{x0} + \sum_{j=1}^{J} \beta_{xj} f_j(t), \tag{3.43}$$

the $\{f_j(t)\}$ being a set or orthogonal temporal basis functions. Our focus is on applications with smooth seasonal (or, as in Section 3.2.5, diurnal) trends. Many air quality parameters display a dominant seasonal trend structure that is not conveniently represented by sums of trigonometric basis functions. In these cases we seek a parsimonious set of nonparametric basis functions $\{f_j(t)\}$ where the first basis function, $f_1(t)$, typically represents the dominant or average trend over the spatial region of interest, and subsequent basis functions, computed to be orthogonal to the first, along with the spatially indexed parameters β_{xj}, permit the shape and amplitude of the spatial structure to vary. We illustrate these empirical trend characteristics and a simple approach to computation of useful basis functions using maximum 8-hour average daily ozone observations at 94 monitoring sites in southern California for the 8-year period 1987 to 1994.

Carrying out spatial analysis, including predictions of air quality concentrations $Z(x_0, t)$ at unmonitored locations x_0, requires spatial prediction of both the trend and the temporal deviations about the trend. The former can be achieved by more-or-less conventional geostatistical kriging of the multivariate spatial data set defined by the spatial data set of vectors of trend coefficients. The latter is addressed here using the Sampson-Guttorp spatial deformation model for nonstationary spatial covariance structure, first introduced in Sampson and Guttorp (1992), but implemented here using the Bayesian framework of Damian et al. (2001, 2003).

3.4.1.1 Computation of temporal trend basis functions from incomplete data using an SVD

Assuming a complete data matrix of T observations on each of N monitoring sites, and letting $\mathbf{Z}$, $\mathbf{M}$, and $\mathbf{E}$ represent the $T \times N$ matrices corresponding to the terms of (3.42), the decomposition can be written

$$\mathbf{Z} = \mathbf{M} + \mathbf{E} \tag{3.44}$$

where

$$\mathbf{M} = \mathbf{F}\,\mathbf{B}, \tag{3.45}$$

$\mathbf{F} = [f_0(t)\ f_1(t)\ \cdots\ f_J(t)]$ being a $T \times J$ matrix with columns the T-vectors of values of the basis functions $f_j(t)$, $j = 1, 2, \ldots, J$, with $f_0(t) \propto 1$. Without loss of generality we can scale these basis functions to norm one, $f_j(t)' f_j(t) = 1$. The matrix $\mathbf{B}$ is the $J \times N$ matrix of trend coefficients for the N sites,

$$\mathbf{B} = \begin{bmatrix} \beta_{01} & \beta_{02} & \cdots & \beta_{0N} \\ \beta_{11} & \beta_{12} & \cdots & \beta_{1N} \\ \vdots & \vdots & \cdots & \vdots \\ \beta_{J1} & \beta_{J2} & \cdots & \beta_{JN} \end{bmatrix}. \tag{3.46}$$

As $\mathbf{M}$ is of rank J, we obtain the most parsimonious set of basis functions for a least squares approximation of the data matrix $\mathbf{Z}$ by taking $\mathbf{F}$ to be the matrix of the first J left singular vectors in the singular value decomposition (SVD) $\mathbf{Z} = \mathbf{U}\mathbf{D}\mathbf{V}'$. To the extent that a large number of sites share similar temporal (seasonal) patterns, the left singular vectors will represent these patterns. There are, however, two problems in proposing these as the empirical orthogonal function basis for this trend modeling problem. First, although the first singular vector does usually represent the dominant shared seasonal pattern, based on observational data with substantial variation about the trend, it turns out to be a noisy representation of this pattern. Second, air quality data matrices always contain substantial numbers of missing observations, so the usual SVD routines cannot be applied directly.

The SVD can be computed using an iterative algorithm in which the left singular vectors are computed by regressions in which smoothness is imposed directly. However, it has proven adequate to simply compute smoothed versions of the left singular vectors and then compute the coefficient matrix $\mathbf{B}$ by ordinary least squares regressions of the columns of $\mathbf{Z}$ on the set of J smoothed columns of $\mathbf{F}$. To deal with missing data, we use a simple "EM-like" iterative algorithm for the SVD. This algorithm can be explained as follows:

1. Specify a dimension (rank), J, for the EOF model.
2. Scale the observations at each monitoring site (columns of $\mathbf{Z}$) to norm (variance) one; call this matrix $\tilde{\mathbf{Z}}$.
3. Fill in the missing observations in the data matrix $\tilde{\mathbf{Z}}$ using elements of an initial rank-one approximation provided by a regression through the origin of each column of $\tilde{\mathbf{Z}}$ on the vector u_1 computed as the average over sites of the columns of $\tilde{\mathbf{Z}}$.
4. Compute the rank-J SVD-approximation of the now complete data matrix $\tilde{\mathbf{Z}}$.
5. Replace the missing values in $\tilde{\mathbf{Z}}$ by the elements of the rank-J SVD approximation.
6. Return to step 3 and iterate to convergence.

Following this calculation we use an ordinary spline smoother to smooth the plots of the left singular vectors against time $t = 1, 2, \ldots, T$. As the imputation of missing values in this algorithm depends explicitly on the assumed rank of the model, one must fit models of varying dimension and then choose a preferred model based on conventional measures like the percent of summed squared variation in $\tilde{\mathbf{Z}}$ explained, along with graphical analysis of the fitted trend models.

3.4.1.2 Spatial deformation modeling of the nonstationary spatial covariance structure of the detrended space-time residuals

Sampson and Guttorp (1994) introduced an approach to nonstationary spatial covariance modeling in which the geographic coordinates are deformed to create a geography (the disperion plane, or D-plane) in which the covariance structure is approximately isotropic. This approach is usually applied to detrended residuals. We assume for simplicity that the temporal structure of the residuals $\hat{\varepsilon}(x, t) = Z(x, t) - \hat{\mu}(x, t)$ is white noise.

We decompose the residuals $\hat{\varepsilon}(x, t) = \nu(x) H(x, t) + E(x, t)$, where $H(x, t)$ is a mean zero, variance one spatial process with covariance structure

$$\text{Cov}(H(x, t), H(y, t)) = \rho_\theta \left(|f(x) - f(y)| \right),$$

and $E(x, t)$ is a white noise process, uncorrelated with $H(x, t)$. The function f is the deformation of the geographic plane, and is fitted using a pair of thin-plate splines (Bookstein, 1989).

Technically we use a Gaussian-based Bayesian approach using MCMC, detailed in Damian et al. (2001, 2003). This has the advantage that we can draw samples from the deformations, and get a good feeling for the uncertainty in the fit.

3.4.1.3 Example: Application to 8-hour maximum average daily
ozone concentrations from southern California

The analysis of 8-hour maximum average daily ozone concentrations from
southern California for the period 1987 to 1994 was one of seven similar analyses of data from regions spanning most of the continental United States.
Ozone seasonal trends are similar nationwide and we hoped to be able to use
a single set of temporal trend basis functions for all regions. The computation
of trend components was based on data from 513 monitoring sites monitoring nearly throughout the year across the country. A large fraction of the
monitors in the United States, especially in the north and northeast, are in
operation only over the primary ozone season from April through October.
Because of the lengthy time series available, and because we were not explicitly concerned with the spatially varying temporal autocorrelation in the
ozone data, we avoided this issue by basing inference on the data obtained by
subsampling every third daily observation, which results in series that have
little autocorrelation (about their trend).

Temporal trend components. Analysis of the ozone concentration data was
judged most appropriate on a square root scale. Figure 3.20 shows the four
temporal basis functions computed from the first four left singular vectors
of the 2912×513 data matrix (considering only those sites sampling essentially year-round). The first function clearly represents the dominant seasonal
ozone cycle with highest concentrations during the sunny summer months.
The shape and amplitude of this seasonal feature varies from year to year, another notable distinction from many seasonal trend models that do not adapt
to fluctuations in trend. The second trend component is necessary for many
sites that display a pair of ozone peaks, one in spring and another later in the
summer. The third and fourth components serve mainly to adjust the exact
shape and locations of the seasonal peaks defined by the first two components. Approximately 67% of the variance in the entire (scaled) 2912×513
data matrix is explained by the first four unsmoothed components.

Figure 3.21 illustrates trends for two monitoring sites in Los Angeles County.
These were computed as linear combinations of the four trend components illustrated in Figure 3.20. We note that there is distinct variation in the shape
of the seasonal trend even over this relatively small spatial region with the first
site, 060370002 (which was inoperative in early 1987 and from 1992 through
1994), showing the dominant seasonal pattern and the second site, 060371902,
showing two reasonably distinct seasonal maxima. Figure 3.22 illustrates the
prediction of the trends at these and two other sites in Los Angeles County, all
of which were left out of the spatial analysis for purposes of cross-validation.
Site 060371902 was selected as an example of the greatest error in the prediction of the trend. The error is primarily in the intercept rather than the
shape of the seasonal trend.

Spatial-deformation and illustration of validation predictions of actual ob-
servations. The coastline, complex topology, and typical weather patterns
combine to effect a complex nonstationary spatial correlation structure in the

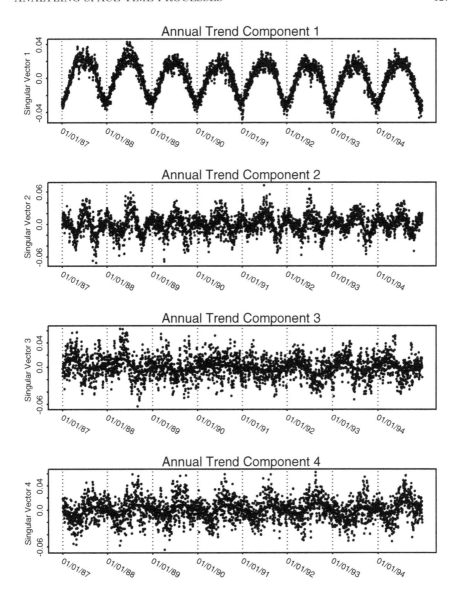

Figure 3.20 First four singular vectors (dots) and smooth trend components derived from the 2912×513 matrix of square root transformed ozone concentrations.

spatio-temporal residuals from the fitted temporal trends. Figure 3.23 depicts the posterior mean estimate of the spatial deformation computed to permit fitting of stationary isotropic correlation models in the deformed coordinate system. The predominant feature of compression along the coastline running NW-SE indicates that spatial correlation is strongest parallel to the coast and

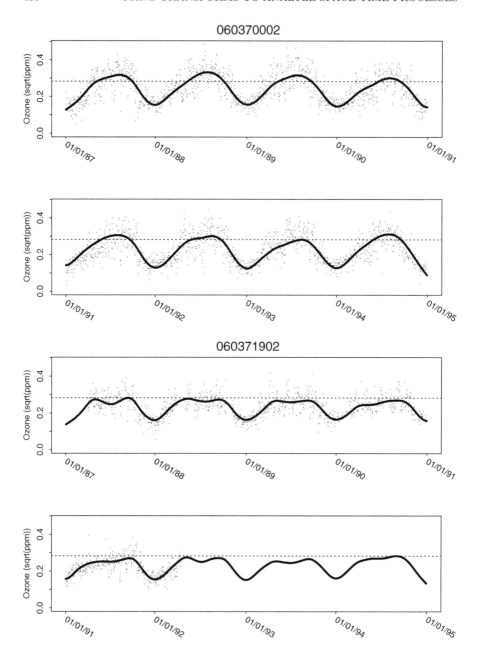

Figure 3.21 Fitted temporal trends for two monitoring sites in Los Angeles County.

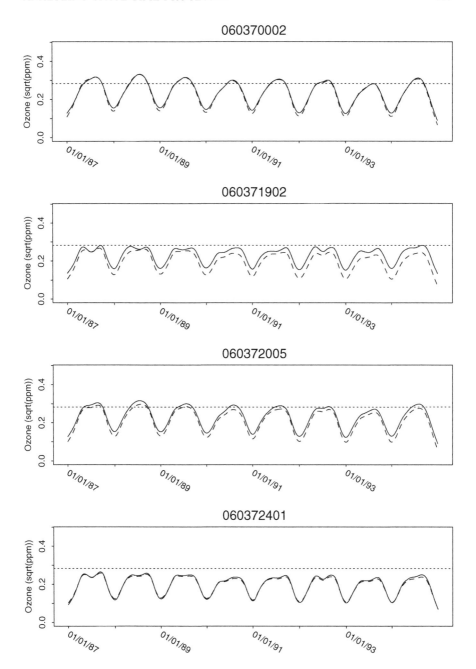

Figure 3.22 Predicted trend curves for four validation monitoring sites in Los Angeles County.

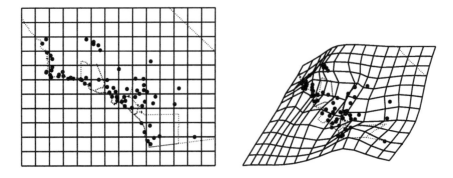

Figure 3.23 Posterior mean spatial deformation representing nonstationary spatial covariance structure.

weaker orthogonal to the coast. Figure 3.24 shows empirical spatial correlations vs. distance in the geographic and new deformed coordinate system. Spatial prediction (kriging) of the trend coefficients combined with prediction of the spatio-temporal residuals produce the cross-validation predictions of the time series for the two Los Angeles County sites, as illustrated in Figure 3.25.

3.4.2 Spectral analysis

3.4.2.1 A spectral representation

A stationary spatial-temporal process $\{Z(x,t) : x \in D \subset \mathbb{R}^d, t \in T \subset \mathbb{R}\}$ has a spectral representation in terms of sine and cosine waves of different

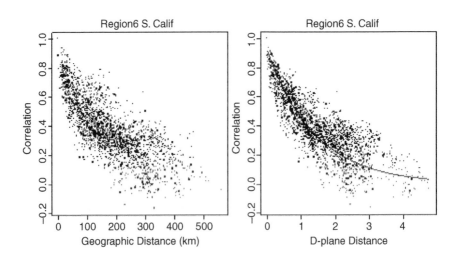

Figure 3.24 Spatial correlations vs. (left) distance in the geographic plane and (right) distance in the deformed coordinate system (D-plane) of Figure 3.23. An estimated exponential spatial correlation function is drawn on the D-plane scatter.

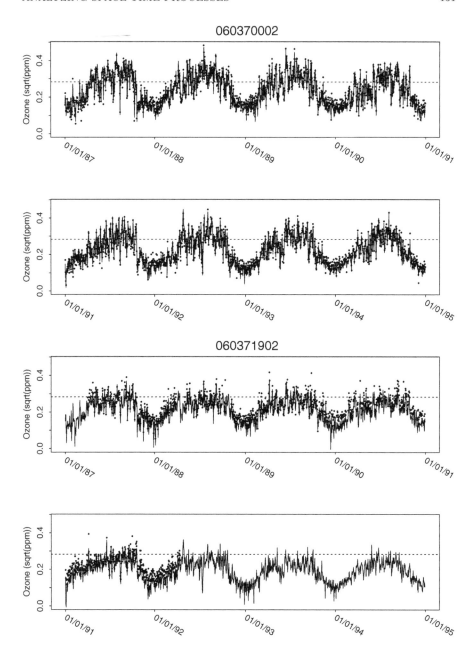

Figure 3.25 Cross-validation predictions of daily (square root) ozone concentrations at two monitoring sites in Los Angeles County.

frequencies (ω, τ), where ω is d-dimensional spatial frequency and τ is temporal frequency. If $Z(x, t)$ is a stationary random field with spatial-temporal covariance $C(x, t)$, then we can represent the process in the form of the following Fourier-Stieltjes integral:

$$Z(\mathbf{x}, t) = \int_{\mathbb{R}^d} \int_{\mathbb{R}} \exp(i\omega^T x + i\tau t) dY(\omega, \tau), \qquad (3.47)$$

where Y is a random function that has uncorrelated increments with complex symmetry except for the constraint, $dY(\omega, \tau) = dY^c(-\omega, -\tau)$, needed to ensure that $Z(\mathbf{x}, t)$ is real-valued. Y^c denotes the complex conjugate of Y. Using the spectral representation of Z and proceeding formally,

$$C(\mathbf{x}, t) = \int_{\mathbb{R}^d} \int_{\mathbb{R}} \exp(i\omega^T x + i\tau t) F(d\omega, d\tau), \qquad (3.48)$$

where the function F is a positive finite measure and is called the spectral measure or spectrum for Z. The spectral measure F is the mean square value of the process Y,

$$E\{|Y(\omega, \tau)|^2\} = F(\omega, \tau).$$

It is easy to see that for any finite positive measure F, the function given in (3.48) is positive-definite. If F has a density with respect to Lebesgue measure, it is the spectral density, f, which is the Fourier transform of the spatial-temporal covariance function:

$$f(\omega, \tau) = \frac{1}{(2\pi)^{d+1}} \int_{\mathbb{R}^d} \int_{\mathbb{R}} \exp(-i\omega^T x - i\tau t) C(\mathbf{x}, t) d\mathbf{x} dt, \qquad (3.49)$$

and the corresponding covariance function is given by

$$C(\mathbf{x}, t) = \int_{\mathbb{R}^d} \int_{\mathbb{R}} \exp(i\omega^T x + i\tau t) f(\omega, \tau) d\omega d\tau. \qquad (3.50)$$

When $f(\omega, \tau) = f^{(1)}(\omega) f^{(2)}(\tau)$, we obtain

$$C(\mathbf{x}, t) = \int_{\mathbb{R}^d} \int_{\mathbb{R}} \exp(i\omega^T x + i\tau t) f^{(1)}(\omega) f^{(2)}(\tau) d\omega d\tau$$

$$= \int_{\mathbb{R}^d} \exp(i\omega^T x) f^{(1)}(\omega) d\omega \int_{\mathbb{R}} \exp(i\tau t) f^{(2)}(\tau) d\tau$$

$$= C^{(1)}(x) C^{(2)}(t),$$

which means the corresponding spatial-temporal covariance is separable.

3.4.2.2 A new class of nonseparable space-time covariances

We propose the following spatial-temporal spectral density (Fuentes et al., 2005) that has a separable model as a particular case,

$$f(\omega, \tau) = \gamma(\alpha^2 \beta^2 + \beta^2 |\omega|^2 + \alpha^2 \tau^2 + \epsilon|\omega|^2 \tau^2)^{-\nu}, \qquad (3.51)$$

where γ, α, and β are positive, $\nu > \frac{d+1}{2}$, and $\epsilon \in [0, 1]$. The function in (3.51) is a valid spectral density. First, $f(\omega, \tau) > 0$ everywhere. Second, $f(\omega, \tau) \leq \gamma(\alpha^2\beta^2 + \beta^2|\omega|^2 + \alpha^2\tau^2)^{-\nu}$, and

$$\int_{\mathbb{R}^d} \int_{\mathbb{R}} \exp(i\omega^T x + i\tau t)\gamma(\alpha^2\beta^2 + \beta^2|\omega|^2 + \alpha^2\tau^2)^{-\nu} d\omega d\tau$$

$$= \frac{\pi^{\frac{d+1}{2}}\gamma}{2^{\nu-\frac{d+1}{2}-1}\Gamma(\nu)\alpha^{2\nu-d}\beta^{2\nu-1}} \left(\alpha\sqrt{(\frac{\beta}{\alpha}t)^2 + |x|^2}\right)^{\nu-\frac{d+1}{2}}$$

$$\times K_{\nu-\frac{d+1}{2}}\left(\alpha\sqrt{(\frac{\beta}{\alpha}t)^2 + |x|^2}\right). \tag{3.52}$$

Therefore, $\int_{\mathbb{R}^d} \int_{\mathbb{R}} \exp(i\omega^T x + i\tau t)f(\omega, \tau)d\omega d\tau$ exists.

In the representation (3.51), the parameter α^{-1} explains the rate of decay of the spatial correlation, β^{-1} explains the rate of decay for the temporal correlation, and γ is a scale parameter. The parameter ν measures the degree of smoothness of the process Z. The parameter ϵ indicates the interaction between the spatial and temporal components. Next, we discuss two particular cases of ϵ.

Two particular cases:

$\epsilon = 1$

When $\epsilon = 1$, the Equation (3.51) can be written as

$$f(\omega, \tau) = \gamma(\alpha^2\beta^2 + \beta^2|\omega|^2 + \alpha^2\tau^2 + |\omega|^2\tau^2)^{-\nu}$$
$$= \gamma(\alpha^2 + |\omega|^2)^{-\nu}(\beta^2 + \tau^2)^{-\nu}.$$

Therefore the corresponding spatial-temporal covariance is separable. Moreover, in the expression of this covariance, both the spatial component and the temporal component are the Matérn type covariances. When $\gamma = \alpha = \beta = d = 1$ and $\nu = 3/2$, a contour plot of the corresponding separable spatial-temporal covariance is given in Figure 3.26. From the plot, there are ridges along the lines where the spatial lag is 0 and the temporal lag is 0.

$\epsilon = 0$

When $\epsilon = 0$,

$$f(\omega, \tau) = \gamma(\alpha^2\beta^2 + \beta^2|\omega|^2 + \alpha^2\tau^2)^{-\nu}. \tag{3.53}$$

The function in (3.53) is an extension of the traditional Matérn spectral density. It treats time as an additional component of space, but it does have a different rate of decay. In the spectral density (3.53), the parameter α^{-1} explains the rate of decay of the spatial correlation. For the temporal correlation, the rate of decay is explained by the parameter β^{-1}. γ is a scale parameter. The parameter ν measures the degree of smoothness of the process Z.

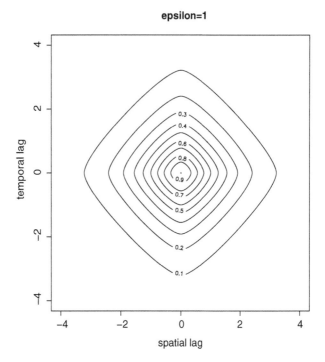

Figure 3.26 The contour plot for a separable spatial-temporal covariance.

The higher the value of ν, the smoother the process Z will be. The corresponding spatial-temporal covariance is given by (3.52), which is a Matérn type covariance. We have

$$C(x, t) = \frac{\sigma^2}{2^{\nu-1}\Gamma(\nu)} \left(\frac{\|(x, \rho t)\|}{r} \right) \mathcal{K}_\nu \left(\frac{\|(x, \rho t)\|}{r} \right), \tag{3.54}$$

where σ^2, r, ρ, and ν are all positive. $\|\cdot\|$ denotes the Euclidean distance. In the representation (3.54), the parameter r measures how the correlation decays with distance; generally this parameter is called *range*. The parameter σ^2 is the variance of the process Z. The parameter $\nu > 0$ measures the degree of smoothness of the process Z. The parameter ρ is a scale factor to take into account the change of units between the spatial and temporal domains. Therefore, this parametric model for C corresponds to a $(d + 1)$-dimensional Matérn type covariance with an extra parameter ρ, which can be explained as a conversion factor between the units in the space and time domains. When $\gamma = \alpha = \beta = d = 1$ and $\nu = 3/2$, a contour plot of corresponding nonseparable spatial-temporal covariance is given in Figure 3.27. It has a very smooth surface.

When $\epsilon \in (0, 1)$, we are not able to write down the exact expression of the spatial-temporal covariance, which correponds to the spectral density in

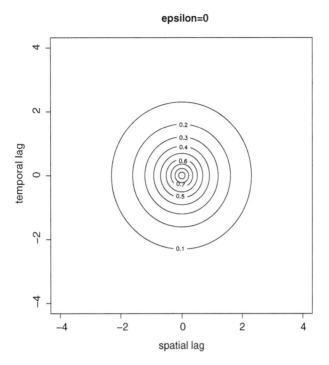

Figure 3.27 The contour plot for a nonseparable spatial-temporal covariance.

(3.51). But we can calculate this numerically because

$$C(\mathbf{x}, t) = \int_{\mathbb{R}^d} \int_{\mathbb{R}} \exp(i\omega^T x + i\tau t) f(\omega, \tau) d\omega d\tau$$

$$= \int_{\mathbb{R}} \exp(i\tau t) g(x, \tau) d\tau, \tag{3.55}$$

where $g(x, \tau) = \int_{\mathbb{R}^d} \exp(i\omega^T x) f(\omega, \tau) d\omega$. The function $g(x, \tau)$ is available from the integration; therefore $C(x, t)$ can be computed by numerically carrying out a one-dimensional Fourier transformation of g. This can be quickly approximated using fast Fourier transform. A separate transform needs to be done for every value of (x, t) of interest, but this is feasible. The expression of g is given by

$$g(x, \tau) = \frac{\pi^{d/2} \gamma}{2^{\nu - \frac{d}{2} - 1} \Gamma(\nu)} (\beta^2 + \epsilon \tau^2)^{-\nu} \left(\frac{|x|}{\theta(\tau)} \right)^{\nu - \frac{d}{2}} \mathcal{K}_{\nu - \frac{d}{2}}(\theta(\tau)|x|),$$

where $\theta(\tau) = \sqrt{\frac{\alpha^2(\beta^2 + \tau^2)}{\beta^2 + \epsilon \tau^2}}$. When $\gamma = \alpha = \beta = d = 1$ and $\nu = 3/2$, contour plots of corresponding nonseparable spatial-temporal covariances with $\epsilon = 0.1, 0.2, 0.3, 0.5, 0.7, 0.8, 0.9$ are given in Figure 3.28. The ridge is getting more

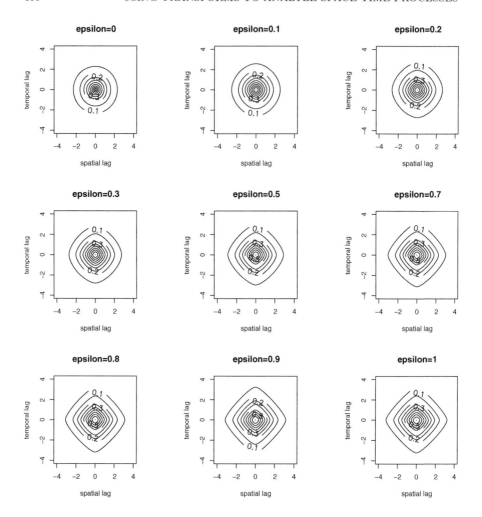

Figure 3.28 Contour plots for some nonseparable spatial-temporal covariances.

obvious when ϵ is getting large, say close to 1. In fact, when ϵ is close to 1, $\epsilon|\omega|^2\tau^2$ is close to $|\omega|^2\tau^2$. Therefore the spectral density in (3.51) is turning close to a separable form from a nonseparable form as $\epsilon \to 1$.

In summary, the new class spectral density in (3.51) is nonseparable for $0 \leq \epsilon < 1$, and separable for $\epsilon = 1$. Therefore, the parameter ϵ plays a role for separability. It controls the interaction between the spatial component and the temporal component. Note that the degree of smoothness is the same for the spatial component and the temporal component. Therefore a more general class is proposed to allow for a different degree of smoothness for space and time:

$$f(\omega,\tau) = \gamma \left\{ c_1(a_1^2 + |\omega|^2)^{\alpha_1} + c_2(a_2^2 + \tau^2)^{\alpha_2} + \epsilon(a_3^2 + |\omega|^2\tau^2)^{\alpha_3} \right\}^{-\nu}, \quad (3.56)$$

where $a_1, a_2, a_3, \alpha_1, \alpha_2, \alpha_3$ and c_1, c_2 are positive; $\epsilon \in [0, 1]$ and $\frac{d}{\alpha_1 \nu} + \frac{1}{\alpha_2 \nu} < 2$. This is a valid spectral density. The spectral density in (3.51) is a special case of this more general class. If $\alpha_1 = \alpha_3 = 1$ and $\frac{d}{\nu} + \frac{1}{\alpha_2 \nu} < 2$ in (3.56), we have

$$
\int_{\mathbb{R}^d} \exp(i\omega^T x) f(\omega, \tau) d\omega
$$

$$
= \frac{\pi^{d/2} \gamma}{2^{\nu - \frac{d}{2} - 1} \Gamma(\nu)} (c_1 + c_3 \tau^2)^{-\nu} \left(\frac{|x|}{\rho(\tau)} \right)^{\nu - \frac{d}{2}} \mathcal{K}_{\nu - \frac{d}{2}} (\rho(\tau) |x|),
$$

(3.57)

where $\rho(\tau) = \left(\frac{c_2(a_2^2 + \tau^2)^{\alpha_2} + c_1 a_1^2 + c_3 a_3^2}{c_1 + c_3 \tau^2} \right)^{1/2}$. So the corresponding spatial-temporal covariance $C(x, t)$ can be quickly approximated using the fast Fourier transformation of (3.58).

3.4.2.3 An example in meteorology

In the analysis presented here we study and model the spatial temporal structure of wind fields using MM5 model output wind fields from July 21, 2002, for the full 24-h period. The complicated flow patterns over the region during this time are evident in Figure 3.29. The arrows indicate the direction from which the winds are coming, while the length of the stem indicates wind speed. An easterly wind (winds from the east, southeast, and northeast) tends to dominate over the majority of the region for this time period. The 3 a.m. (7-h forecast) plot clearly shows an area of confluence (an area where the wind vectors tend to come together) over the Bay. At 9 a.m. (13-h forecast), this area of confluence has been replaced by an area of diffluence (an area where the wind vectors tend to spread apart). Diffluence seems to persist over the majority of the Bay at noon (16-h forecast) and for the rest of the period studied. There is little evidence in these plots to suggest that MM5 was capturing the sea breeze circulation, which observations show to be present.

As a first empirical attempt to deal with the nonstationarity inherent in these kinds of environmental data, we divided the spatial domain into two broad categories: land and water. Five subregions of nonstationarity were found and they are shown in Figure 3.30; these subregions were identified using the test for nonstationary. This final regional arrangement of clusters appears reasonable considering atmospheric and oceanic processes that are occurring in the boundary layer on this day.

We model the nonstationary Z as a mixture of (independent) local stationary space-time processes Z_i for $i = 1, \ldots, k$ that explain the space-time dependence structure in the five subregions of stationarity $S_1, \ldots, S_5$,

$$
Z(x, t) = \sum_{i=1}^{5} K(\mathbf{s} - \mathbf{s}_i) Z_i(x, t), \tag{3.58}
$$

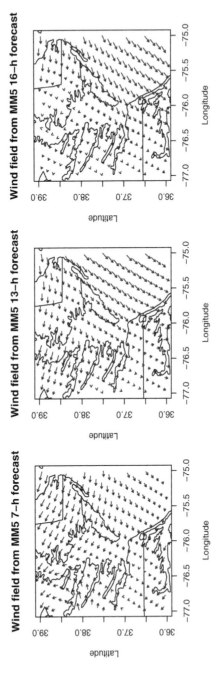

Figure 3.29 (**SEE COLOR INSERT FOLLOWING PAGE 142**) Wind field maps, showing wind direction and speed over the Chesapeake Bay at 3 a.m., 9 a.m., and noon on July 21, 2002.

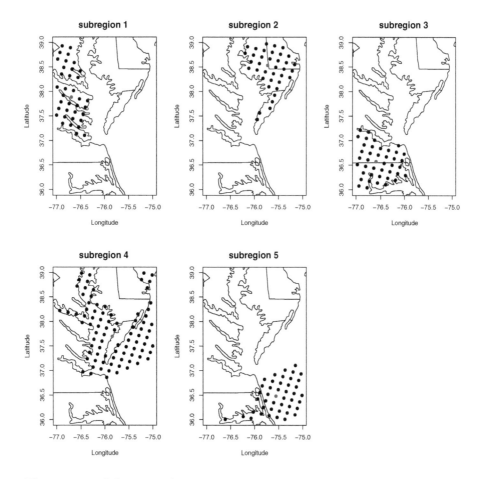

Figure 3.30 Subregions of stationarity.

where $\mathbf{s}_i = (x_i, t_i)$ is the centroid of the i-th subregion, and for $i = 1, \ldots, 5$. Z_i explains the spatial-temporal structure of Z in a subregion of stationarity S_i. The corresponding covariance function for the process Z is

$$\mathrm{cov}\{Z(\mathbf{x}_1, t_1), Z(\mathbf{x}_2, t_2)\} = \sum_{i=1}^{5} K(\mathbf{s}_1 - \mathbf{s}_i) K(\mathbf{s}_2 - \mathbf{s}_i) C_i(\mathbf{x}_1 - \mathbf{x}_2, t_1 - t_2),$$

$$(3.59)$$

where $\mathbf{s}_1 = (\mathbf{x}_1, t_1)$, $\mathbf{s}_2 = (\mathbf{x}_2, t_2)$, and each C_i is a space-time covariance that explains the space-time dependency in a subregion of stationarity S_i.

The weight function in (3.59) is modeled as $K(x - x_i) = \frac{1}{h_i^2} K_0(\frac{x - x_i}{h_i})$, where the location x_i is the centroid of the i-th subregion. The bandwith h_i is defined as half of the maximum distance for the i-th subregion. The function K_0 is modeled as $K_0(\mathbf{u}) = \frac{3}{4}(1 - u_1^2)_+ + \frac{3}{4}(1 - u_2^2)_+$, which is a quadratic weight

Table 3.5   The posterior distribution for ϵ_i

	$P(\epsilon_i = 0)$	$P(\epsilon_i = 1)$
subregion 1	7.62e-48	1
subregion 2	1.03e-91	1
subregion 3	3.76e-94	1
subregion 4	1	0
subregion 5	2.56e-150	1

function for $\mathbf{u} = (u_1, u_2)$. We fit the parameters for the spatial-temporal co-variance of Z using a Bayesian framework. The priors for the sill parameter, spatial range parameter, and temporal range parameter are Inverse Gamma with infinite variance. The prior distribution for the smoothness parameter is a uniform distribution with support $(0, 2]$. The support for the smoothness parameter is a conservative interval based on our previous experience analyzing similar datasets. The prior for ϵ gives all the mass to the values 0 and 1: $p(\epsilon = 0) = .5$, $p(\epsilon = 1) = 1$. The posterior distributions for ϵ_i (Table 3.5) suggest the separability for each subregion except for subregion 4. These seem to indicate that in subregion 4 we have a nonseparable covariance and in the rest of the subregions we have separability.

3.4.2.4 Testing for separability

In this section we introduce a spectral analog of the assumption of separability. Consider $\{Z(\mathbf{s};t) : \mathbf{s} \in D \subset \mathbb{R}^d; t \in \mathbb{Z}\}$ a spatial temporal zero-mean process observed at N space-time coordinates $(\mathbf{s}_1; t_1), \ldots, (\mathbf{s}_N; t_N)$. We start by assuming that the covariance function is stationary in space and time,

$$\mathrm{cov}(Z(\mathbf{s}_1; t_1), Z(\mathbf{s}_2; t_2)) = C(\mathbf{h}; u)$$

for $\mathbf{h} = \mathbf{s}_1 - \mathbf{s}_2$, $u = t_1 - t_2$.

We can write the covariance C in terms of the spectral density g of the spatial temporal process Z,

$$C(\mathbf{h}, u) = \int \int \exp\{i\mathbf{h}^T \omega + iu\tau\} g(\omega; \tau) d\omega d\tau,$$

where

$$g(\omega, \tau) = (2\pi)^{-d-1} \sum_{u=-\infty}^{u=\infty} \int \exp\{-i\mathbf{h}^T \omega - iu\tau\} C(\mathbf{h}; u) d\mathbf{h}$$

$$= (2\pi)^{-d} \int \exp\{-i\mathbf{h}^T \omega\} f(\mathbf{h}; \tau) d\mathbf{h}$$

(3.60)

for any fixed $\mathbf{h}$, $f(\mathbf{h}; \tau)$ is the cross-spectral density function of the time processes $Y_1(t) = Z(\mathbf{s}; t)$, and $Y_2(t) = Z(\mathbf{s} + \mathbf{h}; t)$, and we have

$$f(\mathbf{h}; \tau) = (2\pi)^{-1} \sum_{u=-\infty}^{u=\infty} \exp\{-iu\tau\} C(\mathbf{h}, u). \tag{3.61}$$

If C is a separable covariance, then we can write

$$C(\mathbf{h}, u) = C_1(\mathbf{h}) C_2(u),$$

where C_1 is a positive-definite function in $\mathbb{R}^d$, and C_2 is a positive-definite function in $\mathbb{R}$. Thus, $f(\mathbf{h}; \tau)$ is the product of a function of $\mathbf{h}$ and a function of τ,

$$
\begin{aligned}
f(\mathbf{h}; \tau) &= (2\pi)^{-1} \sum_{u=-\infty}^{u=\infty} \exp\{-iu\tau\} C(\mathbf{h}, u) \\
&= (2\pi)^{-1} \sum_{u=-\infty}^{u=\infty} \exp\{-iu\tau\} C_1(\mathbf{h}) C_2(u) = C_1(\mathbf{h})\kappa(\tau),
\end{aligned}
\tag{3.62}
$$

where κ is an integrable and positive function, and C_1 for each fixed τ is a covariance function of $\mathbf{h}$ and an integrable function of $\mathbf{h}$. We can obtain a nonseparable covariance function by making C_1 depend on τ. Thus, we get

$$C(\mathbf{h}, u) = \int \exp\{iu\tau\} C_1(\mathbf{h}; \tau)\kappa(\tau)d\tau.$$

Cressie and Huang (1999) use this spectral representation to generate parametric models of nonseparable spatio-temporal stationary covariance functions.

Thus, if Z is separable,

$$f(\mathbf{h}; \tau) = C_1(\mathbf{h})\kappa(\tau). \tag{3.63}$$

In principle, f could be a complex function, but when Z is separable and stationary, f is real.

3.4.2.4.1 Test for separability

We propose a test for separability of spatial-temporal processes. The beauty of this method is that the mechanics of the test can be reduced to those of a simple two-way ANOVA procedure. We test for separability by studying if the coherence R is a function of τ.

We consider the standardized asymptotic distribution of the variance stabilizing transformation of R, that is,

$$\phi_{\mathbf{a},\mathbf{b}}(\tau) = \tanh^{-1}(\tilde{R}_{\mathbf{a},\mathbf{b}}(\tau)),$$

and we estimate it with

$$\hat{\phi}_{\mathbf{a},\mathbf{b}}(\tau) = \tanh^{-1}(\hat{R}_{\mathbf{a},\mathbf{b}}(\tau)),$$

where the coherency is estimated by replacing f with $\hat{f}_{\mathbf{ab}}(\omega)$, a tapered second-order periodogram function proposed by Fuentes (2005b):

$$\hat{f}_{\mathbf{ab}}(\omega) = \int_{-\infty}^{\infty} \int_{-\infty}^{\infty} g_\rho(\mathbf{a} - \mathbf{s}) g_\rho(\mathbf{b} - \mathbf{s}) I^*_{\mathbf{a}+\mathbf{s},\mathbf{b}+\mathbf{s}}(\omega) d\mathbf{s}, \qquad (3.64)$$

where

$$I^*_{\mathbf{ab}}(\omega) = 2\pi/T \sum_{t=0}^{T-1} W^{(T)}(\omega - 2\pi t/T) I_{\mathbf{ab}}(2\pi t/T), \qquad (3.65)$$

I is the second-order periodogram, and W and g are two filter functions. Thus, $\hat{f}_{\mathbf{ab}}(\omega)$ can be interpreted as an average of the total energy of the process contained within a band of frequencies in the region of ω and a region in space in the neighborhood of $\mathbf{a}$ and $\mathbf{b}$. We evaluate $\hat{\phi}_{(\mathbf{a}_i,\mathbf{b}_i)}(\tau)$ at k pairs at pairs of locations $\{(\mathbf{a}_i,\mathbf{b}_i)\}_{i=1}^{k}$ and a set of frequencies $\tau_1, \tau_2, \ldots, \tau_n$ that cover the domain. We write

$$\hat{\phi}_{(\mathbf{a}_i,\mathbf{b}_i)}(\tau_j) = \phi_{(\mathbf{a}_i,\mathbf{b}_i)}(\tau_j) + \epsilon((\mathbf{a}_i,\mathbf{b}_i),\tau_j).$$

Asymptotically $E\{\epsilon((\mathbf{a}_i,\mathbf{b}_i),\tau_j)\} = 0$ and $\mathrm{Var}\{\epsilon((\mathbf{a}_i,\mathbf{b}_i),\tau_j)\} = \sigma^2$, where σ^2 is independent of $(\mathbf{a}_i,\mathbf{b}_i)$ and ω_j.

Assuming that $(\mathbf{a}_i,\mathbf{b}_i)$ and τ_j are spaced "sufficiently wide apart," then the $\epsilon((\mathbf{a}_i,\mathbf{b}_i),\tau_j)$ will be approximately uncorrelated; this is based on the asymptotic properties of $\hat{f}_{(\mathbf{a}_i\mathbf{b}_i)}(\tau_j)$ (see Fuentes, 2005b). We write

$$U_{ij} = \hat{\phi}_{(\mathbf{a}_i,\mathbf{b}_i)}(\tau_j), \qquad (3.66)$$

$$m_{ij} = \phi_{(\mathbf{a}_i,\mathbf{b}_i)}(\tau_j),$$

and

$$\epsilon_{ij} = \epsilon((\mathbf{a}_i,\mathbf{b}_i),\tau_j).$$

Then, we have the model

$$U_{ij} = m_{ij} + \epsilon_{ij},$$

that becomes the usual "two-factor analysis of variance" model, and can be rewritten:

$$H_1 : U_{ij} = \mu + \alpha_i + \beta_j + \epsilon_{ij}$$

for $i = 1, \ldots, k$ and $j = 1, \ldots, n$. The parameters $\{\alpha_i\}, \{\beta_j\}$ represent the main effects of the space and frequency factors. We test for separability using the standard techniques to test the model $(\beta_j = 0)$:

$$H_{0a} : U_{ij} = \mu + \alpha_i + \epsilon_{ij}$$

against the more general model H_1.

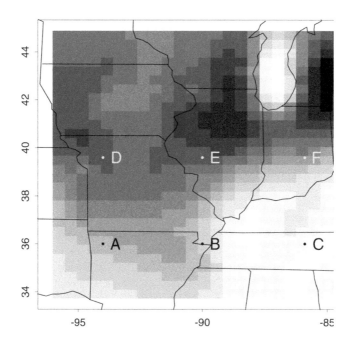

Figure 3.31 This graph shows the ozone deposition flux (kg/hectare) in a region in the Midwest. The values in this graph are the output of the EPA regional scale air quality model (Models-3) on June 2, 1996, at 12 p.m. Central Time. Reprinted from *Journal of Statistical Planning and Inference,* 90, Fuentes, M., Testing for separability of spatial-temporal covariance functions, 183–193, Copyright 2005, with permission from Elsevier.

3.4.2.5 An example in air quality

In this example we analyze the spatial temporal structure of a process Z, which is the hourly averaged ozone fluxes (kg/hectare) in June 1996 (June 2–June 4), using the output of air quality numerical models (Models-3) (see Figure 3.31 and Fuentes, 2005b). We have 72 observations over time. Before applying our tests for stationarity and separability, we need to remove the spatial-temporal trend. The tables in this subsection are from Fuentes (2005b).

To remove the spatial trend, we calculated at each location the ozone anomaly, that is, the corresponding ozone value minus the mean over time (using the 72 observations over time at each location). We removed the temporal trend using a cosine and sine function with a period of 24 hours. The diurnal cycle of the ozone anomalies appeared to be the same everywhere. We implemented our test to the ozone anomalies (after removing this diurnal cycle).

The estimates, $\hat{f}_{\mathbf{ab}}(\omega)$, were obtained using (3.64), the cross-spectral function proposed by Fuentes (2005b), for the pair $\{\mathbf{a}, \mathbf{b}\}$ in D, in which $W(\alpha)$ is given in the example presented in Section 3.2.6.1 with a bandwidth of $2\pi B$ with $B = 1/12$, and $g_\rho(\mathbf{u})$ (as in Section 3.2.6.1) has $\rho = 5.5$ units (1 unit = 36 km), $m = BT$. Thus, to obtain approximately uncorrelated estimates, the

Table 3.6 Analysis of variance (all six sites)

Item	Degrees of freedom	Sum of squares	F Value	Pr(F)
between spatial points	5	10.4784	6.1440	0.0007
between frequencies	5	25.5483	14.9802	0.0000
residuals	25	8.5273		

Source: Reprinted from *Journal of Statistical Planning and Inference*, 90, Fuentes, M., Testing for separability of spatial-temporal covariance functions, 183–193, Copyright 2005, with permission from Elsevier.

frequencies ω_j and pairs $\{(\mathbf{a}_i, \mathbf{b}_i)\}_{i=1}^{k}$ should be chosen so that the spacings between the ω_j are at least $\pi/6$, and the distance between any pairs $(\mathbf{a}_i, \mathbf{b}_i)$ and $(\mathbf{a}_j, \mathbf{b}_j)$ for $i \neq j$ is at least 5.5 grid cells (198 km).

The ω_j were chosen as follows: $\omega_j = \pi j/17$ with $j = 1\,(3)\,16$, corresponding to a uniform spacing of $3\pi/17$ (which just exceeds $\pi/6$). We evaluate $\hat{f}_{\mathbf{ab}}(\omega)$ at the following frequencies: $\omega_1 = \pi/17$, $\omega_2 = 4\pi/17$, $\omega_3 = 7\pi/17$, $\omega_4 = 10\pi/17$, $\omega_5 = 13\pi/17$, and $\omega_6 = 16\pi/17$. We consider six pairs, $\{(\mathbf{a}_i, \mathbf{b}_i)\}_{i=1}^{6}$, such that the distance between pairs is at least 216 km (which just exceeds ρ) for all i, where $\|.\|$ denotes the Euclidean distance. Figure 3.31 shows the locations of $\mathbf{a}_i$ for $i = 1, \ldots, 6$, which correspond to the sites $A, B, \ldots$, and F, respectively. The locations of $\mathbf{b}_i$ for $i = 1, \ldots, 6$, are the other six sites at the same latitude as $A, B, \ldots$, and F but 72 km, 144 km, 180 km, 72 km, 144 km, and 180 km further east, respectively. Table 3.6 shows the results of the test for separability, using these six pairs.

The "between spatial locations" effect is highly significant (p-value < 0.01), confirming that there is clear evidence of lack of separability. The coefficient of determination is .81.

In the next table we study separability in a smaller subregion (the eastern part of our domain) using only pairs $(\mathbf{a}_3, \mathbf{b}_3)$ and $(\mathbf{a}_6, \mathbf{b}_6)$. Table 3.7 shows that the "between spatial locations" effect is not significant, suggesting then that there is no evidence of lack of stationarity (in the eastern part of our domain using pairs $(\mathbf{a}_3, \mathbf{b}_3)$ and $(\mathbf{a}_6, \mathbf{b}_6)$). However, the "between frequencies"

Table 3.7 Analysis of variance (using sites **c** and **f**)

Item	Degrees of freedom	Sum of squares	F Value	Pr(F)
between spatial points	1	0.0129	0.1872	0.6832
between frequencies	5	11.1377	32.1995	0.0008
residuals	5	0.3459		

Source: Reprinted from *Journal of Statistical Planning and Inference*, 90, Fuentes, M., Testing for separability of spatial-temporal covariance functions, 183–193, Copyright 2005, with permission from Elsevier.

Table 3.8 Coherence for pairs $(\mathbf{a}_3, \mathbf{b}_3)$ and $(\mathbf{a}_6, \mathbf{b}_6)$ at six frequencies

Pairs:	ω_1	ω_2	ω_3	ω_4	ω_5	ω_6
$(\mathbf{a}_3, \mathbf{b}_3)$	0.769	0.648	0.891	0.960	0.994	0.999
$(\mathbf{a}_6, \mathbf{b}_6)$	0.931	0.660	0.840	0.978	0.994	0.998

effect is highly significant, confirming that the process is nonseparable. The coefficient of determination is .97. Even in a smaller subregion the assumption of separability is still unrealistic (Table 3.7). The distance between the two components in both pairs is the same, so we could use these two pairs to test for stationarity. It appears that stationarity in that smaller subregion is a reasonable assumption.

Table 3.8 shows the $|\hat{R}_{\mathbf{ab}}(\omega)|^2$ values at each one of the frequencies ω_j for $j = 1, \ldots, 6$ for the pairs $(\mathbf{a}_3, \mathbf{b}_3)$ and $(\mathbf{a}_6, \mathbf{b}_6)$. $|R|^2$ can be interpreted as a coefficient of correlation: values close to 1 indicate high correlation between the two time series. Both pairs seem to have a very similar coherency function, which supports the assumption of stationarity. However, the coherence is clearly changing with frequency. Thus, it seems that locally it might be reasonable to assume stationarity for ozone fields, but separability seems to be unrealistic even for small geographic areas.

3.5 Discussion

We have presented three basic types of transform tools: Fourier transforms, wavelet transforms, and empirical orthogonal functions. In this section we discuss to what types of data these are best applicable, and under what circumstances they should perhaps be avoided. We also make some computational remarks and conclude with some open research questions in this general area.

3.5.1 What are the tools good for?

Fourier and wavelet methods in the spatial domain are very well adapted to gridded data. Generally speaking, Fourier tools are good for data that show some regular behavior over the entire region of observation, while wavelet tools can handle more irregular facets that change over space. There are ways to deal with nongridded data (one approach is mentioned in the subsection on computational aspects), but they are computationally cumbersome. The EOF method is set up to deal with temporally regular but spatially irregular observations, and expands to gridded data using geostatistical tools.

Some parts of the theory are developed using Gaussian assumptions. In the case of Fourier and wavelet transforms, the underlying data need not be Gaussian: the averaging methods used in computing the coefficients will be approximately Gaussian from central limit type results. As always, it is a good idea to try to transform data to approximate symmetry before doing an analysis. However, one then has to be careful when backtransforming, especially as standard errors may be affected by the transformation.

3.5.2 Computational aspects

All the tools used in this chapter do, in the best circumstances, use extremely fast computational algorithms. The fast Fourier transform, the discrete wavelet transform, and the singular value decomposition of a data matrix are all fast, and can handle large data sets. Difficulties come in when circumstances are not the best. For example, if data are on a grid, the wavelet approach to nonstationary covariance estimation is fast; but if they are not, things quickly get complicated. One approach is to fit a simple stationary model to the sample covariances (assuming temporal replications of a spatial field observed at fixed stations), then sample for each time point the conditional distribution of the field on the grid given the fitted covariances. The sample from this distribution is on the grid, so it allows estimation using the multiresolution approach. A new conditional sample is then drawn (using the latest estimate of the spatial covariance model), and the process is iterated until convergence. Similar approaches can be used for Fourier-type analyses.

In the actual analyses of environmental data, be it from monitoring stations, from deterministic models, or from both, an important part of the computation often involves some kind of Markov chain Monte Carlo method, and this is typically where the meat of the computational effort lies.

The trend estimation using wavelets is computationally equivalent to a nonparametric smoother in the iid case with a parametric bootstrap simulation of standard error. The advantage here is that the wavelet estimator is consistent (for polynomial trend and approximately for general smooth trends) even in the case of long-term memory. Also, the smoothers developed for iid situations generally behave poorly on time series data.

3.5.3 Some research questions

The tools described in this chapter do not generally have obvious counterparts for multivariate data sets. There are many possible covariance structures for space-time data (see Chapter 4 in this volume), but the additional complexity of multivariate data is that in reality there are rarely the symmetries (in time) needed for simple Kronecker-type structures. If we think about air quality models, the air chemistry (which actually is rather poorly understood) is definitely not reversible in time. Building some appropriate models for this is an important task.

There is a lot of room for investigating the importance of various assumptions. While, for example, hourly ozone data are clearly not separable, how much damage does it do to make a separability assumption? Our experience from studying nonstationary spatial covariance structures is that spatial estimates generally are very similar to those you would get from a stationary model, but the standard errors can be off by substantial amounts. A well-designed simulation study, looking at many aspects of the assumptions we have been discussing (stationarity in time and/or space, separability, Gaussianity etc.) would be very valuable.

All the authors of this chapter have spent considerable parts of their research careers thinking about nonstationary spatial covariance structures. While many different approaches are available (including the three mentioned in this chapter), there are very few direct comparisons between them. Finding a reasonable suite of data sets to make such a comparison study would be very useful.

References

Abramowitz, M. and Stegun, I.A. (1965). *Handbook of Mathematical Functions,* Dover Publications, New York.

Akaike, H. (1974). A new look at statistical model identification. *IEEE Trans. Auto. Contr.,* AC -19, 716–722.

Arfken, G. and Weber, H.J. (1995). *Mathematical Methods for Physicists,* 4th ed. Academic Press, San Diego, CA.

Beran, J. (1994). *Statistics for Long Memory Processes.* Monographs on Statistics and Applied Probability, Vol. 61, Chapman & Hall. New York.

Bookstein, F.L. (1989). Principal warps — Thin-plate splines and the decomposition of deformations. *IEEE Transactions on Pattern Analysis and Machine Intelligence,* 11(6), 567–585.

Bracewell, R. (1999). *The Fourier Transform and Its Applications,* 3rd ed. McGraw-Hill, New York.

Brillinger, D.R. (1981). *Time Series: Data Analysis and Theory, Expanded Edition,* Holden-Day, San Francisco.

Clark, R.M. (1977). Non-parametric estimation of smooth regresion function. *Journal of the Royal Statistical Society,* 39, 107–113.

Craigmile, P.F., Guttorp, P., and Percival, D.B. (2004). Trend assessment in a long memory dependence model using the discrete wavelet transform. *Environmetrics,* 15, 313–355.

Craigmile, P.F. and Percival, D.B. (2005). Asymptotic decorrelation of between-scale wavelet coefficients. *IEEE Transactions on Information Theory,* 51(3), 1039–1048.

Cole, J., Dunbar, R., McClanahan, T., and Muthiga, N. (2000). Tropical Pacific forcing of decadal variability in the western Indian Ocean over the past two centuries. *Science,* 287, 617–619.

Cramér, H. and Leadbetter, M.R. (1967). *Stationary and Related Stochastic Processes. Sample Function Properties and Their Applications.* Wiley, New York.

Cressie, N. (1985). Fitting variogram models by weighted least squares. *Journal of the International Association for Mathematical Geology,* 17, 563–586.

Cressie, N. and Huang, H. (1999). Classes of nonseparable spatio-temporal stationary covariance functions, *Journal of the American Statistical Association*, 94(448), 1330–1340.

Dahlhaus, R. and Künsch, H. (1987), Edge effects and efficient parameter estimation for stationary random fields. *Biometrika*, 74, 877–882.

Damian, D., Sampson, P.D., and Guttorp, P. (2001). Bayesian estimation of semiparametric non-stationary spatial covariance structures, *Environmetrics*, 12, 161–178.

Damian, D., Sampson, P.D., and Guttorp, P. (2003). Variance modeling for nonstationary spatial processes with temporal replications. *Journal of Geophysical Research-Atmospheres*, 108(D24), Art. No. 8778.

Daubechies, I. (1992). Ten Lectures on Wavelets. *SIAM, CBMS-NSF Conference Series in Applied Mathematics*, Vol. 61.

Fuentes, M. (2001). A new high frequency kriging approach for nonstationary environmental processes. *Environmetrics*, 12, 469–483.

Fuentes, M. (2002). Spectral methods for nonstationary spatial processes. *Biometrika*, 89, 197–210.

Fuentes, M. (2005a). A formal test for nonstationarity of spatial stochastic processes. *Journal of Multivariate Analysis*, 96, 30–54.

Fuentes, M. (2005b). Testing for separability of spatial-temporal covariance functions. *Journal of Statistical Planning and Inference*, 90, 183–193.

Fuentes, M., Chen, L., and Davis, J. (2005). A new class of nonseparable spatial-temporal covariance functions. Tech. report at North Carolina State University.

Fuentes, M. and Smith, R. (2001). A new class of nonstationary models. Tech. report at North Carolina State University, Institute of Statistics Mimeo Series #2534.

Guyon, X. (1982). Parameter estimation for a stationary process on a d-dimensional lattice. *Biometrika*, 69, 95–105.

Guyon, X. (1992). *Champs Aléatoires sur un Réseau,* Masson Paris.

Haas, T.C. (1998), Statistical assessment of spatio-temporal pollutant trends and meteorological transport models. *Atmospheric Environment* 32, 1865–1879.

Handcock, M.S. and Wallis, J.R. (1994). An approach to statistical spatial-temporal modeling of meteorological fields (with discussion). *Journal of the American Statistical Association*, 89, 724–731.

Jenkins, G.M. (1961). General considerations in the estimation of spectra. *Technometrics*, 3, 133–166.

Loéve, M. (1955). *Probability Theory.* Van Nostrand, New York.

Mallat, S. (1989). A theory for multiresolution signal decomposition: The wavelet representation. *IEEE Transactions on Pattern Analysis and Machine Intelligence*, 11, 674–693.

Matérn, B. (1960). *Spatial Variation.* Meddelanden fràn Statens Skogsforskningsinstitut, 49(5), Almaenna Foerlaget, Stockholm. Second edition(1986), Springer-Verlag, Berlin.

Nychka, D., Wikle C., and Royle, J.A. (2002). Multiresolution models for nonstationary spatial covariance functions. *Statistical Modeling*, 2, 315–332.

Pawitan, Y. and O'Sullivan F. (1994). Nonparametric spectral density estimation using penalized Whittle likelihood. *Journal of the American Statistical Association*, 89, 600–610.

Percival, D.B. and Walden, A. T. (2000). *Wavelet Methods for Time Series Analysis.* Cambridge University Press, Cambridge, U.K.

Priestley, M.B. (1981). *Spectral Analysis and Time Series.* London: Academic Press.

Priestley, M.B. and Rao, T.S. (1969). A test for non-stationarity of time-series. *Journal of the Royal Statistical Society,* Series B. 31, 140–149.

R Development Core Team, (2004). *R: A language and environment for statistical computing,* R Foundation for Statistical Computing, Vienna, Austria. http://www.R-project.org.

Renshaw, E. and Ford, E.D. (1983). The interpretation of process from pattern using two-dimensional spectral analysis: methods and problems of interpretation. *Applied Statistics,* 32, 51–63.

Ripley, B.D. (1981). *Spatial Statistics,* New York: John Wiley.

Rosenblatt, M.R. (1985). *Stationary Sequences and Random Fields,* Boston: Birkhäuser.

Sampson, P.D. and Guttorp, P. (1992). Nonparametric estimation of nonstationary spatial covariance structure, *Journal of the American Statistical Association,* 87, 108-119.

Schwarz, G. (1978). Estimating the dimension of a model, *Annals of Statistics,* 6, 461–464.

Stein, M.L. (1995). Fixed-domain asymptotics for spatial periodograms. *Journal of the American Statistical Association,* 90, 1277–1288.

Stein, M.L. (1999). *Interpolation of Spatial Data: Some Theory for Kriging.* New York: Springer-Verlag.

Stein, M.L. and Handcock, M. S. (1989). Some asymptotic properties of kriging when the covariance function is misspecified. *Mathematical Geology,* 21, 171–190.

Stein, M.L., Chi, Z., and Welty, L.J. (2004). Approximating likelihoods for large spatial data sets. *Journal of the Royal Statistical Society, Series B,* 66, 275–296.

Whittle, P. (1954). On stationary processes in the plane. *Biometrika,* 41, 434–449.

Yaglom, A.M. (1987). *Correlation Theory of Stationary and Related Random Functions.* Springer-Verlag, New York.

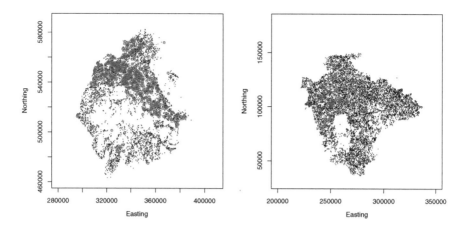

Color Figure 1.4 Locations of at-risk farms (black) and FMD case-farms (red) in Cumbria (left-hand panel) and in Devon (right-hand panel).

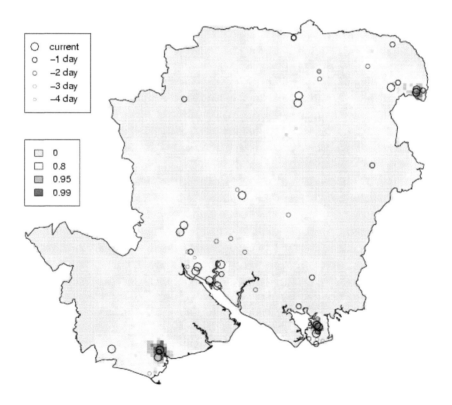

Color Figure 1.17 Predictive exceedance probabilities $p_t(x)$ for the Hampshire gastroenteric disease data on a day in March 2003, using the threshold value $c = 2$. Incident cases over five consecutive days are shown as circles of diminishing size to correspond to the progressive discounting of past data in constructing the predictive distribution of $S(x, t)$.

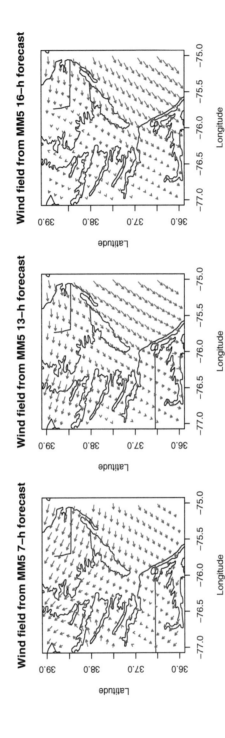

Color Figure 3.29 Wind field maps, showing wind direction and speed over the Chesapeake Bay at 3 a.m., 9 a.m., and noon on July 21, 2002.

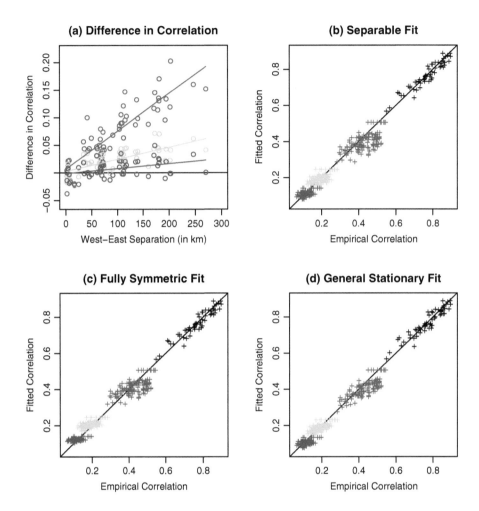

Color Figure 4.3 (a) Difference between the empirical west-to-east and east-to-west correlations for the 55 pairs of distinct stations and temporal lag one day (red), two days (green), and three days (blue), in dependence on the longitudinal (east-west) distance between the stations (in km). Linear least-squares fits are shown as well. (b), (c) and (d) Empirical versus fitted correlations at temporal lag zero (black), one day (red), two days (green), and three days (blue), for the separable model (4.19), the fully symmetric model (4.20), and the general stationary model (4.21), respectively. All displays are based on the training period (1961–1970).

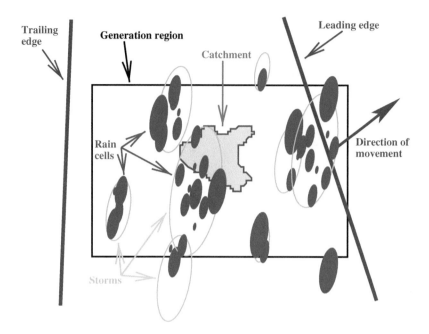

Color Figure 5.8 Schematic diagram of a rain band moving across a catchment. The event interior model is simulated in a large rectangular generation region surrounding the catchment, with toroidal wrap-around at the edges. Storms are only retained if their centres are within the rain band. However, all cells from such storms are retained, even if they fall outside the band.

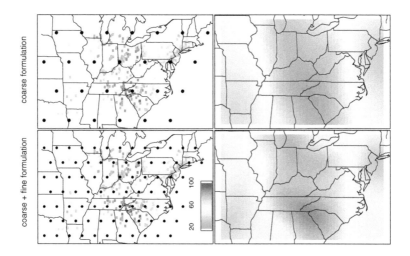

Color Figure 6.31 A multiresolution ozone model. The ozone measurements from a single summer day in the Eastern U.S. (shown in both of the left hand plots) are modeled as the sum of two process convolution models, one coarse and one fine. The knot locations for the coarse model are shown in the top left plot; the knot locations for the fine process are shown in the bottom left plot. The resulting posterior mean for $z_{\text{coarse}}(s)$ is given in the top right plot; that of the multiresolution model $z(s) = z_{\text{coarse}}(s) + z_{\text{fine}}(s)$ is shown in the bottom right plot. The kernels for the two processes are normal, each of whose sd corresponds to their respective knot spacing.

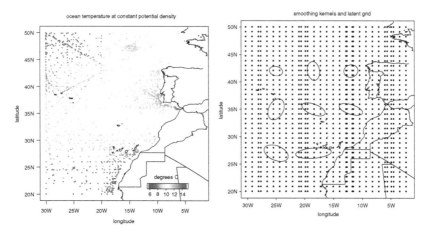

Color Figure 6.39 Left: measured temperatures between 1908 and 1988 along a manifold of constant potential density, which corresponds to depths well below 1000 m. The data consist of 3987 measurements. Right: spatial locations of the latent space-time process $x(s, t)$. The time spacings are every 7 years. Also shown are ellipses that correspond to 1 sd of the spatially varying gaussian smoothing kernels $k_s(\cdot)$.

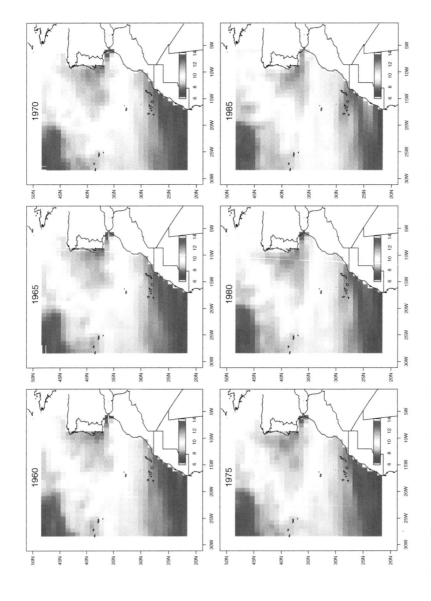

Color Figure 6.40 Posterior mean estimate of the temperature surface $z(s,t)$ at six different times.

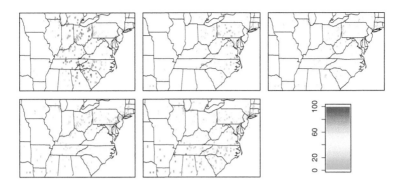

Color Figure 6.42 Ozone concentrations taken for five consecutive summer days over the eastern United States. This is a subset of the 60 days of measurements used in the example.

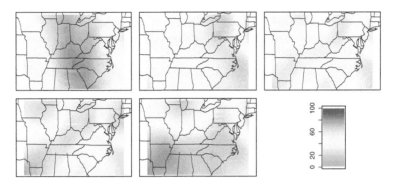

Color Figure 6.43 Posterior mean field for ozone concentration for the 5 days shown in Figure 6.42.

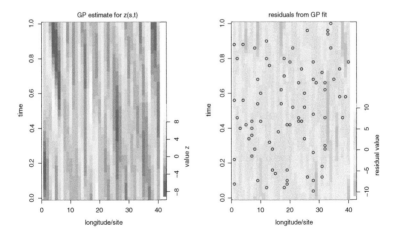

Color Figure 6.46 Left: posterior mean for $z(s,t)$ estimated under the Gaussian process formulation. Right: the residual field obtained by subtracting the posterior mean estimate from the true field shown in Figure 6.45. The circles denote locations where measurements were obtained.

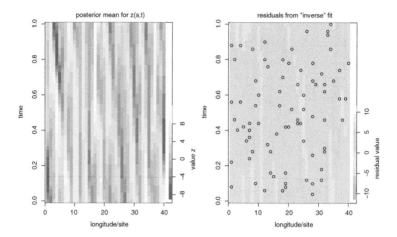

Color Figure 6.48 Left: posterior mean for $z(s,t)$ estimated under the inverse formulation. Right: the residual field obtained by subtracting the posterior mean estimate from the true field shown in Figure 6.45. The circles denote locations where measurements were obtained.

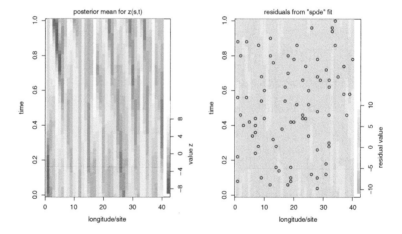

Color Figure 6.51 Left: posterior mean for $z(s,t)$ estimated under the PDE-based MRF formulation. Right: the residual field obtained by subtracting the posterior mean estimate from the true field shown in Figure 6.45. The circles denote locations where measurements were obtained.

CHAPTER 4

Geostatistical Space-Time Models, Stationarity, Separability, and Full Symmetry

Tilmann Gneiting, Marc G. Genton, and Peter Guttorp

Contents

4.1 Introduction

Environmental and geophysical processes such as atmospheric pollutant concentrations, precipitation fields and surface winds are characterized by spatial and temporal variability. In view of the prohibitive costs of spatially and temporally dense monitoring networks, one often aims to develop a statistical model in continuous space and time, based on observations at a limited number of monitoring stations. Examples include environmental monitoring and model assessment for surface ozone levels (Guttorp et al., 1994; Carroll et al., 1997; Meiring et al., 1998; Huang and Hsu, 2004), precipitation forecasts (Amani and Lebel, 1997) and the assessment of wind energy resources (Haslett and Raftery, 1989). Geostatistical approaches model the observations as a partial realization of a spatio-temporal, typically Gaussian random function

$$Z(\boldsymbol{s}, t), \quad (\boldsymbol{s}, t) \in \mathrm{R}^d \times \mathrm{R},$$

which is indexed in space by $\boldsymbol{s} \in \mathrm{R}^d$ and in time by $t \in \mathrm{R}$. Henceforth, we assume that second moments for the random function exist and are finite. Optimal least-squares prediction, or kriging, then relies on the appropriate specification of the space-time covariance structure. Generally, the covariance between $Z(\boldsymbol{s}_1, t_1)$ and $Z(\boldsymbol{s}_2, t_2)$ depends on the space-time coordinates $(\boldsymbol{s}_1, t_1)$ and $(\boldsymbol{s}_2, t_2)$, and no further structure may exist. In practice, however, estimation and modeling call for simplifying assumptions, such as stationarity, separability, and full symmetry.

Specifically, the random field Z is said to have a *separable* covariance if there exist purely spatial and purely temporal covariance functions C_S and C_T, respectively, such that

$$\mathrm{cov}\{Z(\boldsymbol{s}_1, t_1), Z(\boldsymbol{s}_2, t_2)\} = C_\mathrm{S}(\boldsymbol{s}_1, \boldsymbol{s}_2) \cdot C_\mathrm{T}(t_1, t_2) \tag{4.1}$$

for all space-time coordinates $(\boldsymbol{s}_1, t_1)$ and $(\boldsymbol{s}_2, t_2)$ in $\mathrm{R}^d \times \mathrm{R}$. The spatio-temporal covariance structure factors into a purely spatial and a purely temporal component, which allows for computationally efficient estimation and inference. Consequently, separable covariance models have been popular even in situations in which they are not physically justifiable. Many statistical tests for separability have been proposed recently and are based on parametric models (Shitan and Brockwell, 1995; Guo and Billard, 1998; Brown et al., 2000), likelihood ratio tests and subsampling (Mitchell et al., 2005), or spectral methods (Scaccia and Martin, 2005; Fuentes, 2006).

A related notion is that of full symmetry (Gneiting, 2002a; Stein, 2004). The space-time process Z has *fully symmetric* covariance if

$$\mathrm{cov}\{Z(\boldsymbol{s}_1, t_1), Z(\boldsymbol{s}_2, t_2)\} = \mathrm{cov}\{Z(\boldsymbol{s}_1, t_2), Z(\boldsymbol{s}_2, t_1)\} \tag{4.2}$$

for all space-time coordinates $(\boldsymbol{s}_1, t_1)$ and $(\boldsymbol{s}_2, t_2)$ in $\mathrm{R}^d \times \mathrm{R}$. Atmospheric, environmental, and geophysical processes are often under the influence of prevailing air or water flows, resulting in a lack of full symmetry. Transport effects of this type are well-known in the meteorological and hydrological literature and have recently been described by Gneiting (2002a), Stein (2005a), and de Luna and Genton (2005), who considered the Irish wind data of Haslett and Raftery (1989), by Wan et al. (2003) for wind power data, by Huang and Hsu (2004) for surface ozone levels, and by Jun and Stein (2004a, 2004b) for atmospheric sulfate concentrations. Separability forms a special case of full symmetry. Hence, covariance structures that are not fully symmetric are nonseparable, and tests for full symmetry (Scaccia and Martin, 2005; Lu and Zimmerman, 2005) can be used to reject separability.

Frequently, trend removal and space deformation techniques (Haslett and Raftery, 1989; Sampson and Guttorp, 1992) allow for a reduction to a stationary covariance structure. The spatio-temporal random function has *spatially stationary* covariance if $\mathrm{cov}\{Z(s_1, t_1), Z(s_2, t_2)\}$ depends on the observation sites s_1 and s_2 only through the spatial separation vector, $s_1 - s_2$. It has *temporally stationary* covariance if $\mathrm{cov}\{Z(s_1, t_1), Z(s_2, t_2)\}$ depends on the observation times t_1 and t_2 only through the temporal lag, $t_1 - t_2$. If a spatio-temporal process has both spatially and temporally stationary covariance, we say that the process has *stationary* covariance. Under this assumption, there exists a function C defined on $\mathrm{R}^d \times \mathrm{R}$ such that

$$\mathrm{cov}\{Z(s_1, t_1), Z(s_2, t_2)\} = C(s_1 - s_2, t_1 - t_2) \qquad (4.3)$$

for all (s_1, t_1) and (s_2, t_2) in $\mathrm{R}^d \times \mathrm{R}$. We call C the space-time *covariance function* of the process, and its restrictions $C(\cdot, 0)$ and $C(0, \cdot)$ are purely spatial and purely temporal covariance functions, respectively. For tests of stationarity we point to Fuentes (2005) and references therein.

The remainder of the chapter is organized as follows. Section 4.2 reviews the geostatistical approach to space-time modeling and returns to the aforementioned notions of stationarity, separability, and full symmetry. Clearly, these notions apply to correlation structures also and in many cases, such as the case study below, a discussion in terms of correlations is preferable. Section 4.3 turns to recent advances in the literature on stationary space-time covariance functions. This material is largely expository, and much progress has been made since the reviews of Kyriakidis and Journel (1999) and Gneiting and Schlather (2002). Section 4.4 provides a case study based on the Irish wind data of Haslett and Raftery (1989). This rich and well-known data set continues to inspire methodological advances, and recent analyses include the papers by Gneiting (2002a), Stein (2005a, 2005b), and de Luna and Genton (2005). We consider time-forward kriging predictions based on increasingly complex spatio-temporal correlation models, and our experiments suggest that the use of the richer, more realistic models results in improved predictive performance.

4.2 Geostatistical space-time models

4.2.1 Spatio-temporal domains

Geostatistical approaches have been developed to fit random function models in continuous space and time, based on a limited number of spatially and/or temporally dispersed observations. Hence, the natural domain for a geostatistical space-time model is $\mathrm{R}^d \times \mathrm{R}$, where R^d stands for space and R for time. Physically, there is clear-cut separation between the spatial and time dimensions, and a realistic statistical model will take account thereof. This contrasts with a purely mathematical perspective in which $\mathrm{R}^d \times \mathrm{R} = \mathrm{R}^{d+1}$ with no differences between the coordinates. While the latter equality may seem (and indeed is) trivial, it has important implications. In particular, all technical results on spatial covariance functions or on least-squares prediction, or kriging, in Euclidean spaces apply directly to space-time problems, simply by separating a vector into its spatial and temporal components.

Henceforth, we focus on covariance structures for spatio-temporal random functions $Z(s,t)$ where $(s,t) \in \mathrm{R}^d \times \mathrm{R}$. Other spatio-temporal domains are relevant in practice as well. Monitoring data are frequently observed at fixed temporal lags, and it may suffice to model a random function on $\mathrm{R}^d \times \mathrm{Z}$, with time considered discrete. The time autoregressive Gaussian models of Storvik et al. (2002) form a promising tool in this direction. A related result of Stein (2005a) characterizes space-time covariance functions that correspond to a temporal Markov structure. In atmospheric and geophysical applications, the spatial domain of interest is frequently expansive or global and the curvature of the earth needs to be taken into account (Banerjee, 2005). In this type of situation, random functions defined on $\mathrm{S} \times \mathrm{R}$ or $\mathrm{S} \times \mathrm{Z}$ become crucial, where S denotes a sphere in three-dimensional space. Perhaps the simplest way of defining a random field on $\mathrm{S} \times \mathrm{R}$ or $\mathrm{S} \times \mathrm{Z}$ is by defining a random function on $\mathrm{R}^3 \times \mathrm{R}$ and restricting it to the desired domain. Gneiting (1999), Stein (2005a, 2005b), and particularly Jun and Stein (2004b) discuss these and other ways of defining suitable parametric covariance models on global spatial or spatio-temporal domains.

4.2.2 Stationarity, separability, and full symmetry

We consider a generic, typically Gaussian spatio-temporal random function $Z(s,t)$ where $(s,t) \in \mathrm{R}^d \times \mathrm{R}$. The covariance between $Z(s_1,t_1)$ and $Z(s_2,t_2)$ generally depends on the space-time coordinates (s_1,t_1) and (s_2,t_2), and no further structure may exist. In practice, simplifying assumptions are required, such as the aforementioned notions of stationarity, separability, and full symmetry, which are defined in (4.1), (4.2), and (4.3), respectively. In addition, it is sometimes desirable that the space-time process Z has *compactly supported* covariance, so that $\mathrm{cov}\{Z(s_1,t_1), Z(s_2,t_2)\} = 0$ whenever $\|s_1 - s_2\|$ and/or $|t_1 - t_2|$ are sufficiently large. Gneiting (2002b) reviews parametric models of compactly supported, stationary, and isotropic covariance functions in a spatial setting. Unfortunately, the straightforward idea of thresholding covariances to zero using the product with an indicator function, such as the truncation device of Haas (2002, p. 320), yields invalid covariance models. That said, compactly supported covariances are attractive in that they allow for computationally efficient spatio-temporal prediction and simulation. Furrer et al. (2006), for instance, proposed their use for kriging large spatial datasets.

Figure 4.1 summarizes the relationships between the various notions in terms of classes of space-time covariance functions, and an analogous scheme applies to correlation structures. The largest class is that of general, stationary or non-stationary covariance functions. A separable covariance can be stationary or non-stationary, and similarly for fully symmetric covariances. However,

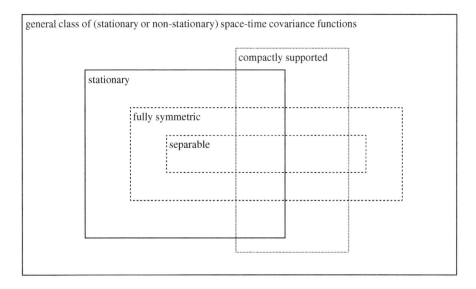

general class of (stationary or non-stationary) space-time covariance functions

compactly supported

stationary

fully symmetric

separable

Figure 4.1 Schematic illustration of the relationships between separable, fully symmetric, stationary, and compactly supported covariances within the general class of (stationary or non-stationary) space-time covariance functions. An analogous scheme applies to correlation structures.

a separable covariance function is always fully symmetric, but not vice versa, and this has implications in testing and model fitting. In particular, to reject separability it suffices to reject full symmetry. We return to these issues in Section 4.4 below, when we discuss modeling strategies using the example of the Irish wind data.

Occasionally, the second-order structure of a spatio-temporal random function is modeled based on variances rather than covariances. Specifically, when viewed as a function of space-time coordinates (s_1, t_1) and (s_2, t_2) in $\mathrm{R}^d \times \mathrm{R}$, the quantity

$$\frac{1}{2} \operatorname{var}(Z(s_1, t_1) - Z(s_2, t_2)) \tag{4.4}$$

is called a *non-stationary variogram*. Separability is not a meaningful assumption for variograms, but we discuss full symmetry and stationarity. The space-time process has a *fully symmetric* variogram structure if

$$\operatorname{var}(Z(s_1, t_1) - Z(s_2, t_2)) = \operatorname{var}(Z(s_1, t_2) - Z(s_2, t_1))$$

for all (s_1, t_1) and (s_2, t_2) in $\mathrm{R}^d \times \mathrm{R}$. The space-time random function has a *spatially intrinsically stationary* variogram if the non-stationary variogram (4.4) depends on the observation sites s_1 and s_2 only through the spatial separation vector, $s_1 - s_2$. Similarly, it has *temporally intrinsically stationary* variogram if (4.4) depends on the observation times t_1 and t_2 only through the temporal lag, $t_1 - t_2$. The process has *intrinsically stationary* variogram if it has both spatially intrinsically stationary and temporally intrinsically stationary variogram. Under this latter assumption, there exists a function γ defined on $\mathrm{R}^d \times \mathrm{R}$ such that

$$\frac{1}{2}\,\mathrm{var}(Z(\boldsymbol{s}_1, t_1) - Z(\boldsymbol{s}_2, t_2)) = \gamma(\boldsymbol{s}_1 - \boldsymbol{s}_2, t_1 - t_2)$$

for all $(\boldsymbol{s}_1, t_1)$ and $(\boldsymbol{s}_2, t_2)$ in $\mathrm{R}^d \times \mathrm{R}$, and we call γ the *stationary variogram* of the process Z. Variograms exist under slightly weaker assumptions than covariances, and we refer to Gneiting et al. (2001) and Ma (2003c) for the various analogies and correspondences between the two classes of dependence measures. The use of variograms has been particularly popular in purely spatial problems. In discussing geostatistical space-time models, we follow the literature and focus our attention on covariances and correlations.

4.2.3 Positive definiteness

A crucial notion for stationary covariance functions is that of positive definiteness. Specifically, if $Z(\boldsymbol{s}, t)$ is a random function in $\mathrm{R}^d \times \mathrm{R}$, k is a positive integer, and $(\boldsymbol{s}_1, t_1), \dots, (\boldsymbol{s}_k, t_k)$ are space-time coordinates in $\mathrm{R}^d \times \mathrm{R}$, the covariance matrix of the random vector $(Z(\boldsymbol{s}_1, t_1), \dots, Z(\boldsymbol{s}_k, t_k))'$ is nonnegative definite. If the process is stationary with covariance function C on $\mathrm{R}^d \times \mathrm{R}$, this matrix can be written as

$$(C(\boldsymbol{s}_i - \boldsymbol{s}_j, t_i - t_j))_{i,j=1,\dots,k}. \tag{4.5}$$

A complex-valued function C on $\mathrm{R}^d \times \mathrm{R}$ is called *positive definite* if the matrix in (4.5) is nonnegative definite for all finite collections of space-time coordinates $(\boldsymbol{s}_1, t_1), \dots, (\boldsymbol{s}_k, t_k)$ in $\mathrm{R}^d \times \mathrm{R}$. Any positive definite function is Hermitian, and a positive definite function is real-valued if and only if it is symmetric. It is well known that the class of stationary covariance functions is identical to the class of symmetric positive definite functions, and we use the two terms interchangeably. When we talk of a positive definite function, we explicitly allow for complex-valued functions.

Unfortunately, it can be quite difficult in general to check whether a function is positive definite, and this forms one of the key difficulties in the construction of parametric space-time covariance models. Until a few years ago, geostatistical space-time models were largely based on stationary, separable covariance functions of the form

$$C(\boldsymbol{h}, u) = C_{\mathrm{S}}(\boldsymbol{h}) \cdot C_{\mathrm{T}}(u), \quad (\boldsymbol{h}, u) \in \mathrm{R}^d \times \mathrm{R} \tag{4.6}$$

where $C_{\mathrm{S}}(\boldsymbol{h})$ and $C_{\mathrm{T}}(u)$ are stationary, purely spatial and purely temporal covariance functions, respectively. A convenient choice is an isotropic model, $C_{\mathrm{S}}(\boldsymbol{h}) = c_0(\|\boldsymbol{h}\|)$, $\boldsymbol{h} \in \mathrm{R}^d$, where c_0 is one of the standard models in geostatistics, such as the powered exponential class (see, for instance, Diggle

Table 4.1 Some parametric classes of isotropic covariance functions. The Whittle-Matérn covariance is defined in terms of the modified Bessel function, K_ν.

Class	Functional form	Parameters
Powered exponential	$c_0(r) = \sigma^2 \exp(-(\theta r)^\gamma)$	$0 < \gamma \le 2;\ \theta > 0;\ \sigma > 0$
Whittle-Matérn	$c_0(r) = \sigma^2 \frac{2^{1-\nu}}{\Gamma(\nu)}\, (\theta r)^\nu K_\nu(\theta r)$	$\nu > 0;\ \theta > 0;\ \sigma > 0$
Cauchy	$c_0(r) = \sigma^2 \left(1 + (\theta r)^\gamma\right)^{-\nu}$	$0 < \gamma \le 2;\ \nu > 0;\ \theta > 0;\ \sigma > 0$

et al., 1998); the Whittle-Matérn class (Whittle, 1954; Matérn, 1986); and the Cauchy class (Gneiting and Schlather, 2004). These parametric models are listed in Table 4.1. Similarly, the temporal covariance function can conveniently be chosen as $C_T(u) = c_0(|u|)$, $u \in R$, where c_0 is another, possibly distinct standard model that guarantees positive definiteness. The product (4.6) yields a positive definite function whenever C_S and C_T are positive definite on R^d and R, respectively, because products — and also sums, convex combinations and limits — of positive definite functions are positive definite.

4.3 Stationary space-time covariance functions

4.3.1 Bochner's Theorem

The celebrated theorem of Bochner (1955) states that a continuous function is positive definite if and only if it is the Fourier transform of a finite nonnegative measure. This allows for the following characterization of stationary space-time covariance functions.

Theorem 4.3.1 (**Bochner**)
Suppose that C is a continuous and symmetric function on $R^d \times R$. Then C is a covariance function if and only if it is of the form

$$C(\boldsymbol{h}, u) = \iint e^{i(\boldsymbol{h}'\boldsymbol{\omega} + u\tau)}\, dF(\boldsymbol{\omega}, \tau), \quad (\boldsymbol{h}, u) \in R^d \times R, \qquad (4.7)$$

where F is a finite, non-negative and symmetric measure on $R^d \times R$.

In other words, the class of stationary space-time covariance functions on $R^d \times R$ is identical to the class of the Fourier transforms of finite, non-negative and symmetric measures on this domain. The measure F in the representation

(4.7) is often called the *spectral measure*. If C is integrable, the spectral measure is absolutely continuous with Lebesgue density

$$f(\boldsymbol{\omega}, \tau) = (2\pi)^{-(d+1)} \int\!\!\int e^{-i(\boldsymbol{h}'\boldsymbol{\omega}+u\tau)} \, C(\boldsymbol{h}, u) \, \mathrm{d}\boldsymbol{h} \, \mathrm{d}u, \quad (\boldsymbol{\omega}, \tau) \in \mathrm{R}^d \times \mathrm{R},$$

and f is called the *spectral density*. If the spectral density exists, the representation (4.7) in Bochner's theorem reduces to

$$C(\boldsymbol{h}, u) = \int\!\!\int e^{i(\boldsymbol{h}'\boldsymbol{\omega}+u\tau)} \, f(\boldsymbol{\omega}, \tau) \, \mathrm{d}\boldsymbol{\omega} \, \mathrm{d}\tau, \quad (\boldsymbol{h}, u) \in \mathrm{R}^d \times \mathrm{R},$$

and C and f can be obtained from each other via the Fourier transform.

A stationary space-time covariance function is separable if there exist stationary, purely spatial, and purely temporal covariance functions C_S and C_T, respectively, such that (4.6) holds or, equivalently, if we can factor the space-time covariance function as

$$C(\boldsymbol{h}, u) = \frac{C(\boldsymbol{h}, 0) \cdot C(\boldsymbol{0}, u)}{C(\boldsymbol{0}, 0)}$$

for all $(\boldsymbol{h}, u) \in \mathrm{R}^d \times \mathrm{R}$ (Mitchell et al., 2005). In spectral terms, a stationary covariance function is separable if and only if the spectral measure factors as a product measure over the spatial and temporal domains, respectively. In particular, if the spectral density exists, it factors as a product over the domains. A stationary space-time covariance function is fully symmetric if

$$C(\boldsymbol{h}, u) = C(\boldsymbol{h}, -u) = C(-\boldsymbol{h}, u) = C(-\boldsymbol{h}, -u)$$

for all $(\boldsymbol{h}, u) \in \mathrm{R}^d \times \mathrm{R}$. In the purely spatial context, this property is also known as axial symmetry (Scaccia and Martin, 2005) or reflection symmetry (Lu and Zimmerman, 2005). For fully symmetric covariances, Bochner's theorem can be specialized as follows.

Theorem 4.3.2
Suppose that C is a continuous function on $\mathrm{R}^d \times \mathrm{R}$. Then C is a stationary, fully symmetric covariance if and only if it is of the form

$$C(\boldsymbol{h}, u) = \int\!\!\int \cos(\boldsymbol{h}'\boldsymbol{\omega}) \cos(u\tau) \, \mathrm{d}F(\boldsymbol{\omega}, \tau), \quad (\boldsymbol{h}, u) \in \mathrm{R}^d \times \mathrm{R}, \qquad (4.8)$$

where F is a finite, non-negative measure on $\mathrm{R}^d \times \mathrm{R}$.

The proof of this result is given in the Appendix. If C is fully symmetric and the spectral density f exists, then f can be chosen as fully symmetric, too, that is,

$$f(\boldsymbol{\omega}, \tau) = f(\boldsymbol{\omega}, -\tau) = f(-\boldsymbol{\omega}, \tau) = f(-\boldsymbol{\omega}, -\tau).$$

for all $(\boldsymbol{\omega}, \tau) \in \mathrm{R}^d \times \mathrm{R}$. A similar characterization applies in terms of general spectral measures. If the space-time covariance function has additional structure, such as spherical symmetry with respect to $\boldsymbol{h} \in \mathrm{R}^d$ for each $u \in \mathrm{R}$, the representation (4.8) can be further specialized. Theorem 2 of Ma (2003a) is a result of this type.

The above results apply to continuous functions. In practice, fitted stationary space-time covariance functions often involve a *nugget effect*, that is, a discontinuity at the origin. In the spatio-temporal context, the nugget effect could be purely spatial, purely temporal or spatio-temporal and takes the general form

$$C(\boldsymbol{h}, u) = a\delta_{(\boldsymbol{h},u)=(0,0)} + b\delta_{\boldsymbol{h}=0} + c\delta_{u=0}, \quad (\boldsymbol{h}, u) \in \mathrm{R}^d \times \mathrm{R}, \qquad (4.9)$$

where a, b, and c are nonnegative constants and δ denotes an indicator function. Products and sums of continuous covariances and functions of the form (4.9) are valid covariances, and for all practical purposes functions of this type exhaust the class of valid stationary covariance functions (Gneiting and Sasvári, 1999). For the Irish wind data which we discuss in Section 4.4 below, we fit a purely spatial nugget effect.

4.3.2 Cressie-Huang representation

The following result of Cressie and Huang (1999) characterizes the class of stationary space-time covariance functions under the additional assumption of integrability.

Theorem 4.3.3 (**Cressie and Huang**)
Suppose that C is a continuous, bounded, integrable, and symmetric function on $R^d \times R$. Then C is a stationary covariance if and only if

$$\rho(\boldsymbol{\omega}, u) = \int e^{-i\boldsymbol{h}'\boldsymbol{\omega}} C(\boldsymbol{h}, u) \, \mathrm{d}\boldsymbol{h}, \quad u \in \mathrm{R}, \qquad (4.10)$$

is positive definite for almost all $\boldsymbol{\omega} \in R^d$.

Cressie and Huang (1999) used Theorem 4.3.3 to construct stationary space-time covariance functions through closed form Fourier inversion in R^d. Specifically, they considered functions of the form

$$C(\boldsymbol{h}, u) = \int e^{i\boldsymbol{h}'\boldsymbol{\omega}} \rho(\boldsymbol{\omega}, u) \, \mathrm{d}\boldsymbol{\omega}, \quad (\boldsymbol{h}, u) \in \mathrm{R}^d \times \mathrm{R},$$

where $\rho(\boldsymbol{\omega}, u)$, $u \in \mathrm{R}$, is a continuous positive definite function for all $\boldsymbol{\omega} \in \mathrm{R}^d$. Gneiting (2002a) gave a criterion that is based on this construction but does not depend on closed form Fourier inversion and does not require integrability. Recall that a continuous function $\varphi(r)$ defined for $r > 0$ or $r \geq 0$ is *completely monotone* if it possesses derivatives $\varphi^{(n)}$ of all orders and $(-1)^n\varphi^{(n)}(r) \geq 0$ for $r > 0$ and $n = 0, 1, 2, \ldots$ Gneiting (2002a) and Ma (2003b) gave various examples of completely monotone functions. In particular, if c_0 is any of the functions listed in Table 4.1, then $\varphi(r) = c_0(r^{1/2})$, $r \geq 0$, is a completely monotone function. Examples of positive functions with a completely monotone derivative include $\psi(r) = (ar^\alpha + 1)^\beta$ and $\psi(r) = \ln(ar^\alpha + 1)$, where $\alpha \in (0, 1]$, $\beta \in (0, 1]$ and $a > 0$.

Theorem 4.3.4 (**Gneiting**)
Suppose that $\varphi(r)$, $r \geq 0$, is a completely monotone function, and that $\psi(r)$, $r \geq 0$, is a positive function with a completely monotone derivative. Then

$$C(\boldsymbol{h}, u) = \frac{1}{\psi(u^2)^{d/2}} \, \varphi\left(\frac{\|\boldsymbol{h}\|^2}{\psi(u^2)}\right), \quad (\boldsymbol{h}, u) \in R^d \times R, \quad (4.11)$$

is a stationary covariance function on $R^d \times R$.

The specific choices $\varphi(r) = \sigma^2 \exp(-cr^\gamma)$ and $\psi(r) = (1 + ar^\alpha)^\beta$ recover Equation (14) of Gneiting (2002a) and yield the parametric family

$$C(\boldsymbol{h}, u) = \frac{\sigma^2}{(1 + a|u|^{2\alpha})^{\beta d/2}} \, \exp\left(-\frac{c\|\boldsymbol{h}\|^{2\gamma}}{(1 + a|u|^{2\alpha})^{\beta\gamma}}\right), \quad (\boldsymbol{h}, u) \in R^d \times R,$$

$$(4.12)$$

of stationary space-time covariance functions. Here, a and c are nonnegative scale parameters of time and space, respectively. The smoothness parameters α and γ and the space-time interaction parameter β take values in $(0, 1]$, and σ^2 is the variance of the spatio-temporal process. The purely spatial covariance function, $C(\boldsymbol{h}, 0)$, is of the powered exponential form, and the purely temporal covariance function, $C(\boldsymbol{0}, u)$, belongs to the Cauchy class.

Clearly, any stationary covariance of the form (4.11) is fully symmetric. Furthermore, under the assumption of full symmetry the test functions $\rho(\boldsymbol{\omega}, u)$ of Theorem 4.3.3 are real-valued and symmetric functions of $u \in R$. If C is not fully symmetric then $\rho(\boldsymbol{\omega}, u)$ is generally complex-valued. For instance, the function $C(h, u) = \exp(-h^2 + hu - u^2)$ has Fourier transform proportional to $\exp(-\frac{1}{3}(\omega^2 + \omega\tau + \tau^2))$ and therefore is a stationary covariance on $R \times R$. The associated test function $\rho(\omega, u)$, $u \in R$, is proportional to $\exp(-\frac{1}{4}(3u^2 + 2i\omega u))$ and positive definite yet generally complex-valued.

4.3.3 Fully symmetric, stationary covariance functions

Non-separable, fully symmetric stationary space-time covariance functions can be constructed as mixtures of separable covariances. The following theorem summarizes relevant results by De Iaco et al. (2002) and Ma (2003b). In view of Theorem 4.3.2, the construction is completely general.

Theorem 4.3.5
Let μ be a finite, nonnegative measure on a non-empty set Θ. Suppose that for each $\theta \in \Theta$, C_S^θ and C_T^θ are stationary covariances on R^d and R, respectively, and suppose that $C_S^\theta(\mathbf{0})\,C_T^\theta(0)$ has finite integral over Θ. Then

$$C(\mathbf{h}, u) = \int C_S^\theta(\mathbf{h})\,C_T^\theta(u)\,\mathrm{d}\mu(\theta), \quad (\mathbf{h}, u) \in R^d \times R, \qquad (4.13)$$

is a stationary covariance function on $R^d \times R$.

Explicit constructions of the form (4.13) have been reported by various authors. Perhaps the simplest special case is the product-sum model of De Iaco et al. (2001),

$$C(\mathbf{h}, u) = a_0 C_S^0(\mathbf{h}) C_T^0(u) + a_1 C_S^1(\mathbf{h}) + a_2 C_T^2(u), \quad (\mathbf{h}, u) \in R^d \times R,$$

where a_0, a_1 and a_2 are nonnegative coefficients and C_S^0, C_S^1 and C_T^0, C_T^2 are stationary, purely spatial and purely temporal covariance functions, respectively. Examples 1 and 2 of De Iaco et al. (2002) consider the particular case of (4.13) in which μ is a gamma distribution on $\Theta = [0, \infty)$ and both C_S^θ and C_T^θ are of powered exponential type. This construction yields the parametric family

$$C(\mathbf{h}, u) = \sigma^2 \left(1 + \left\| \frac{\mathbf{h}}{a} \right\|^\alpha + \left| \frac{u}{b} \right|^\beta \right)^{-\gamma}, \quad (\mathbf{h}, u) \in R^d \times R, \qquad (4.14)$$

of stationary space-time covariance functions, where $\alpha \in (0, 2]$, $\beta \in (0, 2]$, $\gamma > 0$, $a > 0$, $b > 0$, and $\sigma > 0$. De Iaco et al. (2002) gave more stringent parameter ranges which are unnecessarily restrictive. Ma (2003a) reported various interesting examples of parametric space-time covariance functions that are also based on the mixture representation (4.13).

4.3.4 Stationary covariance functions that are not fully symmetric

Environmental, atmospheric, and geophysical processes are often influenced by prevailing winds or ocean currents, which are incompatible with the assumption of full symmetry. In this type of situation, the general idea of a Lagrangian reference frame applies, which can be thought of as attached to and moving with the center of an air or water mass. This approach has been studied in fluid dynamics, meteorology, and hydrology, and we refer to May and Julien (1998) for a comparison of empirical correlations observed in the classical (fixed) Eulerian and in the (moving) Lagrangian reference frame. Specifically, consider a purely spatial random field with stationary covariance function C_S on R^d, and suppose that the entire field moves time-forward with random velocity vector $\mathbf{V} \in R^d$. The resulting spatio-temporal random field has stationary covariance

$$C(\mathbf{h}, u) = \mathrm{E}\, C_S(\mathbf{h} - \mathbf{V}u), \quad (\mathbf{h}, u) \in R^d \times R, \qquad (4.15)$$

where the expectation is taken with respect to the random vector V. Similar constructions of stationary space-time covariance functions have been reported by Cox and Isham (1988) and Ma (2003b), among others. Stationary space-time covariance functions that are not fully symmetric can also be constructed on the basis of diffusion equations or stochastic partial differential equations. These and related approaches have been discussed by Jones and Zhang (1997), Christakos (2000), Brown et al. (2000), Kovolos et al. (2004), Jun and Stein (2004b), and Stein (2005a, 2005b).

For the specification of the random velocity vector V in the Lagrangian covariance (4.15), various choices can be physically motivated and justified. We discuss these in the context of atmospheric transport effects driven by prevailing winds. The simplest case occurs when $V = v$ is constant and represents the mean or prevailing wind. Gupta and Waymire (1987) referred to this as the *frozen field* model. For the Irish wind data of Haslett and Raftery (1989), for instance, v can be identified with the prevailing westerly wind, and we give details in Section 4.4 below. Alternatively, V might attain a small number of distinct values that represent wind regimes, or we might identify the distribution of V with the empirical distribution of velocity vectors, or a smoothed version thereof, as inferred from meteorological records. Finally, the distribution of V could be updated dynamically according to the current state of the atmosphere. This option yields nonstationary, flow-dependent covariance structures similar to those posited by Riishøgaard (1998) and Huang and Hsu (2004). A related approach has been studied under the heading of Lagrangian kriging (Amani and Lebel, 1997), which operates directly within the (moving) Lagrangian reference frame. The recent advent of ensemble Kalman filter techniques in atmospheric and oceanic data assimilation (Evensen, 1994; Houtekamer and Mitchell, 1998; Hamill and Snyder, 2000) marks another promising avenue to nonstationary space-time covariance modeling based on multiple realizations of an underlying spatio-temporal random field.

4.3.5 Taylor's hypothesis

A stationary space-time covariance function C on $\mathrm{R}^d \times \mathrm{R}$ satisfies *Taylor's hypothesis* (Taylor, 1938, p. 478) if there exists a velocity vector $v \in \mathrm{R}^d$ such that

$$C(\mathbf{0}, u) = C(vu, 0), \quad u \in \mathrm{R}. \tag{4.16}$$

Taylor's hypothesis concerns the relationships between the purely spatial and the purely temporal covariance functions only and has found widespread interest in fluid dynamics, meteorology and hydrology. Zawadzki (1973) argued on the basis of empirical correlations that Taylor's hypothesis is plausible for precipitation data and temporal lags less than 40 minutes. Gupta and Waymire (1987) and Cox and Isham (1988) studied the approximate validity of the hypothesis for various space-time covariance models. The following covariance functions, among others, admit Taylor's hypothesis exactly.

1. The covariance function (4.15) for the frozen field model, that is, $C(\boldsymbol{h}, u) = C_S(\boldsymbol{h} - \boldsymbol{v}u)$ where C_S is a stationary covariance on R^d and $\boldsymbol{v} \in \mathrm{R}^d$ is a nonzero velocity vector, satisfies (4.16).
2. Geometrically anisotropic models of the form $C(\boldsymbol{h}, u) = c_0((a^2\|\boldsymbol{h}\|^2 + b^2 u^2)^{1/2})$, where a and b are positive constants and c_0 is any of the parametric models in Table 4.1, admit Taylor's relationship (4.16) with $\boldsymbol{v} = (b/a, 0, \dots, 0)' \in \mathrm{R}^d$.
3. Separable models of the form $C(\boldsymbol{h}, u) = c_0(\|\boldsymbol{h}\|)\, c_0(|u|)$, where c_0 is any of the parametric models in Table 4.1, satisfy (4.16) with $\boldsymbol{v} = (1, 0, \dots, 0)' \in \mathrm{R}^d$.
4. Putting $d = 2$, $\varphi(r) = (1 + cr^\alpha)^{-\nu}$ and $\psi(r) = (1 + ar^\alpha)^\nu$ in (4.11), where $a > 0$, $c > 0$, $\alpha \in (0, 1]$, and $\nu \in (0, 1]$, yields

$$C(\boldsymbol{h}, u) = \sigma^2 (1 + a|u|^{2\alpha})^{-\nu} \left(1 + \frac{c\|\boldsymbol{h}\|^{2\alpha}}{1 + (a|u|^{2\alpha})^{\alpha\nu}} \right)^{-\nu}, \quad (\boldsymbol{h}, u) \in \mathrm{R}^2 \times \mathrm{R}.$$

This function admits Taylor's hypothesis with $\boldsymbol{v} = ((a/c)^{1/(2\alpha)}, 0)'$. Generally, if $d = 2$ the covariance function (4.11) satisfies (4.16) if and only if there exists a nonnegative number b such that the product $\varphi(r)\psi(br)$ does not depend on r.

5. If $\alpha = \beta \in (0, 2]$, the stationary space-time covariance function (4.14) satisfies Taylor's relationship (4.16) with $\boldsymbol{v} = (a/b, 0, \dots, 0)' \in \mathrm{R}^d$.

4.4 Irish wind data

We turn to a case study that considers the Irish wind data of Haslett and Raftery (1989). The dataset consists of time series of daily average wind speed at eleven synoptic meteorological stations in Ireland during the period 1961–1978, as described in Table 4.2. The observations for February 29 were removed in the interest of a convenient handling of the seasonal trend component. We refer to Haslett and Raftery (1989) for background information and to Gneiting (2002a), Stein (2005a, 2005b), and de Luna and Genton (2005) for subsequent analyses of the Irish wind data. To allow for an out-of-sample evaluation of predictive performance, we split the data set into a training period consisting of years 1961–1970 and a test period, comprising 1971–1978. Following the aforementioned authors, we apply the square root transform to the time series of daily average wind speed, fit and extract a common seasonal trend component, and remove the station-specific means, resulting in time series of *velocity measures* at the eleven meteorological stations. We apply the same set of transformations to the training data and to the test data, with the common seasonal trend component and the station-specific means estimated from the training data. The square root transform stabilizes the variance over both stations and time periods and makes the marginal distributions approximately normal.

Table 4.2 Latitude, longitude, and mean wind speed (in m $\cdot$ s^{-1}) at eleven meteorological stations in Ireland, as observed in 1961–1978 and described by Haslett and Raftery (1989)

Station	Latitude	Longitude	Mean wind
Valentia (Val)	51° 56′ N	10° 15′ W	5.48
Belmullet (Bel)	54° 14′ N	10° 00′ W	6.75
Claremorris (Cla)	53° 43′ N	8° 59′ W	4.32
Shannon (Sha)	52° 42′ N	8° 55′ W	5.38
Roche's Point (Roc)	51° 48′ N	8° 15′ W	6.36
Birr (Birr)	53° 05′ N	7° 53′ W	3.65
Mullingar (Mul)	53° 32′ N	7° 22′ W	4.38
Malin Head (Mal)	55° 22′ N	7° 20′ W	8.03
Kilkenny (Kil)	52° 40′ N	7° 16′ W	3.25
Clones (Clo)	54° 11′ N	7° 14′ W	4.48
Dublin (Dub)	53° 26′ N	6° 15′ W	5.05

4.4.1 Exploratory analysis

This section provides an initial study of the spatio-temporal correlation structure for the velocity measures during the training period (1961–1970). Figure 4.2 shows the empirical correlations of contemporary velocity measures at the 55 pairs of distinct meteorological stations as a function of the Euclidean distance between the sites. The correlations decay with distance, and we fit a stationary spatial correlation function of the form

$$C_S(\boldsymbol{h}) = (1 - \nu)\exp(-c\|\boldsymbol{h}\|) + \nu\,\delta_{\boldsymbol{h}=\boldsymbol{0}}, \tag{4.17}$$

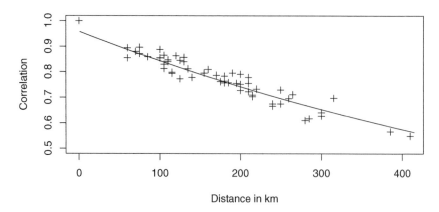

Figure 4.2 Empirical correlations and the fitted spatial correlation function (4.17) for the velocity measures during the training period (1961–1970).

Table 4.3 Empirical correlations between velocity measures at selected meteorological stations in Ireland, as observed in the training period (1961–1970). The table shows the correlations between velocity measures at a temporal lag of one day, with the westerly station leading one day (WE) or lagging one day (EW), respectively.

Westerly station	Easterly station	WE	EW
Valentia	Roche's Point	.48	.35
Belmullet	Clones	.52	.39
Claremorris	Mullingar	.51	.41
Claremorris	Dublin	.50	.36
Shannon	Kilkenny	.51	.39
Mullingar	Dublin	.49	.45

that is, a convex combination of an exponential model and a nugget effect. The weighted least-squares (WLS) estimates for the spatial nugget effect and the scale parameter are $\widehat{\nu} = 0.0415$ and $\widehat{c} = 0.00128$, with distances measured in kilometers. Furthermore, we fit a stationary, purely temporal correlation function of Cauchy type,

$$C_{\mathrm{T}}(u) = (1 + a|u|^{2\alpha})^{-1}, \qquad (4.18)$$

for temporal lags $|u| \leq 3$ days; higher temporal lags are irrelevant given our goal of one-day ahead forecasts at the stations. The WLS estimates for the parameters in (4.18) are $\widehat{a} = 0.972$ and $\widehat{\alpha} = 0.834$, respectively. Plots of the empirical purely temporal correlation functions at the meteorological stations are shown in Figure 4 of Haslett and Raftery (1989).

We now describe some of the interactions between the spatial and the temporal correlation structures. Winds in Ireland are predominantly westerly, so that the velocity measures propagate from west to east. Hence, we expect the empirical correlation between a westerly station today and an easterly station later on to be higher than vice versa. Indeed, Table 4.3 and Figure 4.3(a) illustrate the lack of full symmetry — and thereby the lack of separability — in the correlation structure of the velocity measures. Table 4.3 compares the west-to-east and east-to-west correlations at a temporal lag of one day for the six pairs of meteorological stations with the most dominant longitudinal (east-west) component of the separation vector between the stations. The violation of full symmetry is most pronounced for Valentia and Kilkenny, though, with lag one west-to-east and east-to-west correlations reaching .50 and .30, respectively.

Figure 4.3(a) shows the difference between the empirical west-to-east and east-to-west correlations for all 55 pairs of distinct stations as a function of the longitudinal (east-west) component of the spatial separation vector, for temporal lags of one day (red), two days (green), and three days (blue). Similar graphical displays can be found in de Luna and Genton (2005) and

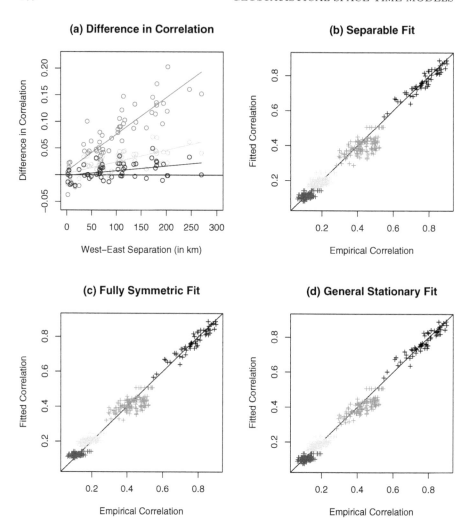

Figure 4.3 (SEE COLOR INSERT FOLLOWING PAGE 142) (a) Difference between the empirical west-to-east and east-to-west correlations for the 55 pairs of distinct stations and temporal lag one day (red), two days (green), and three days (blue), in dependence on the longitudinal (east-west) distance between the stations (in km). Linear least-squares fits are shown as well. (b), (c) and (d) Empirical versus fitted correlations at temporal lag zero (black), one day (red), two days (green), and three days (blue), for the separable model (4.19), the fully symmetric model (4.20), and the general stationary model (4.21), respectively. All displays are based on the training period (1961–1970).

Stein (2005a). The marked difference between the west-to-east and east-to-west correlations persists for temporal lags up to three days. For each lag, the difference increases linearly with the longitudinal distance between the stations.

4.4.2 Fitting a parametric, stationary space-time correlation model

A heuristic approach to fitting a parametric, stationary space-time correlation function can be described as follows. Suppose that a stationary correlation structure is a reasonable approximation for the spatio-temporal data set at hand. To fit a parametric stationary correlation model, we first check (or test) for full symmetry. If full symmetry is not a justifiable assumption, we fit a general stationary model in ways exemplified below. If full symmetry is an appropriate assumption, we check (or test) for separability. If the use of a separable model cannot be justified, we fit a fully symmetric but non-separable model. Otherwise, a separable model is adequate. Note that we fit and discuss correlation structures rather than covariance models.

In the case of the Irish wind data, Table 4.3 and Figure 4.3(a) show that the assumptions of separability and full symmetry are clearly violated. We fit a general stationary correlation model to the velocity measures, based on initial separable and fully symmetric fits. Generally, we consider only those spatio-temporal lags $(\boldsymbol{h}, u)$ in the WLS fits that are relevant to the subsequent prediction experiment; that is, we require that $\|\boldsymbol{h}\| \leq 450$ km and $|u| \leq 3$ days. All fits are based on the training period (1961–1970). The fitted separable model,

$$C_{\text{SEP}}(\boldsymbol{h}, u) = C_{\text{S}}(\boldsymbol{h}) \cdot C_{\text{T}}(u), \tag{4.19}$$

is simply the product of the fitted purely spatial correlation function (4.17) and the fitted purely temporal correlation (4.18), respectively. This separable model can be embedded into the fully symmetric but generally non-separable correlation function

$$C_{\text{FS}}(\boldsymbol{h}, u) = \frac{1 - \nu}{1 + a|u|^{2\alpha}} \left(\exp\left(-\frac{c\|\boldsymbol{h}\|}{(1 + a|u|^{2\alpha})^{\beta/2}} \right) + \frac{\nu}{1 - \nu} \delta_{\boldsymbol{h}=0} \right) \tag{4.20}$$

which derives from (4.12) and reduces to the separable model when $\beta = 0$. With all other parameters fixed at the previous estimates, the WLS estimate of the space-time interaction parameter, $\beta \in [0, 1]$, is $\widehat{\beta} = 0.681$.

To fit a general stationary but not necessarily fully symmetric correlation model, we consider convex combinations,

$$C_{\text{STAT}}(\boldsymbol{h}, u) = (1 - \lambda) C_{\text{FS}}(\boldsymbol{h}, u) + \lambda C_{\text{LGR}}(\boldsymbol{h}, u), \tag{4.21}$$

of the fitted fully symmetric model (4.20) and the compactly supported, Lagrangian correlation function

$$C_{\text{LGR}}(\boldsymbol{h}, u) = \left(1 - \frac{1}{2v} |h_1 - vu| \right)_+. \tag{4.22}$$

Here, the spatial separation vector $\boldsymbol{h} = (h_1, h_2)'$ has longitudinal (east-west) component h_1 and latitudinal (north-south) component h_2, $v \in \mathbb{R}$ is a longitudinal velocity, and we write $p_+ = \max(p, 0)$. Note that (4.22) considers the longitudinal distance between the stations only and forms a special case of the Lagrangian model (4.15). The particular parametric form is motivated by the prevalence of westerly winds over Ireland and the linear relationships in Figure 4.3(a). The random field model underlying (4.21) can be thought of as a superposition of two independent processes with correlation functions C_{FS} and C_{LGR}, respectively, and the Lagrangian component can be interpreted in term of the frozen field model. A less ambitious interpretation is simply in terms of increasing the flexibility of the correlation structure. All other parameters held constant, the WLS estimates in (4.22) are $\widehat{\lambda} = 0.0573$ and $\widehat{v} = 234$ km $\cdot$ d^{-1} or 2.71 m $\cdot$ s^{-1}. To summarize, the fitted general stationary model is

$$C_{\mathrm{STAT}}(\boldsymbol{h}, u) = \frac{(1-\nu)(1-\lambda)}{1+a|u|^{2\alpha}} \left(\exp\left(-\frac{c\|\boldsymbol{h}\|}{(1+a|u|^{2\alpha})^{\beta/2}}\right) + \frac{\nu}{1-\nu}\delta_{\boldsymbol{h}=0} \right)$$
$$+ \lambda\left(1 - \frac{1}{2v}|h_1 - vu|\right)_+$$

with parameter estimates $\widehat{\nu} = 0.0415$, $\widehat{\lambda} = 0.0573$, $\widehat{a} = 0.972$, $\widehat{\alpha} = 0.834$, $\widehat{c} = 0.00128$, $\widehat{\beta} = 0.681$ and $\widehat{v} = 234$, where spatio-temporal lags are measured in kilometers and days, respectively. Full WLS optimization over all correlation parameters simultaneously yields essentially identical estimates. That said, our approach should be understood as illustrative. Ramifications in various directions remain to be explored, such as a more efficient and more appropriate handling of the nugget effect in the general stationary correlation model.

The graphs in Figure 4.3(b)–(d) illustrate the goodness of fit for the separable correlation model (4.19), the fully symmetric model (4.20), and the general stationary model (4.21), respectively. In each of the graphs, the fitted correlations are plotted vs. the empirical ones, for all pairs of stations and for temporal lag zero (black), one day (red), two days (green), and three days (blue). The general stationary correlation model fits better than the fully symmetric model, and the fully symmetric yet non-separable model provides a better fit than the separable model, particularly at temporal lags of one day.

4.4.3 Predictive performance

To assess the performance of predictions based on increasingly complex correlation structures, we consider time-forward predictions of the velocity measures during the test period (1971–1978), based on space-time correlation models fitted to the training data (1961–1970).

Specifically, we consider one-day ahead simple kriging predictions for the velocity measures at the eleven meteorological stations in Ireland. For each station, we obtain $365 \times 9 = 3285$ forecasts during the test period (1971–1978). The predictor variables are the 33 velocity measures observed during the past three days at the eleven stations. At each station, the simple kriging point predictor or point forecast, μ_t, for the velocity measure at time t is given by

$$\mu_t = \mathbf{c_0}' \, \mathbf{C}^{-1} \mathbf{z}_t \qquad (4.23)$$

where $\mathbf{C}$ denotes the 33×33 variance-covariance matrix of the predictor variables, $\mathbf{c_0}$ denotes a vector with the covariances between the predictand and the predictor variables, and $\mathbf{z}_t$ is a vector with the realized values of the predictor variables. The associated simple kriging variance is

$$\sigma_t^2 = \sigma_0^2 - \mathbf{c_0}' \mathbf{C}^{-1} \mathbf{c_0} \qquad (4.24)$$

where σ_0^2 denotes the unconditional variance of the velocity measure at the station at hand (Cressie, 1993, pp. 109–110 and 359; Chilès and Delfiner, 1999, pp. 154–164). Note that σ_t^2 in fact is constant and does not change in time. In addition to the point forecast (4.23), simple kriging provides a probabilistic forecast in the form of a Gaussian predictive distribution, $F_t = \mathcal{N}(\mu_t, \sigma_t^2)$, with predictive mean μ_t and predictive variance σ_t^2, respectively.

In the following we compare out-of-sample simple kriging predictions based on the fitted separable correlation model (4.19), the fully symmetric model (4.20), and the general stationary model (4.21). Specifically, we build $\mathbf{C}$ and $\mathbf{c_0}$ on the basis of the respective correlation model and the station-specific empirical variances of the velocity measures during the training period. Furthermore, we consider simple kriging predictions with $\mathbf{C}$ and $\mathbf{c_0}$ based on the empirical correlations in the training period.

To assess and rank the point forecasts, μ_t, for the velocity measures, x_t, we use the root-mean-square error or RMSE, defined as

$$\text{RMSE} = \left(\frac{1}{3285} \sum_{t=1}^{3285} (\mu_t - x_t)^2 \right)^{1/2} \qquad (4.25)$$

and the mean absolute error or MAE, given by

$$\text{MAE} = \frac{1}{3285} \sum_{t=1}^{3285} |\mu_t - x_t| . \qquad (4.26)$$

To evaluate the predictive distributions, $F_t = \mathcal{N}(\mu_t, \sigma_t^2)$, we consider the logarithmic score and the continuous ranked probability score. The logarithmic score is simply the negative of the logarithm of the predictive density evaluated at the observation. Let $\mathbf{1}(y \geq x)$ denote the function that attains the value 1 when $y \geq x$ and the value 0 otherwise. If F denotes the predictive cumulative distribution function and x materializes, the continuous ranked probability score is defined as

$$\text{crps}(F, x) = \int_{-\infty}^{\infty} (F(y) - \mathbf{1}(y \geq x))^2 \, dy.$$

For point forecasts the continuous ranked probability score reduces to the absolute error. If the predictive distribution is normal with mean μ and variance σ^2, the integral can be evaluated as

$$\mathrm{crps}\left(\mathcal{N}(\mu, \sigma^2), x\right) = \sigma\left(\frac{x - \mu}{\sigma}\left(2\,\Phi\left(\frac{x - \mu}{\sigma}\right) - 1\right) + 2\,\phi\left(\frac{x - \mu}{\sigma}\right) - \frac{1}{\sqrt{\pi}}\right)$$

where ϕ and Φ denote the probability density function and the cumulative distribution function of the standard normal distribution, respectively. We refer to Gneiting and Raftery (2005) and Gneiting et al. (2006) for detailed discussions of the continuous ranked probability score and other devices for the evaluation of probabilistic forecasts. Here, we assess the predictive distributions based on the various correlation structures by comparing the associated LogS and CRPS values, where

$$\mathrm{LogS} = \frac{1}{3285}\sum_{t=1}^{3285}\left(\frac{1}{2}\ln(2\pi\sigma_t^2) + \frac{(x_t - \mu_t)^2}{2\sigma_t^2}\right) \tag{4.27}$$

and

$$\mathrm{CRPS} = \frac{1}{3285}\sum_{t=1}^{3285}\mathrm{crps}\left(\mathcal{N}(\mu_t, \sigma_t^2), x_t\right) \tag{4.28}$$

denote the mean of the logarithmic score and the continuous ranked probability score over the test period, respectively. Clearly, the smaller these values the better.

Tables 4.4, 4.5, 4.6, and 4.7 compare the RMSE, MAE, LogS and CRPS values for the kriging predictions based on the separable, fully symmetric, general stationary, and empirical correlation structures, respectively. In terms of all four performance measures, the simple kriging predictions based on the general stationary correlation model performed better than those based on the fully symmetric model, and the predictions based on the fully symmetric model outperformed those based on the separable model. This suggests that the use of richer, more complex, and more physically realistic correlation and covariance models results in improved predictive performance. Perhaps surprisingly, the predictions based on the empirical space-time correlations performed best. The empirical correlations are computed on 10 years' worth of replicated data and need to be taken at face value — in stark contrast to the classical geostatistical, purely spatial one realization scenario, in which reliable estimates of the correlation and covariance structure are unlikely to be available. However, the simple kriging approach based on the empirical correlations does not allow for predictions of velocity measures away from the meteorological stations. The model-based approaches admit such an extension when combined with a statistical model for the spatially varying variance of the velocity measures.

Table 4.4 Root-mean-square error (RMSE) for one-day ahead simple kriging predictions of the velocity measures at the eleven meteorological stations during the test period (1971–1978)

	Val	Bel	Clo	Sha	Roc	Birr	Mul	Mal	Kil	Clo	Dub
separable	.501	.495	.491	.468	.483	.477	.427	.496	.439	.486	.450
fully symmetric	.501	.495	.492	.468	.479	.476	.424	.492	.436	.484	.445
general stationary	.499	.495	.490	.466	.474	.472	.419	.488	.429	.479	.440
empirical	.500	.494	.486	.454	.465	.462	.414	.479	.414	.466	.427

Table 4.5 Mean absolute error (MAE) for one-day ahead simple kriging predictions of the velocity measures at the eleven meteorological stations during the test period (1971–1978)

	Val	Bel	Clo	Sha	Roc	Birr	Mul	Mal	Kil	Clo	Dub
separable	.398	.395	.389	.372	.387	.375	.340	.399	.347	.385	.359
fully symmetric	.399	.396	.389	.372	.384	.373	.338	.396	.344	.382	.356
general stationary	.397	.395	.387	.369	.379	.370	.334	.393	.339	.377	.351
empirical	.394	.392	.384	.359	.369	.362	.327	.385	.325	.367	.339

Table 4.6 Logarithmic score (LogS) for one-day ahead simple kriging predictions of the velocity measures at the eleven meteorological stations during the test period (1971–1978)

	Val	Bel	Clo	Sha	Roc	Birr	Mul	Mal	Kil	Clo	Dub
separable	.727	.716	.707	.659	.692	.680	.577	.720	.596	.699	.626
fully symmetric	.728	.716	.709	.661	.682	.677	.570	.712	.589	.694	.617
general stationary	.724	.715	.705	.655	.672	.670	.560	.704	.574	.683	.606
empirical	.726	.714	.698	.630	.654	.648	.542	.684	.538	.658	.571

Table 4.7 Continuous ranked probability score (CRPS) for one-day ahead simple kriging predictions of the velocity measures at the eleven meteorological stations during the test period (1971–1978)

	Val	Bel	Clo	Sha	Roc	Birr	Mul	Mal	Kil	Clo	Dub
separable	.282	.279	.276	.264	.273	.268	.241	.281	.247	.273	.254
fully symmetric	.282	.279	.277	.264	.271	.267	.240	.279	.245	.272	.252
general stationary	.281	.279	.275	.262	.267	.265	.237	.276	.241	.269	.249
empirical	.280	.278	.273	.255	.262	.259	.233	.271	.232	.261	.240

Appendix

Proof of Theorem 4.3.2   Suppose that C is a stationary, fully symmetric covariance function. By Bochner's theorem, the representation (4.7) holds. Expand the exponential term under the integral in (4.7) as the product of $\cos(\boldsymbol{h}'\boldsymbol{\omega}) + i\sin(\boldsymbol{h}'\boldsymbol{\omega})$ and $\cos(u\tau) + i\sin(u\tau)$. Since C is real-valued, we can write (4.7) as

$$C(\boldsymbol{h}, u) = \iint \Big(\cos(\boldsymbol{h}'\boldsymbol{\omega}) \cos(u\tau) - \sin(\boldsymbol{h}'\boldsymbol{\omega}) \sin(u\tau) \Big) \, \mathrm{d}F(\boldsymbol{\omega}, \tau),$$
$$(\boldsymbol{h}, u) \in \mathrm{R}^d \times \mathrm{R}. \tag{4.29}$$

Similarly,

$$C(\boldsymbol{h}, -u) = \iint \Big(\cos(\boldsymbol{h}'\boldsymbol{\omega}) \cos(u\tau) + \sin(\boldsymbol{h}'\boldsymbol{\omega}) \sin(u\tau) \Big) \, \mathrm{d}F(\boldsymbol{\omega}, \tau),$$
$$(\boldsymbol{h}, u) \in \mathrm{R}^d \times \mathrm{R}. \tag{4.30}$$

Full symmetry implies that $C(\boldsymbol{h}, u) = C(\boldsymbol{h}, -u)$ and therefore (4.29) and (4.30) yield the representation (4.8), as desired. Conversely, any function C of the form (4.8) is fully symmetric and admits Bochner's representation (4.7).

Acknowledgments

While working on this chapter, Tilmann Gneiting was on sabbatical leave at the Soil Physics Group, Universität Bayreuth, Universitätsstr. 30, 95440 Bayreuth, Germany, and supported in part by Office of Naval Research Grant N00014-01-10745 and National Science Foundation Award 0134264.

References

Amani, A. and Lebel, T. (1997), Lagrangian kriging for the estimation of Sahelian rainfall at small time steps, *Journal of Hydrology*, 192, 125–157.

Banerjee, S. (2005), On geodetic distance computations in spatial modeling, *Biometrics*, 61, 617–625.

Bochner, S. (1955), *Harmonic Analysis and the Theory of Probability*, University of California Press, Berkeley and Los Angeles.

Brown, P.E., Kåresen, K.F., Roberts, G.O., and Tonellato, S. (2000), Blur-generated non-separable space-time models, *Journal of the Royal Statistical Society Ser. B*, 62, 847–860.

Carroll, R.J., Chen, R., George, E.I., Li, T.H., Newton, H.J., Schmiediche, H., and Wang, N. (1997), Ozone exposure and population density in Harris county, Texas (with discussion), *Journal of the American Statistical Association*, 92, 392–415.

Chilès, J.P. and Delfiner, P. (1999), *Geostatistics. Modeling Spatial Uncertainty*, Wiley, New York.

Christakos, A.G. (2000), *Modern Spatiotemporal Geostatistics*, Oxford University Press, Oxford.

Cox, D.R. and Isham, V. (1988), A simple spatial-temporal model of rainfall, *Proceedings of the Royal Society of London Ser. A*, 415, 317–328.

Cressie, N. (1993), *Statistics for Spatial Data*, revised ed., Wiley, New York.

Cressie, N. and Huang, H.-C. (1999), Classes of nonseparable, spatio-temporal stationary covariance functions, *Journal of the American Statistical Association*, 94, 1330–1340; 96, 784.

De Iaco, S., Myers, D.E., and Posa, T. (2001), Space-time analysis using a general product-sum model, *Statistics & Probability Letters*, 52, 21–28.

De Iaco, S., Myers, D.E., and Posa, T. (2002), Nonseparable space-time covariance models: some parametric families, *Mathematical Geology*, 34, 23–42.

de Luna, X. and Genton, M.G. (2005), Predictive spatio-temporal models for spatially sparse environmental data, *Statistica Sinica*, 15, 547–568.

Diggle, P.J., Tawn, J.A., and Moyeed, R.A. (1998), Model-based geostatistics (with discussion), *Applied Statistics*, 47, 299–350.

Evensen, G. (1994), Sequential data assimilation with a nonlinear quasi-geostrophic model using Monte Carlo methods to forecast error statistics, *Journal of Geophysical Research – Oceans*, 99, 10143–10162.

Fuentes, M. (2005), A formal test for nonstationarity of spatial stochastic processes, *Journal of Multivariate Analysis*, 96, 30–54.

Fuentes, M. (2006), Testing for separability of spatial-temporal covariance functions, *Journal of Statistical Planning and Inference*, 136, 447–466.

Furrer, R., Genton, M.G., and Nychka, D. (2006), Covariance tapering for interpolation of large spatial datasets, *Journal of Computational and Graphical Statistics*, in press.

Gneiting, T. (1999), Correlation functions for atmospheric data analysis, *Quarterly Journal of the Royal Meteorological Society*, 125, 2449–2464.

Gneiting, T. (2002a), Nonseparable, stationary covariance functions for space-time data, *Journal of the American Statistical Association*, 97, 590–600.

Gneiting, T. (2002b), Compactly supported correlation functions, *Journal of Multivariate Analysis*, 83, 493–508.

Gneiting, T. and Raftery, A.E. (2005), Strictly proper scoring rules, prediction and estimation, University of Washington, Department of Statistics, Technical Report no. 463R.

Gneiting, T. and Sasvári, Z. (1999), The characterization problem for isotropic covariance functions, *Mathematical Geology*, 31, 105–111.

Gneiting, T. and Schlather, M. (2002), Space-time covariance models, in *Encyclopedia of Environmetrics*, El-Shaarawi, A.H. and Piegorsch, W.W., Eds., Vol. 4, Wiley, Chichester, pp. 2041–2045.

Gneiting, T. and Schlather, M. (2004), Stochastic models that separate fractal dimension and the Hurst effect, *SIAM Review*, 46, 269–282.

Gneiting, T., Sasvári, Z., and Schlather, M. (2001), Analogies and correspondences between variograms and covariance functions, *Advances in Applied Probability*, 33, 617–630.

Gneiting, T., Larson, K., Westrick, K., Genton, M.G., and Aldrich, E. (2006), Calibrated probabilistic forecasting at the Stateline Wind Energy Center: the regime-switching space-time (RST) method, *Journal of the American Statistical Association*, in press.

Guo, J.H. and Billard, L. (1998), Some inference results for causal autoregressive processes on a plane, *Journal of Time Series Analysis*, 19, 681–691.

Gupta, V.K. and Waymire, E. (1987), On Taylor's hypothesis and dissipation in rainfall, *Journal of Geophysical Research*, 92, 9657–9660.

Guttorp, P., Meiring, W., and Sampson, P.D. (1994), A space-time analysis of ground-level ozone, *Environmetrics*, 5, 241–254.

Haas, T.C. (2002), New systems for modeling, estimating, and predicting a multivariate spatiotemporal process, *Environmetrics*, 13, 311–332.

Hamill, T.M. and Snyder, C. (2000), A hybrid ensemble Kalman filter-3D variational analysis scheme, *Monthly Weather Review*, 128, 2905–2919.

Haslett, J. and Raftery, A.E. (1989), Space-time modelling with long-memory dependence: assessing Ireland's wind-power resource (with discussion), *Applied Statistics*, 38, 1–50.

Houtekamer, P.L. and Mitchell, H.L. (1998), Data assimilation using an ensemble Kalman filter technique, *Monthly Weather Review*, 126, 796–811.

Huang, H.-C. and Hsu, N.-J. (2004), Modeling transport effects on ground-level ozone using a non-stationary space-time model, *Environmetrics*, 15, 251–268.

Jones, R. and Zhang, Y. (1997), Models for continuous stationary space-time processes, in *Modelling Longitudinal and Spatially Correlated Data*, Lecture Notes in Statistics, 122, Gregoire, T.G., Brillinger, D.R., Diggle, P.J., Russek-Cohen, E., Warren, W.G., and Wolfinger, R.D., Eds., Springer, New York, pp. 289–298.

Jun, M. and Stein, M.L. (2004a), Statistical comparison of observed and CMAQ modeled daily sulfate levels, *Atmospheric Environment*, 38, 4427–4436.

Jun, M. and Stein, M.L. (2004b), An approach to producing space-time covariance functions on spheres, University of Chicago, Center for Integrating Statistical and Environmental Science, Technical Report No. 18.

Kolovos, A., Christakos, G., Hristopulos, D.T., and Serre, M.L. (2004), Methods for generating non-separable spatiotemporal covariance models with potential environmental applications, *Advances in Water Resources*, 27, 815–830.

Kyriakidis, P.C. and Journel, A.G. (1999), Geostatistical space-time models: a review, *Mathematical Geology*, 31, 651–684.

Lu, N. and Zimmerman, D.L. (2005), Testing for directional symmetry in spatial dependence using the periodogram, *Journal of Statistical Planning and Inference*, 129, 369–385.

Ma, C. (2002), Spatio-temporal covariance functions generated by mixtures, *Mathematical Geology*, 34, 965–975.

Ma, C. (2003a), Spatio-temporal stationary covariance models, *Journal of Multivariate Analysis*, 86, 97–107.

Ma, C. (2003b), Families of spatio-temporal stationary covariance models, *Journal of Statistical Planning and Inference*, 116, 489–501.

Ma, C. (2003c), Nonstationary covariance functions that model space-time interactions, *Statistics & Probability Letters*, 61, 411–419.

Matérn, B. (1986), *Spatial Variation*, 2nd edition, Springer, Berlin.

May, D.R. and Julien, P.Y. (1998), Eulerian and Lagrangian correlation structures of convective rainstorms, *Water Resources Research*, 34, 2671–2683.

Meiring, W., Guttorp, P., and Sampson, P.D. (1998), Space-time estimation of grid-cell hourly ozone levels for assessment of a deterministic model, *Environmental and Ecological Statistics*, 5, 197–222.

Mitchell, M., Genton, M.G., and Gumpertz, M. (2005), Testing for separability of space-time covariances, *Environmetrics*, 16, 819–831.

Riishøjgaard, L.P. (1998), A direct way of specifying flow-dependent background error correlations for meteorological analysis systems. *Tellus Ser. A*, 50, 42–57.

Sampson, P.D. and Guttorp, P. (1992), Nonparametric estimation of nonstationary spatial covariance structure, *Journal of the American Statistical Association*, 87, 108–119.

Scaccia, L. and Martin, R.J. (2005), Testing axial symmetry and separability of lattice processes, *Journal of Statistical Planning and Inference*, 131, 19–39.

Shitan, M. and Brockwell, P. (1995), An asymptotic test for separability of a spatial autoregressive model, *Communications in Statistics, Theory and Methods*, 24, 2027–2040.

Stein, M.L. (2005a), Space-time covariance functions, *Journal of the American Statistical Association*, 100, 310–321.

Stein, M.L. (2005b), Statistical methods for regular monitoring data, *Journal of the Royal Statistical Society Ser. B*, 67, 667–687.

Storvik, G., Frigessi, A., and Hirst, D. (2002), Stationary space-time Gaussian fields and their time autoregressive representation, *Stochastic Modelling*, 2, 139–161.

Taylor, G.I. (1938), The spectrum of turbulence, *Proceedings of the Royal Society of London Ser. A*, 164, 476–490.

Wan, Y., Milligan, M., and Parsons, B. (2003), Output power correlation between adjacent wind power plants, *Journal of Solar Energy Engineering*, 125, 551–555.

Whittle, P. (1954), On stationary processes in the plane, *Biometrika*, 41, 434–449.

Zawadzki, I.I. (1973), Statistical properties of precipitation patterns, *Journal of Applied Meteorology*, 12, 459–472.

CHAPTER 5

Space-Time Modelling of Rainfall for Continuous Simulation

Richard E. Chandler, Valerie Isham, Enrica Bellone, Chi Yang, and Paul Northrop

Contents

5.1 Introduction

In recent years, floods have become a matter of increasing concern in many countries. Flood risk is assessed by hydrologists using rainfall-runoff models, which are deterministic models of the processes by which rainfall input over a particular catchment is converted into surface runoff. Traditionally, the spatial and temporal structure of the rainfall input has been highly simplified and the flood risk assessment has focused on selected individual rainfall events. However, such an approach ignores the antecedent conditions which may be an important determinant of risk. For example, a moderate event following an extended wet spell may pose a higher risk than a severe event at a time when the soil is relatively dry. The risk of extreme flood events can therefore best be assessed by running a rainfall-runoff model continuously over extended time periods, so that the full range of antecedent conditions is represented (Wheater, 2002). Such an exercise requires long records of input rainfall data at a relatively fine temporal resolution (e.g., hourly) and incorporating spatial variation, which can be hydrologically significant even for small catchments (Wheater et al., 2000a).

Unfortunately, such long records of high-resolution rainfall data are not widely available at the present time. This creates a need for models that can be simulated to provide artificial data. Since flood risk assessments are constantly required in new situations (for example, in different catchments), the models must be suitable for routine implementation. Moreover, risk assessments must often be made over substantial periods into the future; thus the models must be nonstationary in time so as to enable the incorporation of climate change scenarios, as well as seasonality and low-frequency climate variability.

This chapter describes work on the modelling of space-time precipitation fields that has been carried out by an interdisciplinary team of statisticians from University College London and hydrologists from Imperial College London. These models are being used to support the continuous simulation of rainfall for use in hydrological design. Their development builds on extensive experience with models for rainfall at a single location. One strand of work was begun by Rodriguez-Iturbe et al. (1987, 1988) and is reviewed in Onof et al., (2000). Another strand builds on the work of Coe and Stern (1982) and Stern and Coe (1984). As with other contributions to this volume (notably that of Diggle), we distinguish between 'mechanistic' and 'descriptive' models for space-time data. In the current context, the former class seeks to represent the physical rainfall process, and in particular its hierarchical structure whereby 'rain cells' cluster within 'storms' within 'rain events' albeit in a highly idealised fashion. These models have a parsimonious parameterisation, with interpretable parameters that relate to physical phenomena. Moreover, they are constructed in continuous space and time so that their spatial and temporal resolution is limited only by that of the data available to fit them. However, apart from seasonality, this first class of models is stationary in both time and space, and is therefore best geared to relatively short temporal durations and spatial extents.

The second class of models is based on generalised linear models (GLMs). These are essentially multivariate time series models for temporally discrete (e.g., daily) rainfall sequences at a countable set of locations. In these models it is easy to incorporate dependence on explanatory variables such as site altitude, time of year, previous days' rainfall, teleconnections such as El Niño and the North Atlantic Oscillation, and climate change scenarios. This dependence can be used to predict the distribution of rainfall at any site, interpolating between locations at which data are available. Thus the GLMs can produce daily sequences at chosen locations that are spatially and temporally nonstationary. However, they are not suitable for use at subdaily time scales, because an excessive number of parameters would be required to describe adequately the complicated structure of rainfall at these scales. Nor do these models produce rainfall fields in continuous space, despite their ability to predict distributions at arbitrary locations.

In applications, the most appropriate modelling approach is dictated not only by the requirements of the situation at hand, but also by the constraints of the available data. Our mechanistic models require fully spatial-temporal data, such as sequences of images from a weather radar station, for parameter estimation. The GLMs, on the other hand, can be fitted to sequences of daily measurements from rain gauges. Gauge data are much more widely available than radar, and generally less problematic in terms of quality. It is helpful to understand some of the issues involved in rainfall measurement before trying to develop and fit models. Accordingly, in Section 5.2 we provide a brief overview of rainfall measurement using rain gauges and radar, and introduce the data used subsequently in our examples. Sections 5.3 and 5.4 then deal respectively with the mechanistic models and the GLMs. In each case we discuss model construction, inference and diagnostics, and provide some specimen results. Despite the focus on rainfall, many of the techniques we describe (particularly those relating to model fitting and inference) are broadly applicable in general spatial-temporal modelling contexts.

As indicated above, both the mechanistic models and the GLMs have strengths and weaknesses. At present, neither class on its own has the capability to provide nonstationary space-time rainfall simulations at resolutions suitable for input to rainfall-runoff models. It is therefore of interest to combine the strengths of the two model classes. In Section 5.5 we propose some ways of achieving this, and discuss some preliminary results. Finally, in Section 5.6 we summarise the current position and outline directions for further work.

5.2 Rainfall data and measurement

In this section we give a brief overview of rainfall measurement techniques, and describe the data used subsequently. This provides some insight into the issues that arise when developing models.

5.2.1 Measurement using rain gauges

The simplest device for measuring rainfall is the manual rain gauge. This is a container that collects water and is emptied by an observer at regular time intervals (usually daily); the rainfall amount is read from a graduated scale. In the U.K., most daily gauges have reported to a resolution of 0.1 mm since the move to a metric system in the mid-1970s. Prior to this, most gauges reported to a resolution of around 0.25 mm (0.01 inches). Although these amounts may appear negligible, the resolution change produces a detectable change in the proportion of days when zero rainfall is recorded, and therefore produces spurious 'trends' in rainfall occurrence unless it is accounted for in the analysis of long rainfall time series. As might be expected, the impact on non-zero rainfall amounts is much less severe (see Section 5.4.5).

Differences in observer practice also create problems, particularly with small rainfall amounts. For example, if there is water in the gauge but the amount is less than the smallest interval on the scale, a 'trace' amount should be recorded — in statistical terms, this corresponds to censored data below the limit of detection. However, in practice, observers often round such values, either down to zero or up to the smallest interval. This can create apparent spatial inconsistencies in patterns of rainfall occurrence; a simple way to resolve these is to zero all values below some appropriate threshold prior to analysis. We tend to use a threshold of 0.5 mm when working with daily gauge data; this is large enough to remove most of the inter-observer discrepancies, but is hydrologically insignificant for most purposes (and certainly for flood risk assessment).

At subdaily time scales, rain gauge measurement is usually automated. Most automatic gauges in use today are of the 'tipping bucket' variety. This records a sequence of times between which a fixed amount of rain (usually 0.2 mm) has fallen; these times correspond to the tipping of a small bucket when it fills with water. By counting the number of bucket tips in any time interval, the corresponding rainfall amount can be estimated. The resulting discretisation limits the temporal resolution of such gauges, particularly during periods of light rain. It is generally accepted that 0.2-mm tipping bucket measurements provide an adequate representation of hourly rainfall in the U.K.; moreover, during periods of significant rainfall they can be used at time scales down to 15 minutes (Wheater et al., 2000a).

A network of gauges provides a set of rainfall time series at discrete spatial locations. For most locations in the U.K. and elsewhere, the spatial density of available daily gauge series is typically less than 0.01 km^{-2}, including records from gauges that are no longer operational. The density of subdaily records is rather lower than this; moreover, there are relatively few long records of subdaily rainfall. As a result, gauge data are not generally suitable for the study of fine-scale spatial and temporal structure in rainfall fields; they do, however, allow the study of systematic regional variation and time trends.

5.2.2 *Measurement using radar*

Radar technology provides a means of obtaining spatial images of rainfall fields. A radar image provides gridded estimates of average rainfall intensity; the pixel size is typically 2×2 km^2 or 5×5 km^2. For the U.K. radar network, it normally takes 1 minute to produce an image which therefore represents, in effect, an instantaneous temporal snapshot of the rainfall field.

Radar measurement of rainfall is indirect: reflected energy within the radar beam is converted to a rainfall intensity using empirical relationships, which are reviewed in Collier (2000). Typically, 'raw' rainfall estimates obtained in this way will differ from rain gauge measurements. There are many reasons for this, in addition to inaccuracies in the empirical relationships used for the conversion. For example, the radar beam measures reflected energy at some height in the atmosphere (increasing with distance from the radar station) whereas the gauges are at ground level; the radar images represent spatial averages whereas the gauges are point values; obstacles may intercept the radar beam causing systematic anomalies; and background noise can cause 'clutter' in radar images. Austin (2001) gives more detail on such issues. In view of these, radar measurements must be calibrated before use in hydrological applications. In the U.K., the standard procedure is that described by Moore et al. (1994).

The first step in this procedure is to remove systematic anomalies. These can be identified by computing summary statistics for each pixel over a long time period (e.g., a year). Anomalous pixels show up as outliers in the distributions of such summary statistics, and can therefore be defined on the basis of suitably chosen percentiles of these distributions. In our work, anomalies are identified from annual images of both mean rainfall intensity and the proportion of time instants with non-zero rainfall; data from anomalous pixels are subsequently treated as missing.

The second step is to remove clutter from each radar image (i.e., to set to zero any isolated pixels that appear to be 'wet'). Having done this, the images are compared with subdaily (typically 15-min) data from a network of rain gauges. The basic procedure is, for each 15-min interval, to multiply the radar images by a spatial calibration surface, obtained by interpolating calibration factors calculated at each gauge. Let $R_g^{(j)}$ be the recorded average rainfall intensity for the j-th gauge during the time interval of interest, and denote by $R_r^{(j)}$ the corresponding value from the radar pixel containing this gauge. Then the calibration factor for the j-th gauge is

$$z_j = \frac{R_g^{(j)} + \epsilon_g}{R_r^{(j)} + \epsilon_r}. \tag{5.1}$$

The non-zero constant ϵ_r is introduced to prevent problems when $R_r^{(j)} = 0$, while ϵ_g allows the calibration factor to be non-zero even when the the gauge records no rain (which may occur due to gauge discretisation errors during periods of light rainfall, as described in Section 5.2.1).

The aim of calibration is to produce radar images that are in reasonable agreement with the gauge data. Unsurprisingly, the procedure is sensitive to the values of ϵ_r and ϵ_g in (5.1). For the radar station used below we find that the choice of $\epsilon_r = \epsilon_g = 0.1$ generally yields calibrated data with acceptable properties, although these values differ from those recommended by Moore et al. (1994). For purposes of flood risk assessment, it is important to reproduce correctly the distributions of rainfall intensity at various space and time scales, particularly in the upper tails. Our experience is that uncalibrated radar data generally overestimate the larger hourly rainfalls, and that the calibration corrects this. There is a tendency for calibrated radar to underestimate annual totals, but the relative differences are generally small (a few percent) and acceptable for our purposes.

Radar data summary statistics are required to fit the stochastic models described in Section 5.3. However, the properties of the data can be affected substantially by the calibration. In particular, each image is calibrated individually, with no temporal smoothing in the computation of the calibration factors defined in (5.1). As a result, we find that calibration often reduces the temporal autocorrelation in a sequence of radar images. This sensitivity of statistical properties to the calibration procedure is a source of some concern and, at present, is a major impediment to the use of models based on radar data in flood risk assessment. Substantial further study is needed to gain a thorough understanding of the issues involved. It is possible that alternative schemes, such as that proposed by Brown et al. (2001) for real-time calibration, may yield improved, and better understood, performance. Nonetheless, the scheme used in this chapter represents existing best practice in the U.K., with some relatively minor modifications for our purposes.

5.2.3 Data used in the examples

In the work reported below, two rainfall data sets are used to illustrate the 'mechanistic' and 'descriptive' modelling approaches, respectively. In Section 5.3 we use data from the Chenies radar, northwest of London. Twelve years of data, from 1990 onwards, are available; 5×5 km^2 images are available at 15-minute intervals to a radius of 210 km from the radar, and 2×2 km^2 images at 5-minute intervals to a radius of 76 km. There are frequent breaks in the record, with durations ranging from a single image to several days. The recorded intensities are discretised so that the smallest non-zero value is 0.03 mm/hr. For calibration, 15-minute data from a network of 122 tipping bucket gauges have been used, although the gauges were not all recording throughout the whole period. The left panel of Figure 5.1 shows the radar location, as well as the extent of the 2 km and 5 km data.

The second data set is used in Section 5.4 to illustrate the fitting and simulation of GLMs. The data are from a network of 34 daily gauges run by

(a)   (b)

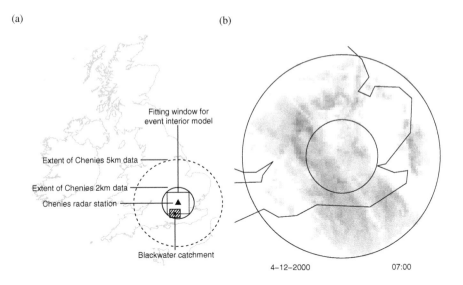

Figure 5.1 (a) Locations of Chenies radar (showing the data window used for model fitting in Section 5.3) and Blackwater catchment (b) example of a 5×5 km^2 image from the Chenies radar.

the U.K. Meteorological Office, in the catchment of the river Blackwater in southern England. The location is also shown in the left panel of Figure 5.1. The area is relatively flat; gauge altitudes range from 10 to 170 m above sea level. The earliest record starts in 1904, with data from some gauges available until 2000. Typically, between six and ten gauges have been operational at any one time since the 1960s. The size of the Blackwater catchment is about 50×40 km^2, which is large by U.K. standards. The gauge density is also quite high, relative to other locations in the U.K. This gives some idea of the spatial scales of interest to hydrologists, and of the density of rain gauge data that may typically be available.

To give some idea of the structure of a rainfall field, the right panel of Figure 5.1 shows a 5 km image from the Chenies radar. This image has not been calibrated, and illustrates some of the artefacts commonly associated with radar data (e.g., spurious radial features and, at the northwestern edge, a failure to detect rainfall because the radar beam is above the cloud level). It also shows the spatial scale of the weather systems that are responsible for much of the rainfall in the U.K. The system itself, which we subsequently refer to as a 'rain event,' is much larger than any catchment (compare with the size of the Blackwater catchment in the left hand panel of Figure 5.1). This suggests that at typical catchment scales, rainfall could be modelled as a sequence of such events, with an appropriate spatial-temporal model for the event interiors. This idea forms the basis for our stochastic modelling.

5.3 Stochastic models for rainfall intensity based on point processes

The models described in this section are for rainfall in continuous space and time, and are constructed using point processes. On the basis of the observations above, separate models are built for the interior of a rain event (Section 5.3.1) and for the timing of a sequence of rain events that will affect a particular catchment (Section 5.3.2). The continuous simulation of the combined model is discussed in Section 5.3.3.

5.3.1 The interior of a single rain event

5.3.1.1 Model definition and properties

Our model for rain event interiors was developed by Northrop (1998), building on previous work by Cox and Isham (1988). In this model, a rain event is a superposition of clusters of 'rain cells,' each cluster being termed a 'storm.' Each rain event has a random velocity V and is associated with an underlying random elliptical 'shape' specified by an orientation Θ and eccentricity E. Storm origins, which are unobserved, occur in a Poisson process of rate λ in space-time; given the shape and velocity of an event, its storms are realised independently and identically about their origins. Each storm has a temporal duration (its 'lifetime') that is exponentially distributed with rate γ. During this lifetime, the storm moves with velocity V and gives birth to rain cell origins at a rate β in time, starting with a cell at the temporal storm origin. Thus the temporal structure of the storm is that of a Bartlett-Lewis Poisson cluster process (Cox and Isham, 1980, Section 3.1); the number, C, of cells per storm has a geometric distribution with mean $\mu_C = 1 + \beta/\gamma$. This is illustrated in the left panel of Figure 5.2.

In space, each cell origin is independently displaced relative to the moving storm origin, according to a bivariate zero mean Gaussian distribution whose contours share the elliptical shape (Θ, E) of the event. This yields a Neyman-Scott cluster structure (Cox and Isham, 1980, Section 3.1). Storms will tend to have a banded appearance when E is close to 1, but will appear more circular for E near 0. The spatial scale A_s of a storm is defined as the variance of its spatial cell displacements in direction Θ, and varies independently between storms. The cells themselves move with velocity V and share the elliptical shape (Θ, E), with random semi-major axis length A_c. Each cell has a random temporal duration D, during which it produces rainfall at a constant random intensity X throughout its spatial extent. Thus the cell can be viewed as a cylinder with elliptical cross-section and height X. A_c, D, and X are realised independently for all cells within an event.

The right panel of Figure 5.2 illustrates the instantaneous spatial structure of a single storm. A rain event is a superposition of many such storms; the total rainfall intensity at any space-time point is the sum of intensities from all cells covering that point. Both storms and cells may (and often will) overlap

(a)

(b)

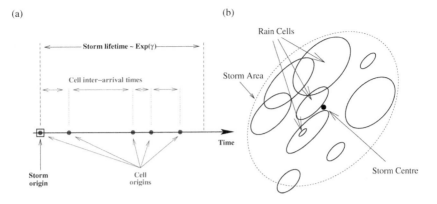

Figure 5.2 Schematic diagram of a storm in the space-time model for the interior of a rain event. (a) Temporal structure: cell origins occur in a Poisson process during the lifetime of a storm, with a cell at the storm origin. (b) Spatial structure: each cell is elliptical, and is displaced from the (moving) storm centre according to a bivariate Gaussian distribution with the same elliptical shape.

in space and time. In this regard, it may be helpful to think of overlapping cells as contributions from different levels in the atmosphere.

The assumption that cell intensities are constant throughout their lifetimes and spatial extents is made primarily for mathematical convenience, although other possibilities can be considered. For example, a multiplicative random noise could be applied to give the cell an irregular profile (Rodriguez-Iturbe et al., 1987), or a truncated Gaussian shape could be used (Northrop and Stone 2005). Detailed assumptions about the shape of a cell are in any case unlikely to be important except at very short time scales.

According to the model description above, the rainfall intensity at spatial location $\boldsymbol{u}$ and time t can be written as

$$Y(\boldsymbol{u}, t) = \int_{\tau=-\infty}^{t} \int_{\boldsymbol{w} \in \mathbb{R}^2} I(\boldsymbol{w}, \tau; \boldsymbol{u}, t) X(\boldsymbol{w}, \tau) \, dN(\boldsymbol{w}, \tau), \qquad (5.2)$$

where N is the cluster process of cell origins, $X(\boldsymbol{w}, \tau)$ denotes the depth of a cell with origin $(\boldsymbol{w}, \tau)$, and $I(\boldsymbol{w}, \tau; \boldsymbol{u}, t)$ is an indicator taking the value 1 if this cell covers $\boldsymbol{u}$ at time t, and zero otherwise. This representation is the key to deriving the statistical properties of the model. For example, taking expectations and taking advantage of the independence of model components, we obtain

$$E[Y(\boldsymbol{u}, t)] = \int_{\tau=-\infty}^{t} \int_{\boldsymbol{w} \in \mathbb{R}^2} E[I(\boldsymbol{w}, \tau; \boldsymbol{u}, t)] \, E[X(\boldsymbol{w}, \tau)] \, E[dN(\boldsymbol{w}, \tau)]$$

$$= \lambda \mu_C E(X) \int_{\tau=-\infty}^{t} \int_{\boldsymbol{w} \in \mathbb{R}^2} E[I(\boldsymbol{w}, \tau; \boldsymbol{u}, t)] \, d\boldsymbol{w} \, d\tau. \qquad (5.3)$$

Here we have used the fact that the space-time cell arrival rate is $\lambda \mu_C$, so that $E[dN(\boldsymbol{w}, \tau)] = \lambda \mu_C d\boldsymbol{w} d\tau$. To evaluate $E[I(\boldsymbol{w}, \tau; \boldsymbol{u}, t)]$, consider first the case when $E = 0$ and $\boldsymbol{V} = 0$, so that the cells are circular and do not move. In this case a cell born at $(\boldsymbol{w}, \tau)$ will cover $(\boldsymbol{u}, t)$ if its duration exceeds $t - \tau$ and its radius exceeds $|\boldsymbol{u} - \boldsymbol{w}|$. Since the duration and radius are independent, we have

$$E[I(\boldsymbol{w}, \tau; \boldsymbol{u}, t)] = \Pr\{D > t - \tau\} \Pr\{A_c > |\boldsymbol{u} - \boldsymbol{w}|\} = \mathcal{S}_D(t - \tau)\mathcal{S}_{A_c}(|\boldsymbol{u} - \boldsymbol{w}|),$$

where $\mathcal{S}(.)$ denotes a survivor function. Substituting this into (5.3), the mean intensity is

$$\lambda \mu_C E(X) \int_{\tau=-\infty}^{t} \mathcal{S}_D(t - \tau)d\tau \int_{\boldsymbol{w} \in \mathbb{R}^2} \mathcal{S}_{A_c}(|\boldsymbol{u} - \boldsymbol{w}|)d\boldsymbol{w}$$

$$= \lambda \mu_C E(X)E(D) \int_{r=0}^{\infty} 2\pi r \mathcal{S}_{A_c}(r)dr = \lambda \mu_C E(X)E(D)E\left(\pi A_c^2\right), \quad (5.4)$$

in agreement with intuition. The $\boldsymbol{w}$-integral here has been evaluated using a transformation to polar coordinates.

In the general case when $\boldsymbol{V}$ and E are non-zero, one can proceed by defining a linear coordinate transformation $\mathcal{L} : (\boldsymbol{u}, t) \rightarrow (\boldsymbol{u}^*, t)$ such that, in the new space-time coordinate system, cells are circles with radius A_c and the velocity is zero. In this case result (5.4) can be applied to the transformed process. The transformation is given by

$$\boldsymbol{u}^* = \begin{pmatrix} 1 & 0 \\ 0 & (1 - E^2)^{-1/2} \end{pmatrix} \begin{pmatrix} \cos \Theta & \sin \Theta \\ -\sin \Theta & \cos \Theta \end{pmatrix} (\boldsymbol{u} - \boldsymbol{V}t). \quad (5.5)$$

Note, however, that (5.5) does not preserve volume unless $E = 0$. Hence the space-time rate of storm arrivals in the transformed (Lagrangian) coordinate system is $\lambda\sqrt{1 - E^2}$, and the general expression for the mean of the rainfall process is

$$E[Y(\boldsymbol{u}, t)] = \lambda \mu_C E(X)E(D)E\left(\pi A_c^2 \sqrt{1 - E^2}\right). \quad (5.6)$$

A Lagrangian transformation is not necessary for the derivation of the mean in (5.6), but it simplifies considerably the derivation of the space-time covariance function $c(\boldsymbol{u}, t) = \text{Cov}(Y(\boldsymbol{0}, 0), Y(\boldsymbol{u}, t))$ for the model, which is defined since the model is spatially and temporally stationary. Specifically, it can be shown that $c(\boldsymbol{u}, t) = \sqrt{1 - E^2} \, c_s(\boldsymbol{u}^*, t)$, where c_s is the covariance function for a model with zero velocity and circular cells. To derive c_s it is also convenient to assume that the cell duration D is exponentially distributed, since in this case the model has a temporal Markov property. We will generally assume that the cell scale A_c has a gamma distribution rather than the simpler exponential form, since the zero mode of the exponential distribution tends to generate large numbers of unrealistically small cells (Northrop, 1996, 1998). Similarly, we assume a gamma distribution for the reciprocal of the storm scale $1/A_s$. The second-order properties of the model require only the corresponding properties of X, so no particular distributional form need be assumed in this case, although a parameter can be saved by assuming, for example, that $\text{var}(X)= \text{E}[(X)]^2$ (as for the exponential distribution). The model is only fitted within a rain event, for which all cells have the same velocity, eccentricity, and orientation, so these can be treated as parameters $(v, e, \theta,$ respectively) with no distributional forms assumed at this point. In this case, derivation of the autocovariance function is simplified as all contributions come either from single cells overlapping both space-time points, or from distinct cells within a single storm. However, allowance for variation between events will be required subsequently, for simulation of long space-time rainfall sequences.

Further details of the model and derivations of its properties are given by Northrop (1998), although that paper contains an error in the evaluation of the space-time covariance function $c_s (\boldsymbol{u}^*, t)$; a correction is given in the Appendix below. A simpler version of the model, in which the rain event is made up of a simple Poisson process of circular cells (so that $\beta = 0$ and $E = 0$), was proposed by Cox and Isham (1988). Similar stochastic models are also discussed by Vedel Jensen et al. in Chapter 2 of this volume.

5.3.1.2 Model fitting

Given adequate radar data, the model described above can be fitted to the interiors of observed rain events. Although it is stationary, features such as seasonal variation can be accommodated by fitting separately to data from each month of the year. As described in Section 5.2.2, radar data can be treated as temporally instantaneous but spatially averaged. If the radar pixels have dimensions $h \times h$ km^2, then the data have the form

$$Y_{i,j}^{(h)}(t) = \frac{1}{h^2} \int_{(i-1)h}^{ih} \int_{(j-1)h}^{jh} Y(\boldsymbol{u}, t) \, du_2 \, du_1 \qquad (5.7)$$

where $\boldsymbol{u} = (u_1, u_2)$. The first- and second-order properties of these variables can be derived straightforwardly from those of $Y(\boldsymbol{u}, t)$.

To fit the model, radar sequences must be selected that are consistent with its assumptions regarding the stationarity of rain event interiors. To achieve this for the Chenies radar described in Section 5.2.3, we inscribe a square within the 2 km data circle; this has dimensions 52×52 pixels (i.e., 104×104 km^2 — see Figure 5.1). Within this 'fitting window,' the proportion of wet pixels in each image (the 'coverage') is calculated to identify periods that may initially be considered as rain events. These periods are then refined, using criteria designed to ensure temporal and spatial stationarity. For example, to filter out periods when an event might be entering or leaving the window, we discard any image with a coverage substantially lower than the maximum achieved. The model is only fitted to sequences for which the various criteria are met throughout, and which last for at least an hour. The latter restriction ensures that both spatial and temporal properties of the data are determined reasonably accurately.

For any given rain event, the model description involves four random variables $(X, D, A_c, \text{and } A_s)$ as well as the parameters $\lambda, \beta, \gamma, e, \theta$, and the two components of the velocity $\boldsymbol{v}$. Thus the model has at least 11 parameters. If two-parameter distributions are used for the cell intensity X, the cell scale A_c, and the storm scale A_s, the number of parameters rises to 14. In the modelling reported below, an exponential distribution is assumed for cell intensities so that $\mathrm{Var}(X) = [\mathrm{E}(X)]^2$. Following exploratory data analysis, the gamma distributions for A_c and A_s are approximated by one-parameter distributions with the scale parameters held fixed. Thus, 11 parameters remain to be determined.

For this type of model, likelihood-based fitting is infeasible because the density of the spatially aggregated data is not available in a computationally useful form. Moreover, the model's constant cell intensities impose short-term deterministic features that are not present in real rainfall. Although such features are hydrologically insignificant, they may adversely affect the performance of methods that attempt to maintain fidelity to all aspects of the data (Rodriguez-Iturbe et al., 1988); likelihood-based fitting may therefore be undesirable. As an alternative therefore, we choose to use a generalised method of moments (GMM), in which parameters are chosen to provide as close as possible a match between selected sample statistics and the corresponding model properties. Specifically, denote by $\boldsymbol{\phi}$ the vector of unknown parameters, and let $\mathbf{T} = (T_1 \ \ldots \ T_k)'$ be a vector of observed summary statistics with expected value $\mathrm{E}(\mathbf{T}) = \boldsymbol{\tau}(\boldsymbol{\phi})$. Then the GMM estimator minimises the objective function

$$S(\boldsymbol{\phi}; \mathbf{T}) = \sum_{i=1}^{k} w_i [T_i - \tau_i(\boldsymbol{\phi})]^2 \tag{5.8}$$

for some set of weights $w_1, \ldots, w_k$. Equivalently, the estimator solves the estimating equation

$$\mathbf{g}(\boldsymbol{\phi}; \mathbf{T}) = \partial S / \partial \boldsymbol{\phi} = \mathbf{0}. \tag{5.9}$$

In principle, the representation (5.9) enables inference (e.g., the construction of confidence intervals for parameters) to be carried out using standard results for estimating equations (see, for example, Davison, 2003, Section 7.2). Specifically, in regular problems any estimator defined via an equation of the form (5.9) is distributed as

$$\hat{\boldsymbol{\phi}} \sim N(\boldsymbol{\phi}_0, \mathbf{H}^{-1} \mathbf{V} \mathbf{H}^{-1}) \tag{5.10}$$

asymptotically, where ϕ_0 is the target value of ϕ (formally defined as the solution to the equation $\mathrm{E}\left[\mathbf{g}\left(\phi; \mathbf{T}\right)\right] = \mathbf{0}$), $\mathbf{H} = \partial \mathbf{g}/\partial \phi$ is the Hessian matrix of the objective function (5.8) at ϕ_0 and $\mathbf{V}$ is the covariance matrix of $\mathbf{g}\left(\phi_0\right)$. Wheater et al. (2005) discuss the estimation of $\mathbf{H}$ and $\mathbf{V}$ in the context of stochastic models for rainfall at a single site.

In the work reported here, models are fitted using the mean $\mathrm{E}(Y^{(h)})$, the variance $\mathrm{var}(Y^{(h)})$, and the space-time autocorrelation function

$$\rho^{(h)}(\boldsymbol{k}, \tau) = \mathrm{Corr}\left(Y_{i,\,j}^{(h)}(t), Y_{i+k_1,\,j+k_2}^{(h)}(t + \tau)\right),$$

where $\boldsymbol{k} = (k_1, k_2)$, for various levels of aggregation (h) and spatial and temporal lags, $\boldsymbol{k}$ and τ respectively. Minimisation of the objective function (5.8) is not feasible analytically, so numerical methods must be used. Some reparameterisation helps improve stability at this point. In particular, β is replaced by the mean number of cells $\mu_C = 1 + \beta/\gamma$; $\mathrm{E}(A_c)$ by the mean cell area $\mu_A = \pi\sqrt{1 - e^2}\,\mathrm{E}(A_c^2)$; and $\mathrm{E}(A_s)$ by the mean storm area μ_S (defined in terms of the area in which approximately 40% of the rain cell origins are expected to fall).

Unfortunately, even after reparameterisation, the numerical minimisation of (5.8) is often unstable for the model considered here. To improve the stability, fitting can be carried out in two stages (see Wheater et al., 2000b, for more details). First, the velocity $\boldsymbol{v}$ is determined from the cross-correlation $\rho^{(h)}(\boldsymbol{k}, \tau)$ using the property that $\rho^{(h)}(\boldsymbol{k}, \tau)$ is maximised when $\boldsymbol{k} = \boldsymbol{v}\tau$. Because the arguments of $\rho^{(h)}$ can take only a discrete set of values, a centroid estimate is obtained. The velocity is then treated as known in the second stage of fitting.

This second stage exploits the fact that the purely spatial autocorrelation function $\rho^{(h)}(\boldsymbol{k}, 0)$ depends strongly on the spatial parameters μ_A, μ_S, e and θ (e, and θ determine the shape of the spatial autocorrelation contours, while μ_A and μ_S control their decay rate). Similarly, for given $\boldsymbol{v}$, the temporal parameters η and γ control the decay of the maximised cross-correlation $\rho^{(h)}(\boldsymbol{v}\tau, \tau)$. The remaining parameters λ, μ_C, and $\mathrm{E}(X)$ influence the mean and variance of the rainfall field. Thus, the second stage of the estimation procedure iterates between the following steps:

1. The parameters μ_A, μ_S, e, and θ are estimated using a generalised method of moments applied to Fisher's logarithmic transformation of $\rho^{(h)}(\boldsymbol{k}, 0)$:

$$F(\rho) = \frac{1}{2} \ln \left(\frac{1 + \rho}{1 - \rho}\right). \tag{5.11}$$

   Only spatial lags $\boldsymbol{k}$ with high $\rho^{(h)}(\boldsymbol{k}, 0)$ (and hence clearly defined elliptical contours) are used.

2. Given the values of the other parameters, closed form expressions for η and γ are found by equating observed and fitted values of $\rho^{(h)}(\lfloor \boldsymbol{v}\tau \rfloor, \tau)$ for two values of τ; $\tau = 10$ and $\tau = 20$ minutes have been found to work well. Here, $\lfloor \boldsymbol{v}\tau \rfloor$ denotes the value of $\boldsymbol{v}\tau$ rounded to the nearest pixel.

3. Given the other parameters, closed form expressions for λ, μ_C, and $E(X)$ are obtained by equating observed and fitted values of $E(Y^{(2)})$, $\text{var}(Y^{(h_1)})$, and $\text{var}(Y^{(h_2)})$; $h_1 = 2, h_2 = 8$ are suitable values.

Unfortunately, it is not obvious that the estimators defined by this two-stage procedure can be regarded as solutions to an estimating equation of the form (5.9); at present therefore, we are not in a position to provide confidence intervals for parameter estimates. The estimating equation theory has been applied to the fitting of stochastic models for purely temporal rainfall, however; Wheater et al. (2005) give an overview and preliminary results.

5.3.1.3 Assessment of the fitted model

Having fitted any model, it is necessary to assess its performance. Several assessment criteria may be considered in the current context. As a basic requirement, the parameters should be identifiable. Moreover, since the parameters all relate to physical quantities such as velocities of rain events and durations of rain cells, their fitted values should be physically sensible. In addition, their seasonal variation should be realistic so that, for example, storms are shorter and more localised in summer than in winter.

Given that the parameters of the fitted model meet these basic requirements, the fitted model can be assessed by its ability to reproduce properties of interest. These might include the first- and second-order properties used in fitting but at other spatial and/or temporal scales. Figure 5.3 gives examples for a model fitted to Chenies data for the interior of the event shown

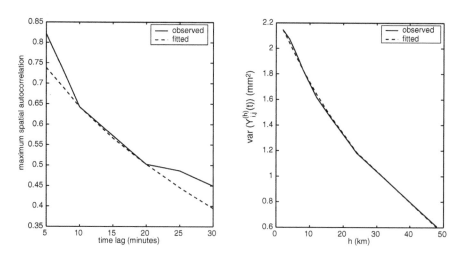

Figure 5.3 Observed and fitted rainfall properties for Chenies event of 4th December 2000, 07:00–08:15. Left: maximal space-time cross correlation $\rho^{(h)}(\lfloor v\tau \rfloor, \tau)$ for 2-km data, at time lags up to 30 minutes. Right: variance $\text{Var}(Y^{(h)})$ of spatially averaged rainfall intensities, at spatial scales from 2×2 km^2 to 48×48 km^2.

in Figure 5.1. Here, it appears that the maximal space-time cross-correlation $\rho^{(h)}\left(\lfloor \boldsymbol{v}\tau \rfloor, \tau\right)$ is not reproduced particularly well except at the temporal lags (10 and 20 min) used to fit the model. On the other hand, the variance of spatially averaged rainfall is reproduced almost exactly over a wide range of scales.

As well as checking the first- and second-order properties of the models, it is of interest to examine other properties not used in fitting. Some of these may have to be determined by simulation because analytic expressions are not available — these include extremes and the coverage. Properties such as the coverage (defined in Section 5.3.1.2), the mean intensity, and the conditional mean intensity (taken over only those pixels that are wet) are average properties and do not provide information about the spatial organisation of the rainfall field. It is therefore informative to take an increasing sequence of thresholds and, for each, to look at these properties for fields of threshold exceedances. This enables an assessment of the models' ability to reproduce localised areas of heavy rainfall. Examples are given in Figure 5.4. Here, several simulations have been used to illustrate the variability inherent in the model. The figure shows that for this particular event, the model reproduces well the spatial organisation of both instantaneous and cumulative rainfall fields.

As well as comparing statistical properties of the fitted model, it is useful to compare the visual appearance of the simulated images with that of the event to which the model was fitted. Figure 5.5 shows part of the observed sequence corresponding to Figures 5.3 and 5.4; Figure 5.6 shows a corresponding specimen simulation from the fitted model.

5.3.2 Durations of events and dry periods

So far, attention has concentrated on models for the interior of a single rain event. For continuous simulation of rainfall sequences it is necessary to model the sequence of events as they move across the fitting window. Here we focus first on the times of event arrivals and departures, and subsequently on the shapes of the events as they affect the window.

On the basis of a 4-year sequence of radar data from a station at Wardon Hill in South West England, Wheater et al. (2000b) distinguished two types of event producing a significant amount of rainfall. The first type extended over the fitting window for at least an hour and could be regarded as spatially and temporally stationary during this period; the second consisted of events that were unsuitable for fitting. By inspection of 5×5 km^2 resolution images such as that in Figure 5.1, it was found that the second type of event generally appeared to have a similar structure to the first, but followed a path that intersected only an edge or corner of the window. A semi-Markov model for the sequence of durations of the two types of events and the dry periods in between them was proposed and fitted.

For the Chenies radar data we use a simpler model for event sequences, in which a single type of event is defined. We assume that the durations of

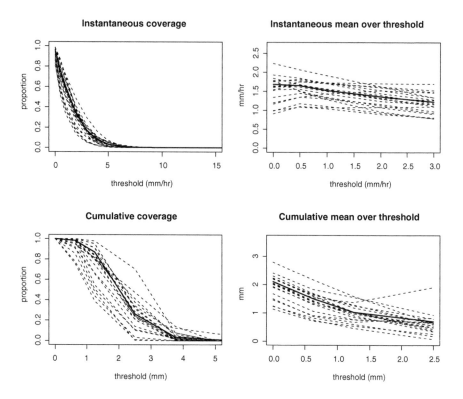

Figure 5.4 Observed (solid) and simulated (dashed) rainfall properties for Chenies event of 4th December 2000, 07:00–08:15. The left-hand plots show the proportion of the area with rainfall above a threshold, as the threshold is raised. The right-hand plots show the mean exceedance over the threshold, for thresholds up to the 90th percentile of the instantaneous intensity distribution (at thresholds higher than this, sampling variability starts to dominate the estimates). The upper plots are for instantaneous rainfall fields; the lower plots are for cumulative rainfall over the 75-minute period.

the events over the window, and of the dry periods that alternate with them, form independent sequences of independent and identically distributed random variables; thus the sequence of wet and dry period durations forms an alternating renewal process (Cox and Isham, 1980, Section 3.2). The approximate independence of successive durations has been verified empirically using scatterplots.

The arrival and departure times of the rain events may be defined as the times at which the coverage over the fitting window crosses a threshold. The choice of this threshold depends on the characteristics of the radar data. If it is too high some events will remain undetected. However, if it is too low periods may be erroneously classified as 'wet' on the basis of radar noise; moreover, a sequence of events following each other rapidly across a catchment may incorrectly be identified as a single event for which the stationarity

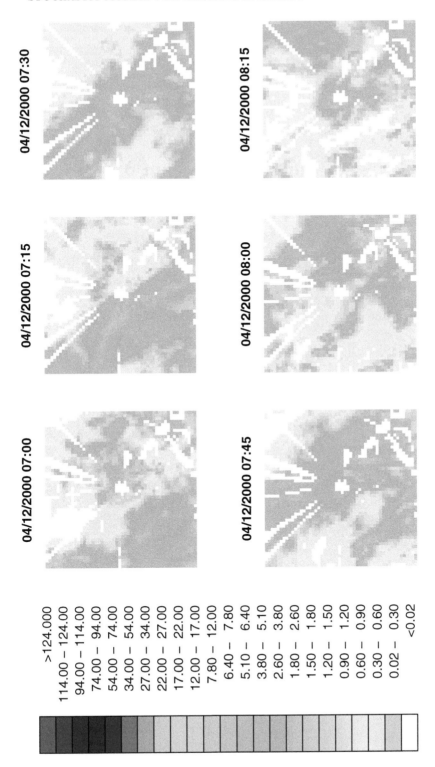

Figure 5.5 Part of the observed sequence of calibrated radar images from Chenies, for the event of 4th December 2000 (see also Figure 5.1). The images show clearly where anomalies have been removed during the calibration process (described in Section 5.2.2). The scale on the left shows rainfall intensities in mm hr^{-1}.

>124.000
114.00 – 124.00
94.00 – 114.00
74.00 – 94.00
54.00 – 74.00
34.00 – 54.00
27.00 – 34.00
22.00 – 27.00
17.00 – 22.00
12.00 – 17.00
7.80 – 12.00
6.40 – 7.80
5.10 – 6.40
3.80 – 5.10
2.60 – 3.80
1.80 – 2.60
1.50 – 1.80
1.20 – 1.50
0.90 – 1.20
0.60 – 0.90
0.30 – 0.60
0.02 – 0.30
<0.02

Figure 5.6 Specimen simulation of the model fitted to the event of 4th December 2000. The time separation of the images here is 15 minutes, as in Figure 5.5.

assumption is questionable. After exploratory data analysis, we set the threshold for Chenies radar data at 15%. Thus the periods termed 'dry' would be more accurately described as having at most very light rainfall. In Wheater et al. (2000b) it was demonstrated, using a rather higher threshold of 25% (reflecting the different characteristics of the data used there), that positive radar values below the threshold (both rainfall and radar noise) contribute a relatively minor proportion of the total rainfall intensity over the window. Nevertheless, further research is needed to determine the extent of the influence of very light rainfall on the runoff properties of a catchment, and to adjust the model to allow for such rainfall if appropriate.

In Section 5.2.3 it was noted that there are many breaks in the Chenies radar sequence. Breaks of several hours' duration are not uncommon, as a result of which many of the wet and dry interval durations are censored. To ignore this (for example, by discarding partial intervals prior to fitting) would lead to biased estimates of the interval properties. Adopting methodology from survival analysis (Davison, 2003, Section 5.4) therefore, we model both types of interval using Weibull distributions, and use likelihood methods to estimate the parameters. The survivor function of the Weibull distribution has the simple algebraic form $\mathcal{S}(x) := \mathrm{P}(X \geq x) = \exp\left[-(x/\alpha)^c\right]$, so that it is straightforward to include the censored data in the likelihood. To incorporate seasonality, separate alternating renewal models are fitted for each month of the year. The fit of the models can be assessed using standard techniques from survival analysis, such as a comparison of empirical and fitted cumulative hazard functions.

5.3.3 Continuous simulation of a sequence of rain events

For the continuous simulation of rainfall over a catchment, the alternating renewal model for wet and dry interval durations must be combined with the stochastic model for event interiors. In this chapter, we consider simulating over the whole of the fitting window. As described above, events arrive and depart when the radar coverage crosses a threshold which, for the Chenies radar, is set at 15%. The basic idea of the continuous simulation is first to simulate the alternating renewal model to give the sequence of event durations and dry periods, then, independently for each wet period, to simulate an appropriate event interior.

By inspection of radar images we find that, at the spatial scales of interest, the edges of weather systems are approximately linear (see Figure 5.1, for example). For continuous simulation over the fitting window, we therefore treat rain events as moving bands of rainfall, defined via the separation and orientations of their leading and trailing edges. These edges define, in some sense, the linear function of eastings and northings that best separates the 'wet' and 'dry' pixels when the event is just entering (or leaving) the window. This suggests that for any observed event, the edge orientations can be obtained by applying Fisher's Linear Discriminant Function (see, for example, Krzanowski

**Chenies 2km radar image, 4th December 2000, 04:45
(coverage = 57%)**

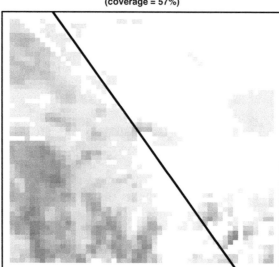

Figure 5.7 The use of Fisher's Linear Discriminant Function to identify the orientation of the leading edge of a rainfall event. This is the same event as that shown in Figure 5.5. The solid line is the linear discriminant; the gradient of this is taken as the orientation of the leading edge.

1988, Section 12.3) to an image near the start or end of the event. An example is given in Figure 5.7. The width of each band is determined so that the rainfall coverage will be above the threshold for the simulated duration of the event.

Different types of event will yield different parameter combinations for the event interior model. Moreover, there is dependence between the model parameters and the event duration; in particular, fast moving events are seldom observed to have long durations. These dependencies should be reflected in simulations. To achieve this, we construct a library of the parameters for every fitting event in the Chenies radar record. For each event, the library includes the 11 parameters of the within-event model, the leading and trailing edge orientation, and the observed duration of the event (i.e., the time for which its coverage was above 15%). To reproduce the inter-parameter dependence during simulation, we propose simply to sample a set of parameter estimates from this library, conditional on the required event duration. This is much easier than trying to fit a joint distribution to all the parameters. Seasonal structure is incorporated by sampling parameter sets on a month-by-month basis. Thus the continuous simulation algorithm is as follows:

1. Generate an event duration by sampling from the fitted Weibull duration distribution for the current month.
2. Sample a parameter set at random from the sub-library of events for the current month with approximately the required duration.

3. Determine the width of a moving rain band that will yield an event duration, over the area of interest, equal to that sampled in step 1. This width depends upon the duration and also upon the event velocity and edge orientations, which are part of the parameter set sampled in step 2.
4. Simulate the event interior model within the rain band; store the locations and properties of any cells that affect the area of interest.
5. Generate the duration of next dry period.
6. Go to Step 1.

In practice, in step 4 the event interior model is simulated within the area of intersection of the moving rain band and a large fixed rectangle (the 'generation region') containing the area of interest, with a toroidal wrap-around to deal with edge effects (this is a standard technique for the simulation of stationary spatial processes — see, for example, Diggle, 2003, Section 1.3). Any storm with a centre outside the rain band is discarded; however, cells outside the band are retained if they belong to a storm inside the band. This prevents the simulated events from having unrealistically sharp boundaries. A schematic diagram of the event interior simulation is given in Figure 5.8.

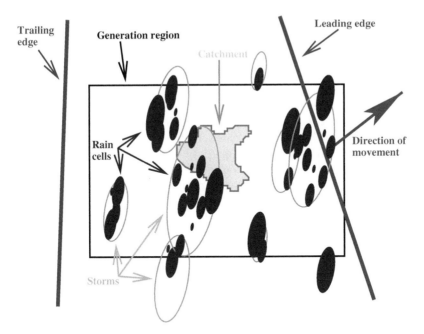

Figure 5.8 (SEE COLOR INSERT FOLLOWING PAGE 142) Schematic diagram of a rain band moving across a catchment. The event interior model is simulated in a large rectangular generation region surrounding the catchment, with toroidal wrap-around at the edges. Storms are retained only if their centres are within the rain band. However, all cells from such storms are retained, even if they fall outside the band.

Further details are given in Wheater et al. (2000b), who also assess the performance of the continuous simulation on the basis of a 4-year sequence of radar data. The longer sequence from Chenies will enable more detailed assessment to be carried out, especially with respect to properties such as extremes.

5.4 Multisite models for daily rainfall

In principle, the stochastic models of Section 5.3 provide a powerful means of generating high-resolution synthetic rainfall data. At present, however, they suffer from two major drawbacks. The first is their reliance on radar data for calibration. Apart from the data quality issues that have already been discussed, radar technology is relatively new so that record lengths are limited (the Chenies sequence used above is the longest currently available in the U.K.). In view of the general consensus that climate varies on decadal or longer time scales (IPCC, 2001, Section 1.2.2), it may be unwise to rely too heavily upon such short runs of data to provide model parameterisations that will hold into the future.

The second drawback of the stochastic models is their stationarity in both space and time: this renders them unsuitable, in the form presented above, for reproducing such hydrologically important features as long-term changes in climate or systematic orographic effects. In this section therefore, we describe an alternative class of models capable of representing spatial and temporal nonstationarity, and also suitable for fitting to daily rain gauge data. As discussed in Section 5.2.1, gauge data are more suited to the study of large-scale variation and trends than to the study of fine-scale spatial and temporal structure. We therefore adopt a descriptive, rather than mechanistic, modelling approach. The specific approach we describe is based on generalised linear models (McCullagh and Nelder, 1989); we give a brief overview of the construction of GLMs for space-time data in Section 5.4.1. Our approach to inference takes time series likelihoods as its starting point; we therefore review some useful results for time series data in Section 5.4.2 before discussing the space-time extensions in Section 5.4.3. Simulation of the models is discussed in Section 5.4.4, and the methodology is illustrated with an example in Section 5.4.5.

5.4.1 Model construction

The use of GLMs to model daily rainfall goes back to the work of Coe and Stern, (1982) and Stern and Coe (1984), who focused on single-site analyses. Their work was extended to a multi-site (i.e., space-time) setting by Chandler and Wheater (2002). In this respect our modelling approach differs slightly from other 'descriptive' models for space-time processes (see, e.g., the contributions given in Chapters 3, 4, and 6 of this volume): it is a spatial extension of a time series model, rather than a temporal extension of a spatial model. The idea is to use relevant covariate information to forecast a probability distribution for each day's rainfall at each site. To handle the mixed discrete–continuous nature of rainfall distributions (there is an atom of probability at zero), it is convenient to proceed in two stages, as follows:

1. Use logistic regression to model the pattern of wet and dry days (this will be referred to as the 'occurrence model'). Specifically, denoting by p_{st} the probability of non-zero rainfall at site s on day t, and by $\mathbf{x}_{st}$ a corresponding vector of covariates, the model is given by

$$\ln\left(\frac{p_{st}}{1 - p_{st}}\right) = \mathbf{x}'_{st}\boldsymbol{\beta} \qquad (5.12)$$

   for a parameter vector $\boldsymbol{\beta}$.

2. Use gamma distributions to model the amount of rain if non-zero (the 'amounts model'). The distributions are assumed to share a common shape parameter, and the means $\{\mu_{st}\}$ are related to covariate vectors $\{\boldsymbol{\xi}_{st}\}$ via a log link:

$$\ln \mu_{st} = \boldsymbol{\xi}'_{st}\boldsymbol{\gamma} \qquad (5.13)$$

   for a parameter vector $\boldsymbol{\gamma}$. A constant shape parameter corresponds to a constant coefficient of variation (McCullagh and Nelder, 1989, Chapter 8); the validity of this can be verified empirically by, for example, calculating coefficients of variation at a site for each month of the year.

Within this framework, temporal and spatial nonstationarities can be accommodated via appropriate choice of covariates. This also allows interpolation at sites for which covariate information is available but rainfall data are not. For example, in Chandler and Wheater (2002) seasonality was represented using sine and cosine functions, while systematic regional variation was included via a combination of site altitude and orthogonal bases for 'eastings' and 'northings' effects. Temporal dependence and persistence were modelled using functions of previous days' rainfalls; and temporal nonstationarity was incorporated through the use of simple trend functions along with the North Atlantic Oscillation (NAO), which is an index of large-scale climate that is known to affect winter rainfall in the area considered. Interaction terms can be used to specify that some covariates may modulate the effects of others — for example, to allow the effect of the NAO to be greater during winter than in summer. The representation of space-time structure in GLMs is discussed in more detail in Yan et al. (2002) and Chandler (2005).

5.4.2 Likelihood inference for time series data

In standard applications, inference for GLMs is usually carried out using likelihood methods. Standard algorithms rely on the responses being conditionally independent given the covariates, so that the joint density of the responses factorises into a product of contributions from each observation. Time series

data can be treated within this framework since, in this case, the joint density
of an observation vector $\mathbf{y} = (y_1 \ \ldots \ y_T)'$ can be written as

$$f_{\mathbf{Y}}(\mathbf{y}|\mathbf{X}) = \prod_{t=1}^{T} f_t(y_t|\mathbf{x}_t, \mathcal{H}_t), \qquad (5.14)$$

where $\mathbf{X}$ is a matrix containing all covariate information and $f_t(y_t|\mathbf{x}_t, \mathcal{H}_t)$
is the density of the tth observation given both its covariates $\mathbf{x}_t$ and its history $\mathcal{H}_t = \{y_1, \ldots, y_{t-1}\}$. Therefore, standard methods can be used for time
series data providing dependence on $\mathcal{H}_t$ is explicitly built into the model. The
simplest way to achieve this is to include functions of past values as extra
covariates, leading to the class of generalised autoregressive models (Fahrmeir
and Tutz, 1994, Section 6.1). We refer to these functions of previous values
as 'autoregressive terms' in the model. A potential disadvantage of this approach is that in most cases, the resulting time series structure is sufficiently
nonlinear that properties of the 'marginal' distributions $\{f_t(y_t|\mathbf{x}_t)\}$ are not
available analytically. This means, for example, that the first few observations
must be discarded from the joint density (5.14), and therefore that these cannot be included in likelihood calculations. Providing T is large, however, this is
not a major concern. A further potential disadvantage is that the inclusion of
previous values can substantively affect the interpretation of other coefficients
in the model, which must therefore be carried out with care. For the present
application, however, the purpose is simulation rather than interpretation, so
we do not consider this aspect further here.

   We now derive a useful result for time series data that provides the foundation for our treatment of space-time data in the next section. Suppose that the
data $\mathbf{y}$ were generated from a joint density of the form (5.14), with parameter
vector $\boldsymbol{\theta} = \boldsymbol{\theta}_0$. The log-likelihood function for $\boldsymbol{\theta}$ can clearly be written as
$\sum_{t=1}^{T} \ln f_t(y_t|\mathbf{x}_t, \mathcal{H}_t; \boldsymbol{\theta})$, and for each $\boldsymbol{\theta}$, this is the realised value of a random
variable

$$\ell(\boldsymbol{\theta}) = \sum_{t=1}^{T} \ln f_t(Y_t|\mathbf{x}_t, \mathcal{H}_t; \boldsymbol{\theta}) = \sum_{t=1}^{T} \ell_t(\boldsymbol{\theta}), \text{ say.} \qquad (5.15)$$

Define the score function

$$\mathbf{U}(\boldsymbol{\theta}) = \frac{\partial \ell}{\partial \boldsymbol{\theta}} = \sum_{t=1}^{T} \frac{\partial \ell_t}{\partial \boldsymbol{\theta}} = \sum_{t=1}^{T} \mathbf{U}_t(\boldsymbol{\theta}), \text{ say,} \qquad (5.16)$$

so that the maximum likelihood estimator of $\boldsymbol{\theta}$ is the solution of the score
equation $\mathbf{U}(\boldsymbol{\theta}) = \mathbf{0}$. For GLMs, the algebraic form of the score contributions
$\{\mathbf{U}_t\}$ makes it clear that they are vectors of weighted residuals (Chandler,
2005); this suggests that they might be uncorrelated. In fact, this lack of
correlation holds quite generally when score contributions are defined in terms
of a factorisation of the form (5.14). This can be shown using martingale
arguments, as in Section 7.2.3 of Davison (2003). Here we sketch an alternative
derivation which, to our knowledge, appears for the first time.

It is well known that under general conditions, $\mathrm{E}\left[\mathbf{U}_t(\boldsymbol{\theta}_0)\right] = 0$. It is, however, instructive to derive this particular result afresh in the context of the factorisation (5.14). We have

$$\mathrm{E}\left[\mathbf{U}_t(\boldsymbol{\theta}_0)\right] = \int_{\mathbf{y}} \left.\frac{\partial \ln f_t}{\partial \boldsymbol{\theta}}\right|_{\boldsymbol{\theta}_0} f_{\mathbf{Y}}(\mathbf{y}|\mathbf{X})\ d\mathbf{y}$$

$$= \int_{\mathbf{y}} \frac{1}{f_t\left(y_t|\mathbf{x}_t, \mathcal{H}_t; \boldsymbol{\theta}_0\right)} \left.\frac{\partial f_t}{\partial \boldsymbol{\theta}}\right|_{\boldsymbol{\theta}_0} \prod_{j=1}^{T} f_j\left(y_j|\mathbf{x}_j, \mathcal{H}_j; \boldsymbol{\theta}_0\right)\ d\mathbf{y}$$

$$= \int_{y_1} f_1 \int_{y_2} f_2 \cdots \int_{y_t} \left.\frac{\partial f_t}{\partial \boldsymbol{\theta}}\right|_{\boldsymbol{\theta}_0} \int_{y_{t+1}} f_{t+1} \cdots \int_{y_T} f_T\ d\mathbf{y}, \quad (5.17)$$

where the $\{f_j\}$ are conditional densities and hence satisfy $\int_{y_j} f_j\left(y_j|\mathbf{x}_j, \mathcal{H}_j; \boldsymbol{\theta}_0\right)$ $dy_j = 1$. In (5.17), the variable y_j appears in the integrand only from the j-th integral onwards. The multiple integral can therefore be evaluated one variable at a time, working from right to left. The integrals from f_T back to f_{t+1} are all equal to 1, so that (5.17) can be terminated at $\int_{y_t} \partial f_t/\partial \boldsymbol{\theta} dy_t$. Providing the range of integration here is not dependent on $\boldsymbol{\theta}$, we can interchange the order of differentiation and integration: $\int_{y_t} \partial f_t/\partial \boldsymbol{\theta} dy_t = \partial/\partial \boldsymbol{\theta} \int_{y_t} f_t dy_t = 0$. Hence $E\left[\mathbf{U}_t(\boldsymbol{\theta}_0)\right] = 0$ for each t.

It follows that

$$\mathrm{Cov}\left[\mathbf{U}_s(\boldsymbol{\theta}_0), \mathbf{U}_t(\boldsymbol{\theta}_0)\right] = \mathrm{E}\left[\mathbf{U}_s(\boldsymbol{\theta}_0)\mathbf{U}_t(\boldsymbol{\theta}_0)\right]$$

$$= \int_{\mathbf{y}} \left.\frac{\partial \ln f_s}{\partial \boldsymbol{\theta}}\right|_{\boldsymbol{\theta}_0} \left.\frac{\partial \ln f_t}{\partial \boldsymbol{\theta}}\right|_{\boldsymbol{\theta}_0} f_{\mathbf{Y}}(\mathbf{y}|\mathbf{X})\ d\mathbf{y}.$$

An identical argument to the one above now yields the result

$$\mathrm{Cov}\left[\mathbf{U}_s(\boldsymbol{\theta}_0), \mathbf{U}_t(\boldsymbol{\theta}_0)\right] = 0 \qquad (s \neq t). \qquad (5.18)$$

It may be worth noting that although the expectations above are with respect to the marginal distribution of $\mathbf{Y}$, analogous results hold for conditional distributions. For example, if $s < t$, then $\mathrm{Cov}\left[\mathbf{U}_s(\boldsymbol{\theta}_0), \mathbf{U}_t(\boldsymbol{\theta}_0)|\mathcal{H}_s\right] = 0$. This is an immediate consequence of the above result, regarding $y_s, \ldots, y_T$ as a sample from the distribution conditioned on $\mathcal{H}_s$ (obtained by omitting the first $(s - 1)$ terms from the factorisation (5.14)).

The derivation above can also be used for vector-valued observations $y_1, \ldots, y_T$; hence the result also holds in this case.

5.4.3 Inference for GLMs in a space-time setting

With space-time data, inter-site dependence usually precludes an interpretable factorisation of the form (5.14), although a partial factorisation can still be obtained (simply replace y_t in (5.14) with $\mathbf{y}_t$, the vector of all observations at time t). An exception occurs if inter-site dependence arises solely through dependence on previous values at all sites, but this is rarely realistic for the type of application considered here. In the absence of an easy solution, one can proceed either by formulating a model that explicitly captures the dependence (as in Diggle et al., 1998, and Hughes et al., 1999, for example) or by fitting models as though sites were independent and making appropriate adjustments in retrospect. Here we adopt the latter approach. The motivation for this is primarily computational since, for routine hydrological application, the fitting of models to large data sets (often running into hundreds of thousands of cases) needs to be accomplished quickly.

To demonstrate the adjustments required for space-time data, suppose as above that $\boldsymbol{\theta} = \boldsymbol{\theta}_0$ is a parameter vector of interest, and let $\ell(\boldsymbol{\theta})$ denote the log-likelihood function computed as though all sites were independent. The 'independence' estimator of $\boldsymbol{\theta}$ solves the equation

$$\mathbf{U}(\boldsymbol{\theta}) = 0, \tag{5.19}$$

where $\mathbf{U}(\boldsymbol{\theta}) = \partial\ell(\boldsymbol{\theta})/\partial\boldsymbol{\theta}$ is the vector of independence scores. From result (5.10) for estimating equations, in regular problems this estimator is distributed as

$$\hat{\boldsymbol{\theta}} \sim N\left(\boldsymbol{\theta}_0, \mathbf{H}^{-1}\mathbf{V}\mathbf{H}^{-1}\right) \tag{5.20}$$

asymptotically, where $\boldsymbol{\theta}_0$ is the target value of $\boldsymbol{\theta}$, $\mathbf{H}$ is the Hessian of the log-likelihood at $\boldsymbol{\theta}_0$ and $\mathbf{V}$ is the covariance matrix of $\mathbf{U}(\boldsymbol{\theta}_0)$. If the observations are independent, $\mathbf{V} = -\mathbf{H}$ (Davison, 2003, Section 4.4), and the covariance matrix reduces to $\mathbf{V}^{-1}$. In general however, this simplification does not arise. For GLMs, the Hessian is estimated using its expected value at $\hat{\boldsymbol{\theta}}$. This expected value is unaffected by dependence between observations, and emerges as a by-product of the numerical algorithm used to maximise the independence log-likelihood (McCullagh and Nelder, 1989, Section 2.5). It remains to estimate $\mathbf{V} = \mathrm{Var}\left[\mathbf{U}(\boldsymbol{\theta}_0)\right]$.

As noted above, space-time data can be regarded as vector-valued time series so that a partial factorisation of the joint density can be obtained. The score vector $\mathbf{U}(\boldsymbol{\theta}_0)$ can therefore be written as a sum of contributions from each time point:

$$\mathbf{U}(\boldsymbol{\theta}_0) = \sum_{t=1}^{T} \mathbf{U}_t(\boldsymbol{\theta}_0). \tag{5.21}$$

According to (5.18), these contributions are uncorrelated; hence

$$\mathbf{V} = \mathrm{Var}\left[\sum_{t=1}^{T} \mathbf{U}_t(\boldsymbol{\theta}_0)\right] = \sum_{t=1}^{T} \mathrm{Var}[\mathbf{U}_t(\boldsymbol{\theta}_0)] = \sum_{t=1}^{T} \mathrm{E}\left[\mathbf{U}_t(\boldsymbol{\theta}_0)\mathbf{U}_t'(\boldsymbol{\theta}_0)\right].$$

This suggests estimating $\mathbf{V}$ using its empirical counterpart

$$\hat{\mathbf{V}} = \sum_{t=1}^{T} \mathbf{U}_t(\hat{\boldsymbol{\theta}})\mathbf{U}_t'(\hat{\boldsymbol{\theta}}). \tag{5.22}$$

Standard asymptotic arguments show that in well-behaved problems, $\hat{\mathbf{V}} - \mathbf{V}$ is $O(T^{1/2})$ in probability, whereas $\mathbf{V}$ itself is $O(T)$. The estimator (5.22) can therefore be plugged into (5.20) to obtain an estimate of the covariance matrix of $\hat{\boldsymbol{\theta}}$. This enables the construction of confidence intervals, as well as formal tests for the inclusion of covariates in a model. Alternative procedures can be based upon an adjusted likelihood ratio statistic, as developed by Bate (2004) and implemented in the rainfall modelling software used to generate the results in the next section (Chandler, 2002). However, formal tests should be interpreted with caution when fitting models to large datasets since even tiny effects can appear statistically significant. It is therefore wise to supplement such tests with carefully chosen residual analyses, as described in Chandler and Wheater (2002) and Yan et al. (2002), for example.

The analysis above corresponds exactly to the use of an 'independence' working correlation structure within the context of generalized estimating equations (GEEs) (Liang and Zeger, 1986). The resulting estimators are consistent with respect to T, but are inefficient relative to estimators based on the correct spatial dependence model. This inefficiency suggests that it may be preferable to specify a more plausible dependence model (for example, using a standard family of spatial correlation functions) within the GEE framework. This has been investigated by Bate (2004), who found that the use of a spatial correlation model affected parameter estimates by far more than their standard errors would suggest. This in turn implies that one of the two procedures is biased. Since the independence estimators are known to be consistent, suspicion naturally falls upon the GEE — particularly in view of its known lack of robustness to a misspecified correlation structure (McDonald, 1993; Crowder, 1995; Sutradhar and Das, 1999) and sensitivity to complex conditional dependence relationships (Sullivan Pepe and Anderson, 1994). Further research is required to examine this in more depth.

The validity of the estimator (5.22) depends ultimately upon that of the decomposition (5.14). For space-time data, and vector-valued time series in general, this requires that the model contains an adequate representation of temporal dependence both within and between sites. This usually requires that at any site, the conditional distributions are modelled as dependent upon its neighbours' past as well as its own.

5.4.4 Multi-site simulation

The previous section provides some justification for fitting marginal time series models, such as (5.12) and (5.13), to multi-site sequences of daily rainfall. However, the aim of the modelling is to produce multi-site rainfall simulations; to be realistic, these simulations must reflect the dependence between sites. To generate a day's rainfall, it is therefore necessary to define a joint distribution for the multi-site pattern of rainfall occurrence, and then the distribution of the rainfall amounts vector at the wet sites. The mean vectors of these joint distributions are obtained by applying (5.12) and (5.13) separately at each site.

We deal with amounts first. It is convenient to proceed via a transformation to marginal normality, since in this case the spatial dependence is characterised by the inter-site correlation structure of the transformed values.

An exact normalising transformation is given by $Z = \Phi^{-1}[F_{st}(Y)]$, where $\Phi[\cdot]$ is the distribution function of the standard normal distribution and F_{st} is the distribution function specified by the GLM at site s and time t. However, the evaluation of this transformation, and its inverse, is relatively computationally expensive for the amounts model here; this is a drawback if large quantities of data are to be generated. As an alternative, note that if Y_{st} has a gamma distribution with mean μ_{st}, then the distribution of the Anscombe residual $(Y_{st}/\mu_{st})^{1/3}$ is extremely close to normal (Terrell, 2003). The mean and variance of this normal distribution can be calculated numerically, as described in Section 3.3.2 of Chandler and Wheater (1998). Therefore rainfall amounts at wet sites can be generated by simulating a vector of correlated Anscombe residuals and inverting the cube root transformation, to recover the corresponding rainfalls with the correct marginal properties.

For the rainfall occurrence model, a transformation to marginal normality is not possible and an alternative approach is required. There is an extensive literature on the modelling of dependent binary data — see, for example, Oman and Zucker (2001); Lunn and Davies (1998); Emrich and Piedmonte (1991); Cox and Wermuth (1996); Hughes et al. (1999). However, at catchment scales the dependence in rainfall fields is often strong since sites tend to be either mostly wet or mostly dry (a consequence of the fact that weather systems tend to affect all sites simultaneously — see Figure 5.1). Many of the available approaches are unable to cope with this degree of dependence; others suffer from computational cost, either in estimating or in simulating the dependence structure. More details can be found in Wheater et al. (2000b, Chapter 4).

The tendency for sites to be mostly wet or mostly dry suggests modelling the number of wet sites directly. Specifically, suppose we wish to simulate, for day t, a vector of dependent binary random variables at S sites, $\mathbf{Y}_t = (Y_{1t} \; \ldots \; Y_{St})'$ say: then we specify a distribution for $Z_t = \sum_{s=1}^{S} Y_{st}$. A flexible family of distributions for discrete random variables, taking values in $\{0, 1, \ldots, S\}$ and for which there is some justification in the current context (see below) is the beta-binomial. Accordingly, we take $Z_t \sim BB(S, \alpha_t, \beta_t,)$, with expected value $S\alpha_t/(\alpha_t + \beta_t) = S\theta_t$ and variance

$$S\alpha_t\beta_t(\alpha_t + \beta_t + S)/(\alpha_t + \beta_t)^2(\alpha_t + \beta_t + 1) = S\theta_t(1 - \theta_t)(\phi_t + S)/(\phi_t + 1)$$

respectively. Here, $\theta_t = S^{-1}E(Z_t) = S^{-1}\sum_{s=1}^{S} E(Y_{st})$ is a mean parameter, which is determined by the occurrence model (5.12). The parameter $\phi_t = \alpha_t + \beta_t$ controls the shape of the beta-binomial distribution, and can be regarded as a measure of dependence. As $\phi_t \to 0$, the distribution becomes increasingly concentrated around 0 and 1 ('perfect dependence'), with U-shaped distributions possible for $\phi_t < 2$ ('strong dependence'); and as $\phi_t \to \infty$ the distribution tends to the binomial (independence). In the first instance it is convenient, and not implausible, to assume that $\phi_t = \phi$ is constant for all t, so that θ_t is the only time-varying parameter of the distribution. Spatial dependence in rainfall occurrence is therefore modelled using a single parameter, which can be estimated from the second-order properties of an observed collection $\{Z_t\}$ as described by Yang et al. (2005). An algorithm for simulating vectors $\mathbf{Y}_t$, in such a way as to respect both the distribution of Z_t and the marginal probabilities given by the logistic regression model, is given in the Appendix of Chandler (2002).

In standard applications, the beta-binomial distribution arises as that of a binomial (S, p) random variable, where p is itself a random variable distributed according to a beta distribution with parameters α_t and β_t. Thus the following simple mechanism would give rise to a beta-binomial distribution for the number of wet sites in a fixed region sampled at S locations: a proportion p_t of the region is wet, where p_t is a beta-distributed random variable. Given p_t, individual locations are wet or dry independently of each other. Although this mechanism is clearly idealised, it provides a useful insight into the model. In particular, it suggests that the model may fail if sites are too close together, since in this case the assumption of conditional independence given p_t is unlikely to hold even approximately.

A further feature of the dependence structures outlined here is that they enable conditional distributions to be calculated for rainfall at unobserved locations, given the available observations. Again, details can be found in Chandler (2002). These conditional distributions can be used to impute missing values in an observed sequence. This is useful when making comparisons between observed and modelled properties since, by carrying out multiple imputations of the missing data, it is possible to construct simulated uncertainty envelopes for the observed properties themselves. This is illustrated in the next section.

5.4.5 Performance assessment

We now illustrate the use of GLMs to generate dependent daily sequences exhibiting spatial and temporal nonstationarities, using the Blackwater data described in Section 5.2.3. The example is discussed in more detail in Yang et al. (2005). To remove problems with the recording of small rainfall amounts (see Section 5.2.1), we fit GLMs to thresholded data:

$$Y_{st}^* = \max(Y_{st} - \tau, 0), \qquad (5.23)$$

where the threshold $\tau > 0$ is set at 0.5 mm. The fitted models are used to simulate daily sequences of thresholded values, and the thresholding removed by computing

$$\tilde{Y}_{st} = \begin{cases} Y_{st}^* + \tau & Y_{st}^* > 0 \\ 0 & \text{otherwise.} \end{cases} \qquad (5.24)$$

To represent rainfall occurrence, a logistic model of the form (5.12) has been fitted to the thresholded data from all 34 sites. The model contains 19 terms. Systematic regional effects are represented using site altitude, as well as Legendre polynomial transformations of site eastings and northings. Seasonality is represented using sine and cosine terms; temporal dependence and persistence are included by using indicators for previous days' rainfall, and the change in recording resolution in the 1970s is modelled via an indicator variable defined as

$$I_{s,t} = \begin{cases} 1 & \text{for all observations before 1975} \\ 0 & \text{for all observations from 1975 onwards.} \end{cases} \qquad (5.25)$$

In addition to this basic structure, the model contains terms representing the effects of the North Atlantic Oscillation (NAO — see Hurrell, 1995) and Arctic Oscillation (AO — Thompson and Wallace, 1998). These indices of large-scale climate are known to affect weather patterns in the northern hemisphere; the NAO is particularly associated with winter storminess in northwest Europe. Therefore the models used here include a seasonally varying NAO effect (defined via interaction terms), along with a constant AO effect.

The predictors in the amounts model (5.13) are similar to those in the occurrence model, but fewer in number (only ten are included). The reduction is mainly due to simpler site effects, implying that rainfall amounts are less affected by spatial variations within the region than rainfall occurrence. Additionally, the amounts model contains no terms involving adjustment indicators of the form (5.25), since these terms were not significant.

Residual plots for both occurrence and amounts models are given by Yang et al. (2005) and indicate a satisfactory fit. However, the nonlinearity of the models ensures that residuals can be used only to check conditional structure, and simulation is required to examine the marginal properties of the models at each site. When simulating, we use a constant inter-site correlation (estimated as 0.786) between Anscombe residuals for rainfall amounts, and a beta-binomial structure with estimated dependence parameter $\hat{\phi} = 0.359$ for occurrence. This value of ϕ corresponds to strong inter-site dependence, as discussed in Section 5.4.4.

To investigate the properties of the models, simulated sequences have been compared with observations at a variety of spatial scales. For this exercise, ten sites were selected that have few missing values during the period 1961 to 1999, and the simulations were carried out over this period. Prior to simulation, ten sets of imputations were carried out, replacing any missing observations by simulated values conditional on all the observations at the selected sites. One hundred sets of simulated data were then generated at the same ten sites to simulate the modelled dynamics of the rainfall processes during the 1961–1999 period. Each simulation was initialised using the historical data for December 1960, and was conditioned on the historical NAO and AO series. Summary statistics were calculated for each simulation and each set of imputed data, and the results compared. Statistics were calculated separately for each month of the year. As well as examining individual sites, statistics were calculated for daily time series obtained by averaging groups of gauges. These average time series can be regarded as estimates of areal mean rainfall at scales up to 2000 km^2, and provide a means of assessing the appropriateness of the spatial dependence structures used in the simulations.

Figure 5.9 shows a variety of summary statistics for a series obtained by averaging data from three gauges that are within about 5 km of each other.

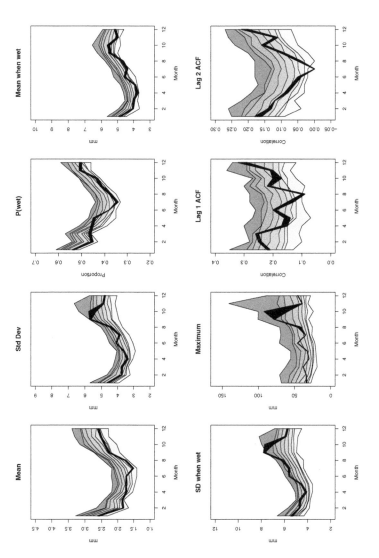

Figure 5.9 Observed and simulated monthly summary statistics for the daily average time series from a group of three sites. Row-wise from top: mean, standard deviation, proportion of wet days (i.e., proportion of days with non-zero rainfall after thresholding), conditional mean and standard deviation (computed for wet days only), maximum, and autocorrelations at lags 1 and 2. The 'observed' lines show the envelopes obtained from ten imputations of missing data. The largest and smallest simulated values of each property are shown, along with the 5th, 10th, 25th, 50th, 75th, 90th, and 95th percentiles of the simulated distributions.

These results are typical, and complement those given in Yang et al. (2005). The ranges of the imputation-based statistics provide some indication of the observational uncertainty due to missing data, which is clearly small here. The statistic showing the worst disagreement between observations and simulations is the conditional mean, for which the simulated distributions are slightly too high. Elsewhere there are some isolated discrepancies, but overall the observed structure is well reflected in the simulations. Note in particular that the simulated distributions of monthly maxima appear plausible; this suggests that they are suitable for use in applications where spatial extremes are of interest.

Simulations can be used to assess the models' ability to reproduce almost any property of interest. Yang et al. (2005) showed that the models are able to reproduce seasonal rainfall totals at spatial scales ranging from a single point to the average of all ten gauges in the 40×50-km^2 region, and that the beta-binomial dependence model does well in reproducing the distributions of numbers of wet sites. Further, simulated extreme rainfalls are generally in agreement with those derived from conventional extreme value analyses of the observations. Similar results have been obtained with other data sets. Ongoing research seeks to extend the work described here by incorporating covariate information from physically-based regional climate models; this effectively allows simulations to be conditioned upon future greenhouse gas emission scenarios. A major issue to be tackled in this respect, however, is the representation of uncertainty in the climate models themselves.

5.5 Continuous simulations that are nonstationary in space and time

We have described two classes of models for spatial-temporal rainfall. The stochastic models of Section 5.3 have parameters that are easily physically interpretable. They are built in continuous space and time, so that their simulation provides high-resolution data. The GLMs described in Section 5.4 are built in discrete space and time although, given appropriate covariates, they can be interpolated to any specific spatial location. These models have the considerable advantage that they can incorporate both spatial heterogeneity, which is hydrologically significant even in small catchments, and temporal effects including long-range climate cycles and possible scenarios for climate change. For input into rainfall-runoff models, it is highly desirable to have the advantages of both types of model, combining high resolution with spatial and temporal nonstationarities. In this section we discuss how this may be achieved. The basic idea is to use GLM simulations to determine the statistical properties of nonstationary multi-site daily sequences, and then to condition the continuous simulation of the stochastic model to respect these properties. We discuss the incorporation of spatial and temporal nonstationarity separately, in Sections 5.5.1 and 5.5.2, respectively. In Section 5.5.3 we describe some other possible approaches.

5.5.1 Incorporating spatial nonstationarity

Perhaps the easiest way to incorporate spatial nonstationarity into the stochastic model, at least as a first approximation, is to generate a stationary simulation and then rescale each image to reflect the spatial structure given by the GLM. An algorithm for achieving this is as follows:

1. Simulate the GLM many times, to give daily values at a set of sites (e.g., a subset of the centres of the radar pixels) for the time period of interest.
2. For each month in the simulation period, calculate the mean daily rainfall at each site by averaging over simulations.
3. Smooth these averages in space, to give an estimated mean field for each month of the simulation period.
4. Calculate the ratio of the estimated mean field to the mean for the stochastic model (which is spatially constant). The latter must also be estimated by simulation, since it does not appear feasible to study analytically the contribution of events arriving and departing from the region of interest.
5. Generate a realisation of the stochastic model, and rescale each image using the ratios calculated in step 4.

This scheme will incorporate the spatial and temporal nonstationarities built in to the GLM; however, the rescaling will not affect the wet and dry space-time regions of the stochastic simulation. At catchment scales this is probably acceptable as far as spatial heterogeneity is concerned. However, it is implausible that, for example, the distribution of dry interval durations in a particular season will remain constant in the presence of appreciable climate change. Therefore a different approach will probably be needed to cope with time trends.

5.5.2 Incorporating temporal nonstationarity

A more sophisticated variant of the 'multiplicative scaling' approach is to calculate a variety of summary statistics from the GLM simulations and then choose stochastic model parameters that reproduce these statistics while preserving subdaily structure. We outline briefly here two alternative strategies that can be used.

The first strategy is, at any given time point, to simulate wet and dry period durations, along with event interiors, using parameters corresponding to parts of the historical record with daily properties similar to those forecast by the GLM. The resulting realisations will then need to be rescaled, as described above, to allow for spatial heterogeneity. The motivation for this is that under a changing climate scenario we may expect seasonal characteristics of rainfall to alter because of changes in the relative frequencies of frontal and convective rainfall so that, for example, the typical rainfall of May 2020 might resemble that of June 2000; such seasonal shifts can be detected in GLM simulations. The statistics used for this comparison should be chosen to reflect the relative frequencies of different types of rainfall — for example, the means, proportions of dry days, and cross-correlations and durations of dry intervals. The scheme is similar in spirit to the approach sometimes used for prediction in complex dynamical systems (Casdagli, 1992; Isham, 1993), whereby predictions are made by identifying historical sequences similar to the current system state.

An alternative approach is to note that many authors (e.g., Smithers et al., 2002) have reported stable empirical relationships between rainfall properties at different time scales. These relationships can be used to reconstruct sub-daily properties from the daily properties simulated by the GLM, and the stochastic model can then be refitted to these reconstructed properties. This allows for spatial and temporal nonstationarity in the actual events simulated. Some progress has been made in this direction, in the case of modelling rainfall at a single site (Sheikh, 2004).

Under both of these strategies, the GLM determines the amounts of rainfall and their spatial heterogeneity. In this case, radar data from the exact region of interest are not required. The radar data are used only to determine the within-event spatial structure and the timing of events; since these features are unlikely to vary rapidly in space, radar data from a nearby region can be used.

5.5.3 Alternative approaches

The strategies outlined above represent what we currently regard as the most promising means of incorporating spatial and temporal nonstationarity into the stochastic model simulations. In earlier work (Wheater et al., 2000b), an alternative approach has been explored for the incorporation of temporal nonstationarity into simulations of a stochastic model for rainfall at a single site. The idea here is, at the beginning of each day of simulation, to use the GLM to forecast a distribution for the day's rainfall and then to simulate from the stochastic model in such a way as to respect this forecast distribution. This is achieved using a variant of rejection sampling (Von Neumann, 1951; Devroye, 1986), in which realisations of the stochastic model are accepted or rejected with some appropriately chosen probability, which depends on the densities of the daily rainfall total under both the GLM and the stochastic model. In practice, however, even for the single site case it is extremely difficult to compute the density under the stochastic model, particularly when storms and cells are already active at the start of the day. At present, the extension to spatial-temporal simulation appears prohibitively difficult.

5.6 Discussion and further work

The simulation of spatial-temporal rainfall fields poses many challenging statistical questions. Some of these, such as how to cope with dependence in space and time when fitting models, are common to many applied areas; others, such as how to handle the peculiarities of radar data, are more specific to this particular application. We have attempted here to give an overview of the issues involved, and to review some of the techniques that we have found to be useful in addressing them. It is worth noting that our models are not designed for forecasting: the aim is to produce long sequences with the right distributional properties. At present, our view is that for this particular application, the most promising way forward is to combine the stochastic spatial-temporal models with the GLMs; this combines the high space-time resolution of the stochastic models with the nonstationary simulation capability of the GLMs and, moreover, enables the best possible use to be made of limited amounts of radar data in conjunction with widely available daily gauge records. It is of particular interest to incorporate physically based climate change scenarios into the rainfall simulations. This can be achieved by conditioning the GLMs on the output of numerical climate models; research in this area is ongoing.

In the work presented above, we have presented some suggestions for checking and assessing the performance of the various models; these suggestions all involve examining the models' ability to reproduce features of observed rainfall fields. However, since the aim of the work is to provide sequences that can be used as input to rainfall-runoff models, the ultimate test is whether the rainfall inputs produce realistic runoff sequences. Joint testing of rainfall and rainfall-runoff models is currently under way.

There are many unresolved issues, in particular regarding the use of radar data. Apart from difficulties with quality control and calibration, radar data are spatially averaged, whereas rain gauge data represent values at a point. Perfect agreement is therefore not to be expected. To a certain extent, by combining the GLMs with the stochastic models we avoid this issue since the radar data are used only to give the spatial structure, rather than the actual rainfall values. Nonetheless, it is of interest to understand in more detail the relationship between point and spatially averaged rainfall, which depends on the fine-scale spatial structure of rainfall fields. Recent advances in instrumentation offer the potential to study this structure for the first time; this will be the subject of further research.

Acknowledgments

This research has been funded partly by DEFRA (R&D project FD2105), and partly by the European Community's Human Potential Programme under contract HPRN-CT-2000-00100, DYNSTOCH. We are grateful to many colleagues who have contributed to the work: in particular, Howard Wheater and Christian Onof. The Chenies radar and supporting rain gauge data were provided by Martin Crees of the Environment Agency, who also gave invaluable guidance on their use.

Appendix: Space-time covariance function for the event interior model

The properties of the event interior model (Section 5.3.1.1) are discussed in detail in Northrop (1998). However, that paper contains an error in the evaluation of the space-time covariance function $c_s(\boldsymbol{u}, t)$ of a model with zero velocity and circular cells. We take this opportunity to provide a correction, although a full derivation of the covariance function is too lengthy to include here.

The problem is in the evaluation of the temporal integrals

$$I = \int_{\tau_1=-\infty}^{0} \mathcal{F}_D(-\tau_1) \int_{\tau_2=-\infty}^{t} \mathcal{F}_D(t - \tau_2) e^{-\gamma|\tau_2 - \tau_1|} \, d\tau_2 \, d\tau_1$$

in Northrop's equation (2.10). Here, $\mathcal{F}_D(u) = e^{-\eta u}$ is the survivor function of the exponential cell duration distribution. The explicit form for I is $[\gamma e^{-\eta t} - \eta e^{-\gamma t}] / [\eta(\gamma^2 - \eta^2)]$, rather than $e^{-\gamma t}/[\eta(\gamma + \eta)]$ as claimed. Northrop's (1998) expression for the covariance function is of the form $Ae^{-\eta t} + Be^{-\gamma t}$ (see the bottom of his page 1883); this can be corrected by replacing the $e^{-\gamma t}$ in the second term with

$$[\gamma e^{-\eta t} - \eta e^{-\gamma t}]/(\gamma - \eta).$$

References

Austin, G. (2001). Weather radar: theory and practice. In Cluckie, I. and Griffith, R., Editors, *Radar Hydrology for Real Time Flood Forecasting*, pages 33–38. European Commission. ISBN: 92-894-1640-8.

Bate, S. (2004). Generalized Linear Models for Large Dependent Data Sets. Ph.D. thesis, Department of Statistical Science, University College London.

Brown, P., Diggle, P., Lord, M., and Young, P. (2001). Space-time calibration of radar-rainfall data. *J. Roy. Stat. Soc.*, C50:221–241.

Casdagli, M. (1992). Chaos and deterministic versus stochastic and non-linear modelling. *J. Roy. Stat. Soc.*, B54:303–328.

Chandler, R.E. (2002). GLIMCLIM: Generalized linear modelling for daily climate time series (software and user guide). Technical report, no. 227, Department of Statistical Science, University College London, London WC1E 6BT. http://www.ucl.ac.uk/stats/research/Resrpts/abstracts.html.

Chandler, R.E. (2005). On the use of generalized linear models for interpreting climate variability. *Environmetrics*, 16(7):699–715.

Chandler, R.E. and Wheater, H.S. (1998). Climate change detection using Generalized Linear Models for rainfall — a case study from the West of Ireland. II. Modelling of rainfall amounts on wet days. Technical report, no. 195, Department of Statistical Science, University College London. http://www.ucl.ac.uk/stats/research/Resrpts/abstracts.html.

Chandler, R.E. and Wheater, H.S. (2002). Analysis of rainfall variability using Generalized Linear Models — a case study from the West of Ireland. *Water Resources Research*, 38, No.10:doi:10.1029/2001WR000906.

Coe, R. and Stern, R.D. (1982). Fitting models to daily rainfall. *J. Appl. Meteorol.*, 21:1024–1031.

Collier, C. (2000). Developments in radar and remote-sensing methods for measuring and forecasting rainfall. *Proc. R. Soc. Lond.*, A360:1345–1361.

Cox, D.R. and Isham, V. (1980). *Point Processes*. Chapman & Hall, London.

Cox, D.R. and Isham, V. (1988). A simple spatial-temporal model of rainfall. *Proc. R. Soc. Lond.*, A415:317–328.

Cox, D.R. and Wermuth, N. (1996). *Multivariate Dependencies: Models, Analysis and Interpretation*. Chapman & Hall, London.

Crowder, M. (1995). On the use of a working correlation matrix in using generalised linear models for repeated measures. *Biometrika*, 82:407–410.

Davison, A.C. (2003). *Statistical Models*. Cambridge University Press, Cambridge.

Devroye, L. (1986). *Non-uniform random variate generation*, Springer-Verlag, New York.

Diggle, P., Tawn, J., and Moyeed, R. (1998). Model-based geostatistics. *Appl. Statist.*, 47:299–326.

Diggle, P.J. (2003). *Statistical Analysis of Spatial Point Patterns (second edition)*. Arnold.

Emrich, L.J. and Piedmonte, M.R. (1991). A method for generating high-dimensional multivariate binary variates. *Amer. Statistician*, 45, (4):302–304.

Fahrmeir, L. and Tutz, G. (1994). *Multivariate Statistical Modelling Based on Generalized Linear Models*. Springer-Verlag, New York.

Hughes, J.P., Guttorp, P., and Charles, S. (1999). A nonhomogeneous hidden Markov model for precipitation. *Appl. Statist.*, 48:15–30.

Hurrell, J.W. (1995). Decadal trends in the North Atlantic Oscillation: regional temperatures and precipitation. *Science*, 269:676–679.

IPCC (2001). *Climate Change 2001 — The Scientific Basis*. Cambridge University Press, Cambridge. Report of the Intergovernmental Panel on Climate Change.

Isham, V. (1993). Statistical aspects of chaos: A review. In O.E. Barndorff-Nielsen, J.J. and Kendall, W., Editors, *Chaos and Networks: Statistical and Probabilistic Aspects*, pages 124–200. Chapman & Hall, London.

Krzanowski, W. (1988). *Principles of Multivariate Analysis*. Oxford University Press.

Liang, K.-Y. and Zeger, S. (1986). Longitudinal data analysis using generalized linear models. *Biometrika*, 73(1):13–22.

Lunn, A. and Davies, S. (1998). A note on generating correlated binary variables. *Biometrika*, 85(2):487–490.

McCullagh, P. and Nelder, J. (1989). *Generalized Linear Models (second edition)*. Chapman & Hall, London.

McDonald, B. (1993). Estimating logistic regression parameters for bivariate binary data. *J. R. Statist. Soc., Series B*, 55:391–397.

Moore, R.J., May, B.C., Jones, D.A., and Black, K.B. (1994). Local calibration of weather radar over London. In *Advances in Radar Hydrology*, pages 186–195. European Commission.

Northrop, P.J. (1996). Modelling and Statistical Analysis of Spatial-Temporal Rainfall Fields. Ph.D. thesis, Department of Statistical Science, University College London.

Northrop, P.J. (1998). A clustered spatial-temporal model of rainfall. *Proc. Roy. Soc. London.*, A454:1875–1888.

Northrop, P.J. and Stone, T.M. (2005). A point process model for rainfall with truncated Gaussian rain cells. Technical report No. 251, Department of Statistical Science, University College, London, `http://www.ucl.ac.uk/stats/research/Resrpts/abstracts.html`.

Oman, S. and Zucker, D. (2001). Modelling and generating correlated binary variables. *Biometrika*, 88(1):287–290.

Onof, C., Chandler, R.E., Kakou, A., Northrop, P., Wheater, H.S., and Isham, V. (2000). Rainfall modelling using Poisson-cluster processes: a review of developments. *Stoch. Env., Res. & Risk Ass.*, 14:384–411.

Rodriguez-Iturbe, I., Cox, D.R., and Isham, V. (1987). Some models for rainfall based on stochastic point processes. *Proc. R. Soc. Lond.*, A410:269–288.

Rodriguez-Iturbe, I., Cox, D.R., and Isham, V. (1988). A point process model for rainfall: further developments. *Proc. R. Soc. Lond.*, A417:283–298.

Sheikh, I. (2004). Stochastic Rainfall Modelling. M.Sc. dissertation, Department of Statistical Science, University College London.

Smithers, J.C., Pegram, G.G.S., and Schulze, R.E. (2002). Design rainfall estimation in South Africa using Bartlett-Lewis rectangular pulse rainfall models. *J. Hydrol.*, 258:83–99.

Stern, R.D. and Coe, R. (1984). A model fitting analysis of rainfall data (with discussion). *J. Roy. Stat. Soc.*, A147:1–34.

Sullivan Pepe, M. and Anderson, G. (1994). A cautionary note on inference for marginal regression models with longitudinal data and general correlated response data. *Commun. Statist. Simula.*, 23:939–951.

Sutradhar, B. and Das, K. (1999). On the efficiency of regression estimators in generalised linear models for longitudinal data. *Biometrika*, 86:459–465.

Terrell, G. (2003). The Wilson-Hilferty transformation is locally saddlepoint. *Biometrika*, 90(2):445–453.

Thompson, D.W.J. and Wallace, J.M. (1998). The Arctic Oscillation signature in the wintertime geopotential height and temperature fields. *Geophys. Res. Lett.*, 25(9):1297–1300.

Von Neumann, J. (1951). Various techniques used in connection with random digits, 'Monte Carlo Method.' *U.S. Natl. Bur. Stand. Appl. Math. Ser.*, 12:36–38.

Wheater, H.S. (2002). Progress in and prospects for fluvial flood modelling. *Proc. R. Soc. Lond.*, A360:1409–1432.

Wheater, H.S., Chandler, R.E., Onof, C.J., Isham, V.S., Bellone, E., Yang, C., Lekkas, D., Lourmas, G., and Segond, M.-L. (2005). Spatial-temporal rainfall modelling for flood risk estimation. *Stoch. Env. Res. & Risk Ass.*, 19:403–416.

Wheater, H.S., Isham, V.S., Cox, D.R., Chandler, R.E., Kakou, A., Northrop, P.J., Oh, L., Onof, C., and Rodriguez-Iturbe, I. (2000a). Spatial-temporal rainfall fields: modelling and statistical aspects. *Hydrological and Earth Systems Science*, 4:581–601.

Wheater, H.S., Isham, V.S., Onof, C., Chandler, R.E., Northrop, P.J., Guiblin, P., Bate, S.M., Cox, D.R., and Koutsoyiannis, D. (2000b). Generation of spatially consistent rainfall data. Report to the Ministry of Agriculture, Fisheries and Food (2 volumes). Also available as Research Report No. 204, Department of Statistical Science, University College London `http://www.ucl.ac.uk/stats/research/Resrpts/abstracts.html`.

Yan, Z., Bate, S., Chandler, R.E., Isham, V.S., and Wheater, H.S. (2002). An analysis of daily maximum windspeed in northwestern Europe using Generalized Linear Models. *J. Climate*, 15(15):2073–2088.

Yang, C., Chandler, R.E., Isham, V.S., and Wheater, H.S. (2005). Spatial-temporal rainfall simulation using generalized linear models. *Water Resources Research*, 41, doi:10.1029/2004WR003739.

CHAPTER 6

A Primer on Space-Time Modeling from a Bayesian Perspective

David Higdon

Contents

6.1 Introduction

The following notes are designed to get the interested reader up and running — developing his or her own spatial or space-time models — as quickly as possible. For better or worse, this means many details regarding alternative modeling, model checking, and computation are omitted for the sake of keeping this primer streamlined. The goal of this primer is to convey an overall sense of what is involved in developing space-time models. Far more detailed accounts of spatial and space-time modeling can be found in standard texts such as Cressie (1991), Stein (1999), Wackernagel (1995), Chilés and Delfiner (1999), or Banerjee et al. (2003).

In this primer, the emphasis is on Gaussian spatial and space-time models. These models are quite flexible and can be adapted to a wide variety of applications, even where the observed data are markedly non-Gaussian as will be shown in Section 6.4. Due to my own bias and experience, purely spatial models are first developed in Sections 6.2 through 6.4. These models are then extended to the space-time domain in Sections 6.5 and 6.6. Throughout, the computation for the actual applications is carried out using Markov chain Monte Carlo (MCMC), which was first implemented on a spatial system in 1953. I will do my best to show the natural links between spatial systems and MCMC. I also caution the reader, that for the sake of reducing the number of details that must be covered, MCMC is essentially the only method used in this primer. Though this technique is widely applicable, in many situations there may be more efficient computational approaches that I neglect here.

6.2 Gaussian computation

The spatial models laid out in this primer make heavy use of Gaussian systems. This section goes over some basics for simulation and conditioning with multivariate normal models.

Figure 6.1 shows a realization $z(s)$ of a Gaussian process model on $s_1, \ldots, s_n$

$$z = \begin{pmatrix} z(s_1) \\ \vdots \\ z(s_n) \end{pmatrix} \sim N\left(\begin{pmatrix} 0 \\ \vdots \\ 0 \end{pmatrix}, \begin{pmatrix} \Sigma \end{pmatrix} \right)$$

where $\Sigma_{ij} = \exp\{-||s_i - s_j||^2\}$ and $||s_i - s_j||$ denotes the distance between locations s_i and s_j. Here z has a multivariate normal probability density function given by

$$\pi(z) = (2\pi)^{-\frac{n}{2}} |\Sigma|^{-\frac{1}{2}} \exp\{-\tfrac{1}{2} z^T \Sigma^{-1} z\}.$$

If the spatial locations $\{s_1, \ldots, s_n\}$ are taken to be $\{0, 1, \ldots, 7\}$, the realizations correspond to the circles plotted in Figure 6.1. Alternatively, the spatial locations $\{s_1, \ldots, s_n\}$ could define a very dense grid of points between 0 and 7.

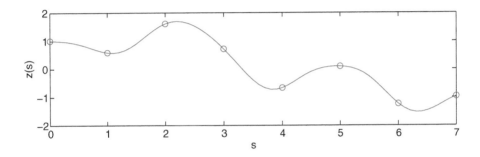

Figure 6.1 A realization from a 1-d Gaussian process (GP) with covariance given by $\Sigma_{ij} = \exp\{-||s_i - s_j||^2\}$.

In this case, the realization corresponds to the line in Figure 6.1. Additional realizations of $z(s)$ are shown in Figure 6.2.

The covariance matrix Σ for the process $z(s)$ is built according to the rule $\Sigma_{ij} = \exp\{-||s_i - s_j||^2\}$. In general, not just any rule will do — a covariance rule needs to give a valid covariance matrix (symmetric and positive definite) for any set of points within the spatial domain. For a catalog of possible covariance rules, consult previously mentioned texts on kriging and spatial modeling. Later in this section, the covariance rules $\Sigma_{ij} = \exp\{-||s_i - s_j||^2\}$ and $\Sigma_{ij} = \exp\{-||s_i - s_j||^1\}$ are compared. Generally, the smoothness and the strength of dependence in the realizations $z(s)$ are influenced by the choice of covariance rule. Note that practical covariance rules will need to acount for scaling in distance as well as variance. Thus a more general covariance rule might look like

$$\Sigma_{ij}(\sigma_z^2, r) = \sigma_z^2 \exp\left\{-\left(||s_i - s_j||/r\right)^2\right\}$$

where σ_z^2 controls the marginal variance of $z(s)$ and r scales distance. However, because the focus of this chapter through Section 6.3.2 is on $z(s)$, the scaling parameters are left at 1 for now.

6.2.1 Generating multivariate normal realizations

Almost any computer package will generate independent, univariate normal draws:

$$u \sim N(0, I_n).$$

A standard property of normals —

$$\text{if } u \sim N(\mu, \Sigma), \quad \text{then } z = Ku \sim N(K\mu, K\Sigma K^T)$$

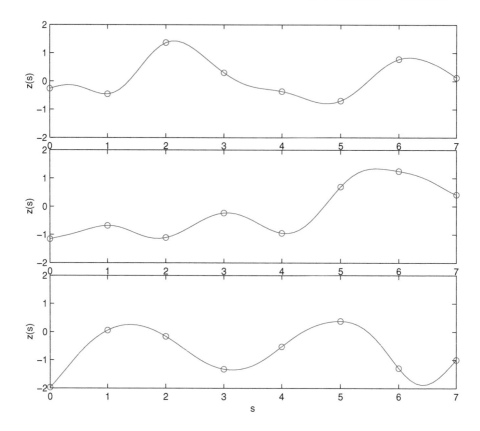

Figure 6.2 Several independent realizations from a 1-d Gaussian process with covariance given by $\Sigma_{ij} = \exp\{-||s_i - s_j||^2\}$.

can be used to construct correlated realizations z from iid draws held in vector u. The following recipe can be used to generate $z \sim N(0, \Sigma)$:

1. compute square root matrix L such that $LL^T = \Sigma$;
2. generate $u \sim N(0, I_n)$;
3. set $z = Lu \sim N(0, LI_nL^T = \Sigma)$.

This simple recipe gives some insight into how one might represent the process z in terms of basis vectors:

- The columns of L are effectively basis vectors for representing realizations:

$$z = \sum_{i=1}^{m} \ell_i u_i$$

where ℓ_i is the ith column of L;

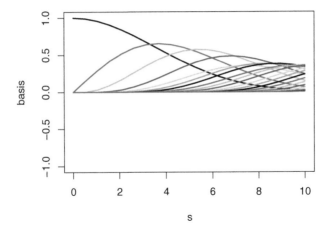

Figure 6.3 Basis vectors resulting from the Cholesky factorization of the covariance matrix obtained by applying the covariance rule $\Sigma_{ij} = \exp\{-||(s_i - s_j)/5||^2\}$ to 20 equally spaced points between 0 and 10.

- The choice of L is not unique — alternative choices for L lead to different basis representations for z;

- L need not be square. If L is a $n \times m$ matrix, then the columns of L give a lower-dimensional basis representation for z if $m < n$, and an overspecified representation if $m > n$

To get an idea of basis representations various square root factorizations produce, we consider a simple 1-d GP $z(s)$ at 20 spatial locations $\{s_1, \ldots, s_{20}\}$ that are equally spaced between 0 and 10 in ascending order. The covariance rule is given by $\Sigma_{ij} = \exp\{-||(s_i - s_j)/5||^2\}$. Hence the 20-vector $z = (z(s_1), \ldots, z(s_{20}))^T$ has a normal distribution with mean 0 and covariance Σ.

The most common method for constructing a square root L of a covariance matrix Σ is with the standard Cholesky factorization (Press et al., 2002, Ch 2.9). The resulting columns of the square root matrix L are shown in Figure 6.3. Because the standard Cholesky factorization builds the basis vectors sequentially, without regard to the spatial locations $s_1, \ldots, s_{20}$ implicit in the vector z, the resulting basis representation is rather inefficient since each of the basis vectors have elements of appreciable size.

This is due to the ordering of the components of z that the resulting Cholesky decomposition of Σ is relatively inefficient. If one reorders the vector z so that its most correlated elements are not always adjacent, the resulting basis vectors given by the Cholesky decomposition will look quite different. A particularly efficient ordering is obtained by using a Cholesky factorization with pivoting (Dongarra et al., 1978). The resulting basis vectors are shown

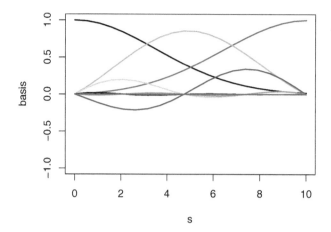

Figure 6.4 Basis vectors resulting from the Cholesky factorization with pivoting of the covariance matrix obtained by applying the covariance rule $\Sigma_{ij} = \exp\{-\|(s_i - s_j)/5\|^2\}$ to 20 equally spaced points between 0 and 10.

in Figure 6.4. In this example, nearly all of the variation in z is explained by the first 5 basis vectors.

A third (of many more) alternatives can be obtained by taking the singular value decomposition (SVD) of Σ for which $\Sigma = UDU^T$ where U is an orthonormal matrix and D is diagonal with non-negative elements. The square root matrix can then be given by $L = UD^{\frac{1}{2}}$. The columns of the resulting matrix L are shown in Figure 6.5. As with the Cholesky factorization with

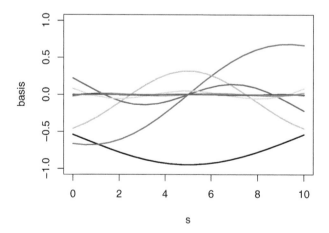

Figure 6.5 Basis vectors resulting from the SVD factorization with pivoting of the covariance matrix obtained by applying the covariance rule $\Sigma_{ij} = \exp\{-\|(s_i - s_j)/5\|^2\}$ to 20 equally spaced points between 0 and 10.

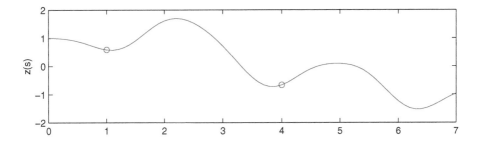

Figure 6.6 From the Gaussian process, we observe the values $z(s_2)$ and $z(s_5)$ and now want to determine the conditional distribution of the entire process $z(s)$.

pivoting, this approach gives an efficient basis representation of z. Sections 6.4 and 6.5 consider modeling approaches that rely on efficient basis representations of z.

6.2.2 Conditional distributions

In nearly any spatial application, some aspect of the problem involves inferring about $z(s)$, over a spatial domain $\mathcal{S}$. To be concrete, suppose $z(s)$, $s \in \mathcal{S}$ is a mean 0 GP with covariance rule $\Sigma_{ij} = \exp\{-\|(s_i - s_j)\|^2\}$, and $\mathcal{S}$ denotes $m = 8$ spatial locations $\{s_1, \dots, s_m\} = \{0, 1, \dots, 7\}$. Now we observe $z(s_i)$ at $n = 2$ spatial locations $s_2 = 1$ and $s_5 = 4$ as shown in Figure 6.6. Now, $z(s_1)$ and $z(s_5)$ are known. It is the conditional distribution of the remaining sites in $\mathcal{S}$ that is required.

Standard normal computations will give us the conditional distribution of $z(s)$, $s \in \{\mathcal{S} \setminus \{s_1, s_5\}\}$. After reordering, we have

$$
\begin{pmatrix}
z(s_2) \\
z(s_5) \\
\hline
z(s_1) \\
z(s_3) \\
z(s_4) \\
z(s_6) \\
z(s_7) \\
z(s_8)
\end{pmatrix}
\sim N \left(
\begin{pmatrix}
0 \\
0 \\
\hline
0 \\
0 \\
0 \\
0 \\
0 \\
0
\end{pmatrix},
\left(
\begin{array}{cc|cccc}
1 & .0001 & .3679 & \cdots & & 0 \\
.0001 & 1 & 0 & \cdots & & .0001 \\
\hline
.3679 & 0 & 1 & \cdots & & 0 \\
\cdots & \cdots & \vdots & \ddots & & \vdots \\
0 & .0001 & 0 & \cdots & & 1
\end{array}
\right)
\right).
$$

Generally, for a mean 0 normally distributed vector, if

$$
\begin{pmatrix} z_1 \\ z_2 \end{pmatrix} \sim N \left(\begin{pmatrix} 0 \\ 0 \end{pmatrix}, \begin{pmatrix} \Sigma_{11} & \Sigma_{12} \\ \Sigma_{21} & \Sigma_{22} \end{pmatrix} \right)
$$

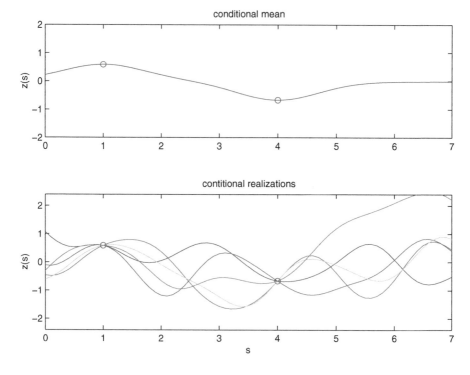

Figure 6.7 The conditional mean of $z(s)$ after observing $z(s)$ at the two spatial locations $s = 1$ and $s = 4$.

then

$$z_2|z_1 \sim N\left(\Sigma_{21}\Sigma_{11}^{-1}z_1, \Sigma_{22} - \Sigma_{21}\Sigma_{11}^{-1}\Sigma_{12}\right). \tag{6.1}$$

See Anderson (1984), Section 2.5, for the derivation. So if we take the first two components of the 8-vector to be z_1, and the remaining six to be z_2, we have the conditional mean of $z_2|z_1 = \mu_{2|1}$ given by $\Sigma_{21}\Sigma_{11}^{-1}z_1$ and the conditional covariance matrix of $z_2|z_1 = \Sigma_{2|1}$ given by $\Sigma_{22} - \Sigma_{21}\Sigma_{11}^{-1}\Sigma_{12}$.

Instead of specifying $\mathcal{S}$ to be 8 points equally spaced between 0 and 7, we could define $\mathcal{S}$ to be a large number of points equally spaced between 0 and 7. Equation (6.1) above just as easily gives the conditional mean and variance of $z(s)$ over this $\mathcal{S}$. Of course, drawing from (6.1) may become computationally demanding if the number of prediction sites is very large. Figure 6.7 shows the conditional mean of $z(s)$ after conditioning on the two points $z(s = 1)$ and $z(s = 4)$. The figure also shows four draws from the GP $z(s)$ conditional on these two observations.

Equation (6.1) is one of the foundations for working with spatial models. Figures 6.8 and 6.9 show draws from the conditional process over the one-dimensional space $\mathcal{S} = [0, 15]$ after conditioning on the same four points. In each plot, a different covariance rule is used. The table below shows the covariance rule used for each plot in Figures 6.8 and 6.9.

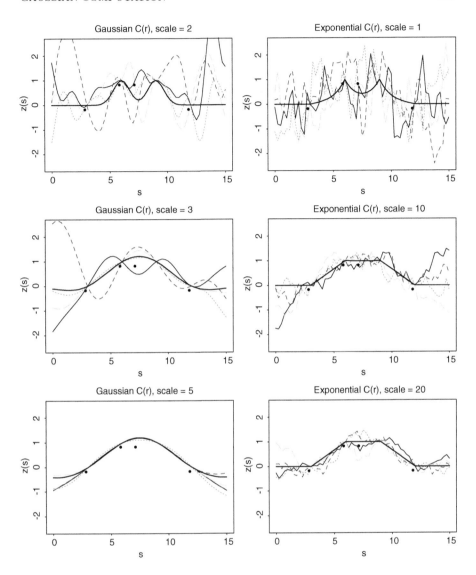

Figure 6.8 Conditional realizations of $z(s)$ under different covariance models after observing the four data points given by the black dots. The conditional mean of $z(s)$ is given by the thick black line; four conditional realizations are given by the dashed lines.

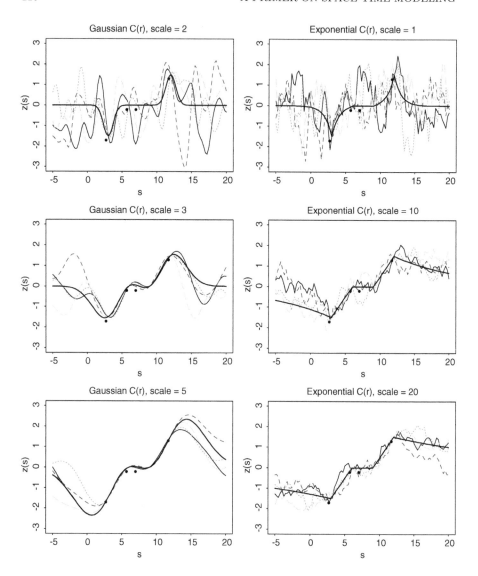

Figure 6.9 Conditional realizations of $z(s)$ under different covariance models after observing the four data points given by the black dots. The conditional mean of $z(s)$ is given by the thick black line; four conditional realizations are given by the dashed lines.

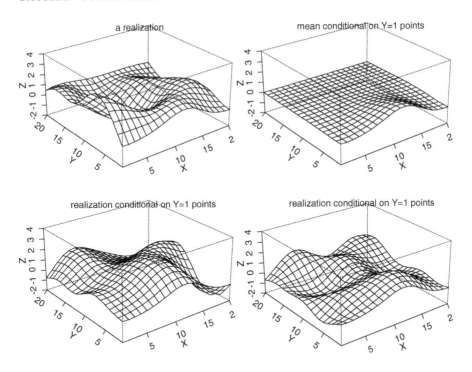

Figure 6.10 Top left: a realization from a 2-d Gaussian process with mean zero and covariance function $C(r) = \exp\{-(r/2)^2\}$. Top right: the conditional mean of the process after conditioning on the points for which $y = 1$. Bottom row: realizations of the Gaussian process after conditioning on the edge points at $y = 1$.

Covariance rules/functions for Figures 6.8 and 6.9

left plots	right plots
$\Sigma_{ij} = \exp\{-|(s_i - s_j)/2|^2\}$	$\Sigma_{ij} = \exp\{-|(s_i - s_j)/1|^1\}$
$\Sigma_{ij} = \exp\{-|(s_i - s_j)/3|^2\}$	$\Sigma_{ij} = \exp\{-|(s_i - s_j)/10|^1\}$
$\Sigma_{ij} = \exp\{-|(s_i - s_j)/5|^2\}$	$\Sigma_{ij} = \exp\{-|(s_i - s_j)/20|^1\}$

Note the left-hand column of plots correspond to GP's with a *Gaussian* covariance function; the right-hand column of plots correspond to GP's with an *exponential* covariance function. Realizations from this exponential process are rougher as compared to the realizations from GP's with the Gaussian covariance rule.

Equation (6.1) is just as applicable to higher dimensional GP's. As an example, the top left plot in Figure 6.10 shows a realization from a GP over the 2-d lattice $\mathcal{S} = \{1, 2, \ldots, 20\}^2$ with covariance given by $\Sigma_{ij} = \exp\{-|d_{ij}/2|^2\}$. Here d_{ij} is the Euclidean distance between points $s_i = (s_{i1}, s_{i2})$ and $s_j = (s_{j1}, s_{j2})$ in $\mathcal{S}$. If we condition on the 20 points corresponding to $s_2 = 1$,

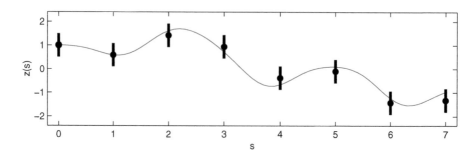

Figure 6.11 Data from a noisy observation of the continuous Gaussian process $z(s)$ at locations $s_i = i$, $i = 0, \ldots, 7$. The dots denote the observations and the bars show $\pm 2\sigma_y$.

the top right plot shows the conditional mean of $z(s)$ and the bottom two figures each show a realization of the process conditional on the 20 edge values $z(s_1, s_2 = 1)$.

6.2.3 Soft conditioning

In almost any actual application, the underlying spatial process $z(s)$ is never measured with complete accuracy at any spatial location. Hence we need to be able to characterize our uncertainty about $z(s)$ given noisy observations. This chapter approaches the problem from a Bayesian perspective.

For this example, we consider the mean 0 GP $z(s)$ over the eight locations $S = \{0, 1, \ldots, 7\}$ with covariance rule $\Sigma_{ij} = \exp\{-|s_i - s_j|^2\}$. We take the observed data y now to be a noisy version of z over each point in S

$$y(s_i) = z(s_i) + \epsilon(s_i), \quad i = 1, \ldots, n,$$

with

$$\epsilon(s_i) \overset{iid}{\sim} N(0, \sigma_y^2), \quad i = 1, \ldots, n.$$

The data are shown in Figure 6.11 with error bars that correspond to $\pm 2\sigma_y$. Also shown is the true underlying process $z(s)$ about which we wish to infer.

Under the Bayesian paradigm, inference about $z(s)$ is based on its posterior distribution given the $n = 8$ noisy observations y. The likelihood — or sampling model — for the observed data is

$$L(y|z) \propto |\Sigma_y|^{-\frac{1}{2}} \exp\left\{-\tfrac{1}{2}(y - z)^T \Sigma_y^{-1}(y - z)\right\},$$

where $\Sigma_y = \sigma_y^2 I_n$ and I_n denotes the $n \times n$ identity matrix. The GP prior model for $z(s)$ is

$$\pi(z) \propto |\Sigma_z|^{-\frac{1}{2}} \exp\left\{-\tfrac{1}{2} z^T \Sigma_z^{-1} z\right\},$$

where Σ_z is obtained by applying the covariance rule to the $n = 8$ spatial locations in $\mathcal{S}$. The resulting posterior density for z given y is proportional to the product of the likelihood and prior

$$\pi(z|y) \propto L(y|z) \times \pi(z) \tag{6.2}$$

$$\propto \exp\left\{ -\tfrac{1}{2}z^T\left(\Sigma_y^{-1} + \Sigma_z^{-1}\right)z + z^T\Sigma_y^{-1}y + f(y)\right\}.$$

The unnormalized density above is that of a normal

$$z|y \sim N\left(V\Sigma_y^{-1}y, V\right)$$

where $V = (\Sigma_y^{-1} + \Sigma_z^{-1})^{-1}$. Note the posterior mean

$$V\Sigma_y^{-1}y = \left(\Sigma_y^{-1} + \Sigma_z^{-1}\right)^{-1}\Sigma_y^{-1}y$$

is a precision weighted average of the prior mean (0) and the data. The identity

$$\left(\Sigma_y^{-1} + \Sigma_z^{-1}\right)^{-1}\Sigma_y^{-1} = \Sigma_z(\Sigma_y + \Sigma_z)^{-1}$$

can be helpful for computations.

Often, the noisy observations at $\{s_1, \dots, s_n\}$ will be used to obtain the posterior distribution for $z(s)$ at a collection of m unobserved sights $\{s_1^*, \dots, s_m^*\}$. The previous recipe for obtaining $\pi(z|y)$ can easily be modified to accommodate this situation.

We define

$$y^d = (y(s_1), \dots, y(s_n))^T$$
$$z^d = (z(s_1), \dots, z(s_n))^T$$
$$y^* = (y(s_1^*), \dots, y(s_m^*))^T$$
$$z^* = (z(s_1^*), \dots, z(s_m^*))^T$$
$$y = (y^d; y^*)$$
$$z = (z^d; z^*)$$

so that y holds a $n + m$ vector of "observations" (the last m components of y are only notional), and z holds $z(s)$ restricted to the observation and prediction locations.

The sampling model for y is the same as before:

$$L(y|z) \propto |\Sigma_y|^{-\frac{1}{2}} \exp\left\{ -\tfrac{1}{2}(y - z)^T\Sigma_y^-(y - z)\right\},$$

except now

$$\Sigma_y^- = \begin{pmatrix} \frac{1}{\sigma_y^2}I_n & 0 \\ 0 & 0_{m \times m} \end{pmatrix},$$

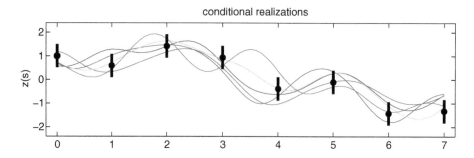

Figure 6.12 Conditional realizations of $z(s)$ after the noisy observations given in Figure 6.11.

and $|\Sigma_y|$ is defined to be σ^{2n}. Here zero precisions (infinite variances) are specified for the components of y that are unobserved. Also, the GP prior is exactly as before for z,

$$\pi(z) \propto |\Sigma_z|^{-\frac{1}{2}} \exp\left\{ -\tfrac{1}{2} z^T \Sigma_z^{-1} z \right\},$$

but now Σ_z is the $(n+m) \times (n+m)$ covariance matrix obtained by applying the covariance rule to the augmented set of spatial locations — both observation and prediction locations.

The resulting posterior distribution for $z = (z^d, z^*)$ is then

$$z|y \sim N\left(V \Sigma_y^- y, V\right)$$

where

$$V = \left(\Sigma_y^- + \Sigma_z^{-1}\right)^{-1}.$$

Note the posterior distribution for z depends only on Σ_y through its inverse which is well defined. Figure 6.12 shows realizations from $\pi(z|y)$ where the prediction sites are a fine grid of locations between 0 and 7.

6.2.3.1 An example

As an example we have $n = 120$ spatially located log-dioxin concentration measurements taken from a 100 m × 200 m contaminated site in Missouri, U.S.A. (Ryti, 1993) shown in the left plot of Figure 6.13.

We wish to predict the concentrations over a 15 × 30 grid denoted by the points in the left plot of the figure. We specify a mean 0 GP prior for $z(s)$ — the log-concentration surface with covariance rule

$$\Sigma_{ij} = \exp\{-\|(s_i - s_j)/15\|^2\}.$$

The posterior mean for the $m = 15 \cdot 30$ prediction sites is shown in the center plot. The prediction standard deviations, given by the square root of the diagonal of the last $m \times m$ submatrix of the posterior covariance matrix V, is shown in the right plot.

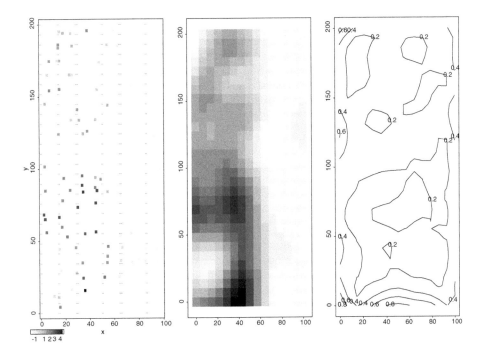

Figure 6.13 Left: log-dioxin concentration measurements from the Piazza Road pilot study. Middle: posterior mean for the concentration at the grid sites denoted by dots in the left-hand plot. Right: pointwise posterior standard deviation of $z(s)$ given the measurements.

Here we specified the prior mean of $z(s)$ as well as the form of the covariance rule. An important component of spatial modeling — ignored here, for now — is determining these features of the spatial model. The geostatistical references mentioned in Section 6.1 have much more to say on this topic.

6.3 Gaussian Markov random fields and Bayesian computation

The posterior distribution of (6.2) depends on precision (inverse covariance) matrices in the prior and likelihood, and can be trivially rewritten

$$\pi(z|y) \propto \exp\left\{ -\tfrac{1}{2}z^T(W_y + W_z)z + z^TW_yy + f(y)\right\}$$

where $W_y = \Sigma_y^{-1}$ and $W_z = \Sigma_z^{-1}$. Hence,

$$z|y \sim N\left(\Lambda^{-1}W_yy, \Lambda^{-1}\right),$$

where $\Lambda = W_y + W_z$.

Rather than specifying a prior dependence structure for the spatial process $z(s)$ through a covariance rule, one can specify the dependence through the precision matrix W_z. As we will see in this section, there is a natural connection between W_z and the conditional dependence structure of $z(s)$. These *Gaussian Markov random field* (GMRF) specifications can be very effective for large systems, such as image models, because they are amenable to efficient computational approaches. The strong links between GMRFs and MCMC make it natural to introduce Gibbs sampling in this section as well.

A GMRF is defined over a fixed set of sites $s_1, \ldots, s_n$. The vector $z = (z(s_1), \ldots, z(s_n))^T$ is the spatial process to be modeled with a GMRF specification. We now list some notation and facts regarding a GMRF specification for $z(s)$ — a far more detailed description of GMRFs can be found in Rue and Held (2005):

- z_i is the value of $z(s)$ at site s_i.
- z_{-i} is the $n-1$-vector $(z_1, \ldots, z_{i-1}, z_{i+1}, \ldots, z_n)^T$.
- $i \sim j$ means s_i and s_j are neighbors.
- $z_{\partial i} = \{z_j : i \sim j\}$ (i.e., all z_j for which s_j and s_i are neighbors).
- $\pi(z_i | z_{-i})$ is called the *full conditional* distribution of z_i (i.e., its distribution given all other components of z).
- $\pi(z_i | z_{-i}) = \pi(z_i | z_{\partial i})$ (i.e., the full conditional for z_i depends only on its neighbors).

A Markov random field (MRF) is defined by specifying a neighborhood system and a set of n conditional densities $\{\pi(z_i | z_{\partial i}), i = 1, \ldots, n\}$. Figure 6.14 gives examples of neighborhood specifications on regular lattices. Note that not just any set of n full conditionals will yield a valid joint distribution $\pi(z)$.

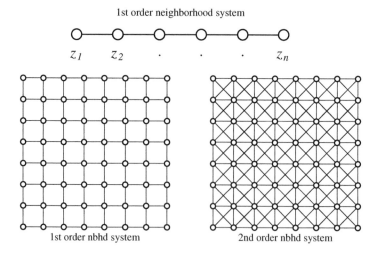

1st order neighborhood system

$z_1 \quad z_2 \quad \cdot \quad \cdot \quad \cdot \quad z_n$

1st order nbhd system   2nd order nbhd system

Figure 6.14 Top: a first-order (nearest neighbor) neighborhood system for a regular 1-d lattice. Bottom left: a first-order neighborhood system for a regular 2-d lattice. Bottom right: a second order (nearest two neighbors) neighborhood system for a regular 2-d lattice.

This section uses only Gaussian MRFs whose full conditionals are determined by the precision matrix W_z. Section 6.6 gives an example of a non-Gaussian MRF.

6.3.1 Locally linear Gaussian MRFs

For a given neighborhood system, define

$$z_i | z_{\partial i} \sim N(\bar{z}_{\partial i}, 1/n_i)$$

where n_i is the number of neighbors belonging to z_i and $\bar{z}_{\partial i}$ is the average of the z_js neighboring z_i. The full conditional density is then

$$\pi(z_i | z_{\partial i}) \propto \exp\left\{ -\frac{1}{2} \sum_{j \in \partial i} (z_i - z_j)^2 \right\}$$

which implies a joint Gaussian distribution for z:

$$\pi(z) \propto |W_z|^{\frac{1}{2}} \exp\left\{ -\frac{1}{2} z^T W_z z \right\} \tag{6.3}$$

$$\propto |W_z|^{\frac{1}{2}} \exp\left\{ -\frac{1}{2} \sum_{i \sim j} (z_i - z_j)^2 \right\}$$

where $|W_z|$ is defined to be the product of the non-zero eigenvalues of W_z, the ij component of W_z is given by

$$W_{zij} = \begin{cases} n_i & \text{if } i = j \\ -1 & \text{if } i \sim j \\ 0 & \text{otherwise} \end{cases}$$

and the sum $\sum_{i \sim j} (z_i - z_j)^2$ is over each edge pair in the neighborhood system. This locally linear MRF model for $z(s)$ gives an alternative to the covariance-based models described in the previous section. Again, any practical application of a MRF model will require that $\pi(x)$ include a parameter to scale $z(s)$. This is addressed at the end of Section 6.3.2 and beyond.

As an example, if we take z to be the $n = 5$-site, first-order system below

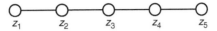

then the precision matrix W_z has the form

$$W_z = \begin{pmatrix} 1 & -1 & \circ & \circ & \circ \\ -1 & 2 & -1 & \circ & \circ \\ \circ & -1 & 2 & -1 & \circ \\ \circ & \circ & -1 & 2 & -1 \\ \circ & \circ & \circ & -1 & 1 \end{pmatrix}.$$

As a 1-d example, suppose there is an underlying true process $z = (z_1, z_2, \ldots, z_{25})^T$ at spatial sites $\{1, 2, \ldots, 25\}$ and we have noisy observations $y = (y_1, y_2, \ldots, y_{25})^T$ at each site. This gives the sampling model

$$L(y|z) \propto |W_y|^{\frac{1}{2}} \exp\left\{ -\tfrac{1}{2}(y-z)^T W_y(y-z) \right\}$$

where $W_y = \tfrac{1}{4}I$, corresponding to noise variance of 4. We specify a first-order, locally linear GMRF for z so that the prior has the form

$$\pi(z) \propto |W_z|^{\frac{1}{2}} \exp\left\{ -\tfrac{1}{2}z^T W_z z \right\}.$$

Hence, the resulting posterior distribution for z given y is multivariate Gaussian

$$z|y \sim N(\Lambda^{-1} W_y y, \Lambda^{-1}),$$

where $\Lambda = W_y + W_z$. Figure 6.15 shows the true underlying process z along with the noisy observations y in the top left frame. Also shown in the figure are the posterior mean of z, some realizations from the posterior, and pointwise 90% intervals for z at each site.

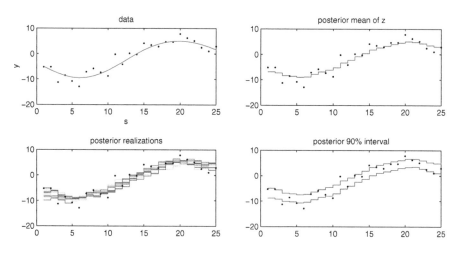

Figure 6.15 Data and posterior quantities for z which is given a first-order, locally linear, Gaussian MRF prior (6.3). The data are independently scattered about the true process (top left plot) according to an $N(0, 4)$ distribution.

The locally linear specification for a 2-d lattice is also quite simple. For the $n = 9$-site, 2-d lattice below

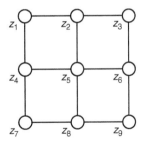

the locally linear specification leads to a 9×9 precision matrix of the form

$$
W_z =
\begin{pmatrix}
2 & -1 & \circ & -1 & \circ & \circ & \circ & \circ & \circ \\
-1 & 3 & -1 & \circ & -1 & \circ & \circ & \circ & \circ \\
\circ & -1 & 2 & \circ & \circ & -1 & \circ & \circ & \circ \\
-1 & \circ & \circ & 3 & -1 & \circ & -1 & \circ & \circ \\
\circ & -1 & \circ & -1 & 4 & -1 & \circ & -1 & \circ \\
\circ & \circ & -1 & \circ & -1 & 3 & \circ & \circ & -1 \\
\circ & \circ & \circ & -1 & \circ & \circ & 2 & -1 & \circ \\
\circ & \circ & \circ & \circ & -1 & \circ & -1 & 3 & -1 \\
\circ & \circ & \circ & \circ & \circ & -1 & \circ & -1 & 2
\end{pmatrix} .
$$

A 2-d example is shown in Figure 6.16 below. We generate data y over a 20×20 lattice that is the true process z (shown in the left plot of Figure 6.16) plus independent $N(0, 4)$ errors.

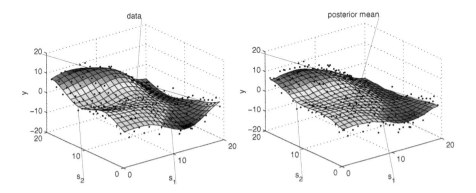

Figure 6.16 Left: data and and true underlying process z over a 20×20 lattice. Right: posterior mean for z resulting from the 2-d, first-order, locally linear, Gaussian MRF prior on z (6.3). As with Figure 6.15, the data are independently scattered about the true process (left plot) according to an $N(0, 4)$ distribution.

6.3.2 General Gaussian MRFs

The locally linear GMRF specification is applicable in a wide range of applications. However, the details of a particular application may require more structure than this prior can deliver. A more general form for a GMRF is

$$\pi(z) \propto |W|^{\frac{1}{2}} \exp\left\{ -\tfrac{1}{2} z^T W z \right\},$$

where W is any symmetric, positive (semi) definite matrix. This model has full conditional distributions

$$z_i | z_{-i} \sim N\left(-\frac{1}{W_{ii}} \sum_{\substack{j \neq i}}^{n} W_{ij} z_j, \frac{1}{W_{ii}} \right).$$

Specifying W determines the neighborhood structure of the system — $W_{ij} \neq 0$ ⇔ sites i and j are neighbors. For details regarding how one might construct alternative precision matrices, see Besag et al. (1995), Besag and Kooperberg (1995), and Rue and Tjelmeland (2002).

It is worth pointing out that the locally linear GMRF specification results in an *intrinsic* prior distribution for z — that is, the model is not proper because $\pi(z) = \pi(z + c\mathbf{1})$ for any constant c (of course, the resulting posterior is proper). A proper alternative to the locally linear GMRF is one for which the locally linear precision W_z is replaced by $W_z + \delta I$. Figure 6.17 shows realizations from the stationary ($\delta = 0.1$) and intrinsic versions of the locally linear prior. These realizations are scaled so that the increments

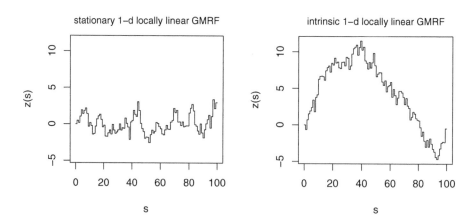

Figure 6.17 Realizations from stationary (left) and intrinsic (right) locally linear Gaussian MRF priors. The stationary prior has mean 0 and precision proportional to $W + 0.1 \cdot I$. The draws are conditional on $z_1 = 0$. The stationary realization is scaled so that the expected squared increment size $E(z_i - z_{i+1})^2$ is the same in both plots.

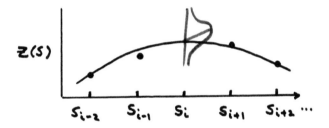

Figure 6.18 The locally quadratic Gaussian MRF specification defines the precision matrix W so that $E(z_i|z_{\partial i})$ is the fit of the parabola estimated from the four pairs (s_j, z_j), $j = i - 2, i - 1, i + 1, i + 2$.

between adjacent sizes are similar. Although the two realizations look similar if one only compares increment magnitude, their global nature is clearly quite different.

As an example of a possible GMRF specification, we take the locally quadratic prior. Here the elements of W are chosen so that $E(z_i|z_{\partial i})$ is the best quadratic fit at s_i given $z_{\partial i}$ (see Figure 6.18). The four pairs of points (s_{i-2}, z_{i-2}), (s_{i-1}, z_{i-1}), (s_{i+1}, z_{i+1}), and (s_{i+2}, z_{i+2}) are used to estimate a parabola of the form $\hat{z}(s) = b_1 s + b_2 s^2$. The fit at site s_i is just a linear combination of the four neighboring z_j's. The resulting precision matrix is then band diagonal with 6's along the diagonal, -4's on the one-off diagonals, and 1's on the two off diagonals. The precision matrix needs to be modified to account for edge sites. Note that the locally linear GMRF specification defines W so that $E(z_i|z_{\partial i})$ is the fit of the line estimated from the two pairs (s_j, z_j), $j = i - 1, i + 1$. Later in this section some comparisons between posteriors obtained under the locally quadratic and locally linear prior formulations are made.

Finally, up to now, the GMRF priors depend only on the precision matrix W_z. In any real application, it is useful to allow for a parameter λ_z to scale this precision matrix. From now on, a GMRF prior for z will typically include the precision scalar so that

$$\pi(z|\lambda_z) \propto \lambda_z^{\frac{n}{2}} \exp\left\{ -\tfrac{1}{2}\lambda_z z^T W_z z\right\}. \tag{6.4}$$

The previous examples were fortuitously scaled so that $\lambda_z = 1$ was a good choice. The precision parameter λ_z controls the regularity of the process $z(s)$. Note the exponent in the term $\lambda_z^{\frac{n}{2}}$ in (6.4) is used here. More generally, it may be preferable to use rank(W) rather than n because W is often not of full rank. We use n throughout this primer. See Knorr-Held (2003) or Hodges et al. (2003) for additional discussion.

6.3.3 Computing with large systems

Recall that in each of the examples in the previous section the resulting posterior was of the form:

$$z|y \sim N(\Lambda^{-1} W_y y, \Lambda^{-1}), \qquad (6.5)$$

where $\Lambda = W_y + \lambda_z W_z$ and λ_z scales the spatial process z. In cases where n is quite large, computing the posterior mean $\Lambda^{-1} W_y y$ can be prohibitive. Here the local nature of the MRF specification can be quite helpful.

The full conditionals of the density implied by (6.5) are

$$\pi(z_i|z_{\partial i}, y) \propto \exp\left\{ -\frac{1}{2}\lambda_y(y_i - z_i)^2 - \frac{1}{2}\lambda_z W_{ii} z_i^2 - \lambda_z \sum_{j \in \partial i} W_{ij} z_i z_j \right\},$$

which implies

$$z_i|z_{\partial i}, y \sim N\left(\frac{\lambda_y y_i + \lambda_z \sum_{j \in \partial i} W_{ij} z_j}{\lambda_y + \lambda_z W_{ii}}, \frac{1}{\lambda_y + \lambda_z W_{ii}} \right).$$

In the case of the locally linear GMRF, this simplifies to

$$z_i|z_{\partial i}, y \sim N\left(\frac{\lambda_y y_i + n_i \lambda_z \bar{z}_{\partial i}}{\lambda_y + n_i \lambda_z}, \frac{1}{\lambda_y + n_i \lambda_z} \right),$$

where

$$n_i = \# \text{ of neighbors belonging to } s_i$$
$$\bar{z}_{\partial i} = \frac{1}{n_i} \sum_{j \in \partial i} z_j.$$

6.3.3.1 Iterated conditional modes

The method of iterated conditional modes (ICM) (Besag, 1986) uses the univariate full conditional distributions in an iterative scheme to determine the mode of the posterior distribution $\pi(z|y)$. The algorithm iteratively replaces each z_i by the mode of its full conditional:

for $t = 1, \ldots, \mathsf{nscan}$ {
    for $i = 1, \ldots, n$ {

$$\mathsf{set}\ z_i = \frac{\lambda_y y_i + n_i \lambda_z \bar{z}_{\partial i}}{\lambda_y + n_i \lambda_z}$$

    }
}.

Here, the newly updated z_js should be used when computing $\bar{z}_{\partial i}$.

In general, the eventual value $\hat{z}_{ICM}$ will correspond to a local mode of the posterior density $\pi(z|y)$. If $\pi(z|y)$ is Gaussian (as it is here), $\hat{z}_{ICM}$ will

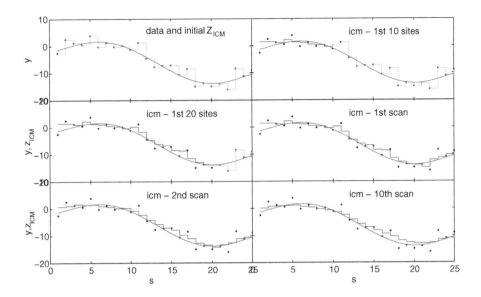

Figure 6.19 ICM applied to a 1-d example where $y = z + \epsilon$ and z is given a locally linear GMRF prior so that the resulting posterior distribution is given by (6.5). By the tenth scan through the data, the ICM solution is reached.

correspond to the posterior mean, which is the global optimum of $\pi(z|y)$. Hence, ICM iteratively solves the posterior mean equation $\hat{z}_{ICM} = \Lambda^{-1} W_y y$.

Figure 6.19 shows ICM being carried out on a 1-d example. The spatial process z is initialized at the observed data y. By the tenth scan through the vector z, the posterior mean solution is reached.

6.3.3.2 Gibbs sampling

While ICM allows one to compute the posterior mean, Markov chain Monte Carlo (MCMC) gives an approach for generating realizations from the posterior distribution. With realizations, one can explore many features of $\pi(z|y)$ — in particular, uncertainties regarding z can be assessed.

The *Gibbs sampler* is a MCMC scheme that iteratively replaces each z_i by a draw from its full conditional:

for $t = 1, \dots, \text{nscan}$ {
    for $i = 1, \dots, n$ {

$$\text{draw } z_i \sim N\left(\frac{\lambda_y y_i + n_i \lambda_z \bar{z}_{\partial i}}{\lambda_y + n_i \lambda_z}, \frac{1}{\lambda_y + n_i \lambda_z} \right)$$

    }
}.

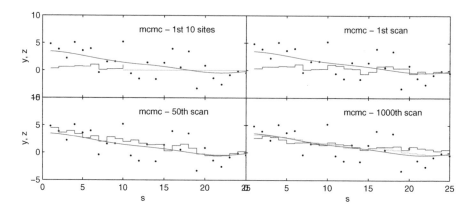

Figure 6.20 Gibbs sampling applied to the posterior distribution (6.5) resulting from the data y (denoted by the dots) and a locally linear Gaussian MRF specification for the underlying spatial process z. The data are generated from the model $y = z + \epsilon$, where the true value for z is given by the smooth line. z is initialized at 0. The light, stepped line in the bottom right plot shows the posterior mean estimated from the 1000 MCMC realizations.

Figure 6.20 shows a 1-d example of sampling from the posterior resulting from a locally linear GMRF prior for z over 25 regularly spaced sites. The unknown spatial process z is initialized at 0. The MCMC continues, giving a dependent sequence of T realizations $z^1, \ldots, z^T$ that can be treated as draws from the posterior distribution $\pi(z|y)$. Of course, the early realizations will be influenced by the initialization of z. Hence it is sensible to discard these early MCMC draws until the sequence has "forgotten" the initialization value of z.

6.3.4 Accounting for unknown model parameters

In most applications, there are various nuisance parameters that need to be incorporated into the model. In the examples of this section, such parameters might include the precision of the observation measurements and the scaling of the GMRF prior for $z(s)$.

Assume the data y are formed corresponding to the model

$$y = z + \epsilon \quad \text{where } \epsilon \sim N\left(0, \frac{1}{\lambda_y} I_n\right).$$

Hence, the likelihood

$$L(y|z, \lambda_y) \propto \lambda_y^{\frac{n}{2}} \exp\left\{-\tfrac{1}{2}\lambda_y (y - z)^T (y - z)\right\}$$

has the precision parameter λ_y that describes the size of the observation errors. Similarly, the GMRF prior for z has the form

$$\pi(z|\lambda_z) \propto \lambda_z^{\frac{n}{2}} \exp\{-\tfrac{1}{2}\lambda_z z^T W_z z\}$$

where the precision λ_z controls the regularity in the spatial process z.

To account for uncertainty regarding these precision parameters, uninformative, conjugate priors are specified them:

$$\pi(\lambda_y) \propto \lambda_y^{a_y-1} \exp\{-b_y\lambda_y\};$$
$$\pi(\lambda_z) \propto \lambda_z^{a_z-1} \exp\{-b_z\lambda_z\}.$$

Typically, $\pi(\lambda_y)$ and $\pi(\lambda_z)$ are chosen to be rather uninformative. The choice of $a_y = a_z = 1$ and $b_y = b_z = 0.0001$ will usually work well. If λ is expected to be very large, then a smaller value b will be necessary. Note that using a value of a_z very close to 0 places a lot of prior mass for λ_z near 0. This can lead to spurious posterior results (Gelman, 2005).

The resulting posterior distribution is proportional to the product of the likelihood and prior

$$\pi(z, \lambda_y, \lambda_z|y) \propto L(y|z, \lambda_y) \times \pi(z|\lambda_z) \times \pi(\lambda_y) \times \pi(\lambda_z) \qquad (6.6)$$
$$\propto \lambda_y^{\frac{n}{2}} \exp\{-\tfrac{1}{2}\lambda_y(y - z)^T(y - z)\}$$
$$\times \lambda_z^{\frac{n}{2}} \exp\{-\tfrac{1}{2}\lambda_z z^T W_z z\}$$
$$\times \lambda_y^{a_y-1} \exp\{-b_y\lambda_y\} \times \lambda_z^{a_z-1} \exp\{-b_z\lambda_z\},$$

which describes uncertainty regarding $(z, \lambda_y, \lambda_z)$.

The full conditional densities of $\pi(z, \lambda_y, \lambda_z|y)$ are then

$$\pi(z|\lambda_y, \lambda_z, y) \propto \exp\left\{ -\tfrac{1}{2}\lambda_y(y - z)^T(y - z) - \tfrac{1}{2}\lambda_z z^T W_z z\right\}$$
$$\pi(\lambda_y|z, \lambda_z, y) \propto \lambda_y^{n/2+a_y-1} \exp\left\{ -[b_y + .5(y - z)^T(y - z)]\lambda_y\right\}$$
$$\pi(\lambda_z|z, \lambda_y, y) \propto \lambda_z^{n/2+a_z-1} \exp\left\{ -(b_z + .5z^T W_z z)\lambda_z\right\}.$$

Each of these full conditionals corresponds to a standard distribution — normal for z and gamma for the precisions.

$$z|\lambda_y, \lambda_z, y \sim N(\Lambda^{-1}\lambda_y y, \Lambda^{-1}), \quad \text{where } \Lambda = \lambda_y I_n + \lambda_z W_z$$
$$\lambda_y|z, \lambda_z, y \sim \Gamma(n/2 + a_y, b_y + .5(y - z)^T(y - z))$$
$$\lambda_z|z, \lambda_y, y \sim \Gamma(n/2 + a_z, b_z + .5z^T W_z z)$$

For z, one can also determine the full conditionals corresponding to each component:

$$z_i|z_{-i}, \lambda_y, \lambda_z, y \sim N\left(\frac{\lambda_y y_i + n_i \lambda_z \bar{z}_{\partial i}}{\lambda_y + n_i \lambda_z}, \frac{1}{\lambda_y + n_i \lambda_z}\right), \quad i = 1, \ldots, n.$$

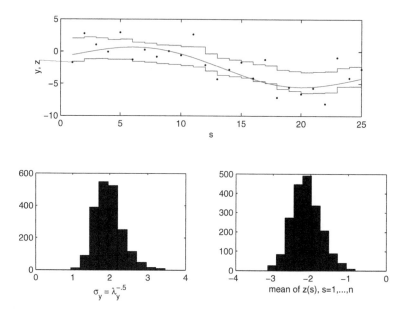

Figure 6.21 Summaries of the Gibbs sampling output produced from sampling the posterior distribution (6.6). Top: pointwise 90% credible intervals for z. Bottom left: a histogram of draws of the error standard deviation obtained by simply taking $\lambda_y^{-1/2}$ at each iteration. Bottom right: the posterior distribution of the average of z.

The Gibbs sampler implementation of MCMC for $\pi(z, \lambda_y, \lambda_z|y)$ simply initializes $(z, \lambda_y, \lambda_z)$ and then iteratively samples from each of the full conditionals. For the results shown in Figure 6.21, z was initialized at $\frac{1}{2}y + \frac{1}{2}\bar{y}$, and the two precisions were set to draws from their full conditionals given the initial value for the vector z. The Gibbs sampler was run for $T = 2000$ scans through each of the parameters, and the first 500 realizations were discarded to remove the effect of the initialization.

Figure 6.21 shows some summaries of the MCMC output produced by this Gibbs sampler: pointwise 90% credible intervals for each component of z; the posterior distribution for $\sigma_y = \lambda_y^{-\frac{1}{2}}$; and the posterior distribution of $\bar{z} = \frac{1}{n}\sum_{i=1}^{n} z_i$. Marginal distributions are simply estimated with histograms of the MCMC output. This example highlights how easy it is to infer about functions of z. Here we obtained the posterior distribution of the spatial mean of z. One could just as easily infer about other quantities such as the max of z, or the spatial location where the max of z is reached.

Figures 6.22 and 6.23 show posterior 80% credible intervals for z under two different formulations — the locally linear and the locally quadratic GMRFs. The spatial process resides on the integers from 1 to 50. Data were generated

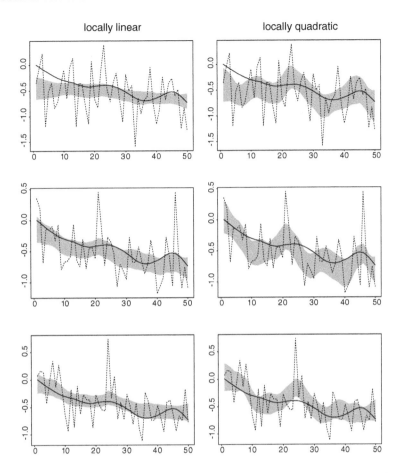

Figure 6.22 Posterior reconstructions under the locally linear and locally quadratic GMRF formulations over sites $\{1, 2, \ldots, 50\}$. In each row data (dotted lines) are generated according to the model $y = z + \epsilon$ where the true z is given by the solid line. Pointwise posterior 80% intervals are computed for z and shown by the gray regions in the plots. The left column shows credible intervals resulting from the locally linear formulation. The right column shows credible intervals resulting from the locally quadratic formulation.

from the model $y = z + \epsilon$ where the true value of z is given by the solid line in the figures, and $\epsilon \sim N(0, \frac{1}{10}I_{50})$. The precision parameters λ_y and λ_z are treated as unknown and assigned $\Gamma(1, .0001)$ priors. Each row in the figures corresponds to a different realization for ϵ. The dotted lines connect the resulting data points. Generally, the locally linear formulation does a better job reconstructing z and more accurately estimates the error precision λ_y given the data in Figure 6.22. For the sine wave data in Figure 6.23, the locally quadratic prior leads to more faithful reconstructions.

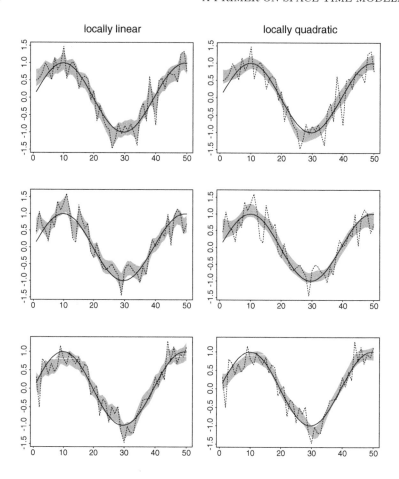

Figure 6.23 Posterior reconstructions under the locally linear and locally quadratic GMRF formulations over sites $\{1, 2, \ldots, 50\}$. In each row data (dotted lines) are generated according to the model $y = z + \epsilon$, where the true z is given by the solid line. Pointwise posterior 80% intervals are computed for z and shown by the gray regions in the plots. The left column shows credible intervals resulting from the locally linear formulation. The right column shows credible intervals resulting from the locally quadratic formulation.

6.4 Convolution-based spatial models

GMRF models work well for image and lattice data; however, when data are irregularly spaced, a continuous model for the spatial process $z(s)$ is usually preferable. In this section, convolution — or, equivalently, kernel — models are introduced. These models construct a continuous spatial model $z(s)$ by smoothing out a simple, regularly spaced latent process. In some cases, a GMRF model is used for this latent process.

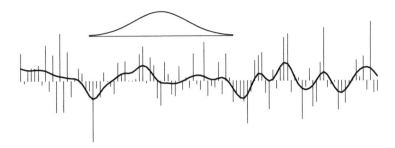

Figure 6.24 A one-dimensional Gaussian process obtained from smoothed white noise.

The convolution process $z(s)$ is determined by specifying a latent process $x(s)$ and a smoothing kernel $k(s)$. We restrict the latent process $x(s)$ to be nonzero at the fixed spatial sites $\omega_1, \ldots, \omega_m$, also in $\mathcal{S}$, and define $x = (x_1, \ldots, x_m)^T$ where $x_j = x(\omega_j)$, $j = 1, \ldots, m$. For now, the x_j's are modeled as independent draws from a $N(0, 1/\lambda_x)$ distribution. The resulting continuous Gaussian process is then

$$z(s) = \int_{\mathcal{S}} k(u - s) dx(u)$$
$$= \sum_{j=1}^{m} k(\omega_j - s) x_j \qquad (6.7)$$

where $k(\omega_j - \cdot)$ is a kernel centered at ω_j. This convolution model is depicted in Figure 6.24. Typically, $k(s)$ is taken to be a normal kernel centered at 0, with an sd σ_k.

As long as $k(s)$ is symmetric, the same representation for $z(s)$ could be explained as a basis construction. Define the m basis functions $k_1(s), \ldots, k_m(s)$, where $k_j(s) = k(s - \omega_j)$ as shown in Figure 6.25. Here the $k_j(s)$'s are normal densities centered at spatial locations ω_j; the ω_j's are shown by the dots in Figure 6.25. The standard deviation of the normal density σ_k is set equal to the spacing between adjacent ω_j's. The continuous spatial process $z(s)$, $s \in \mathcal{S}$, with $\mathcal{S} = [0, 10]$ is defined as

$$z(s) = \sum_{j=1}^{m} k_j(s) x_j \quad \text{where } x \sim N(0, I_m). \qquad (6.8)$$

Note that the basis representation in (6.8) is the foundation of a number of kernel and basis approaches for fitting data. See Hastie et al. (2001) for an overview of these methods.

If our interest is restricted to a finite set of n spatial locations $s_1, \ldots, s_n$, the n-vector $z = (z(s_1), \ldots, z(s_n))^T$ is given by

$$z = Kx \quad \text{where} \quad K_{ij} = k_j(s_i). \qquad (6.9)$$

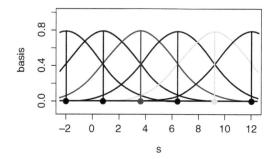

Figure 6.25 Basis functions to construct a 1-d spatial process. Here the $m = 6$ basis functions $k_j(s)$ are normal kernels centered at spatial locations w_j given by the dots. The sd of the kernels is equal to the spacing of the w_j's.

Frequently, applications require the value of $z(s)$ over a grid of spatial locations. Hence, the discrete representation above can be quite useful.

With either interpretation, convolution or basis, the spatial process $z(s)$ is determined by the distribution of x, the latent sites $w_1, \ldots, w_m$, and the kernel $k(s)$. Figure 6.26 shows how $z(s)$ is constructed from a random draw of the $m = 6$-vector x according to (6.7).

Continuing with this 1-d example, if we restrict $z(s)$ to a fine grid of n spatial locations $s_1, \ldots, s_n$ between 0 and 10, the discrete representation (6.9) leads to the result that

$$z \sim N(0, KK^T).$$

Here, K is a $n \times m$ matrix, so that $\Sigma_z = KK^T$ is $n \times n$. Figure 6.27 shows the resulting covariance matrix for z as the number of points m in the latent

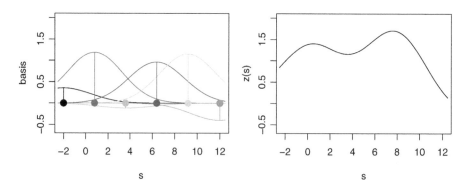

Figure 6.26 A spatial process realization $z(s)$ constructed from a basis representation. The basis kernels, weighted by a standard normal draw, are shown in the left plot. The resulting process $z(s)$ is the sum of these weighted kernels.

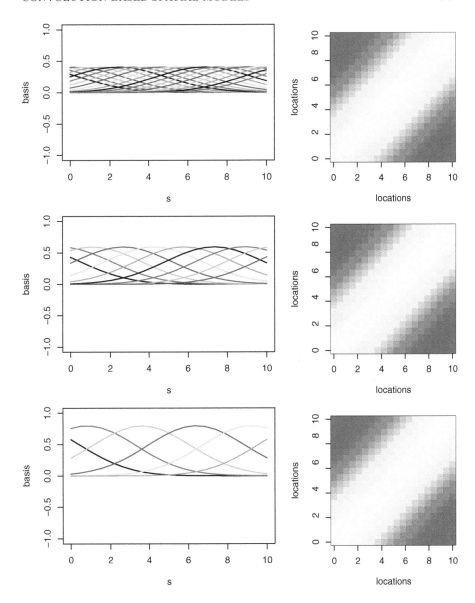

Figure 6.27 Basis kernels and covariance for the induced spatial process $z(s)$ as the dimension m of the latent process x varies. Here the covariance of z is nearly identical for each value of m. Top: $m = 20$; middle: $m = 10$; bottom: $m = 6$.

process x varies. As m increases, the kernel is scaled down to keep the marginal variance of $z(s)$ constant. Hence if $k(s)$ denotes the kernel when $m = 6$, we use $\frac{6}{10}k(s)$ when $m = 10$ and $\frac{6}{20}k(s)$ when $m = 20$. In the top plot, $\omega_1, \ldots, \omega_m$ consist of 20 equally spaced locations between -2 and 12. In the middle plot, $\omega_1, \ldots, \omega_m$ consist of 10 equally spaced locations between -2 and 12; in the bottom plot, it is $m = 6$. The resulting covariance in each case is nearly identical. This means that for any of these constructions, the induced model for the spatial process $z(s)$ is just about the same. Hence, for this particular choice of kernel, m need only be as large as 6.

This convolution — or kernel — based construction of a spatial process $z(s)$ is only appropriate when a smooth representation of $z(s)$ is OK. Using a very peaked kernel ($k(s) = |s|^{-.5} \exp\{-|s|\}$, for example) will require a very dense set of support locations $\omega_1, \ldots, \omega_m$, which will make this approach impractical. Smooth kernels that have been used include the normal density and the tricube (Cleveland, 1979), which has the feature of bounded support. For a given kernel width, one would like the spacing of the ω_j's to be as spread out as possible for computational savings. However, making the spacings too large can lead to unwanted "dead regions" in $z(s)$ where the process moves to 0. For the normal kernel in one or two dimensions, spacing the ω_j's no wider than the kernel sd works fine. Finally, the width of the smoothing kernel $k(s)$ controls the nature of $z(s)$ in exactly the same way the distance scaling in the covariance rule does in Section 6.2.

6.4.1 Modeling and estimation

The appeal of these convolution representations for $z(s)$ is their flexibility as a component of more general model formulations. As an example, consider recording data $y = (y(s_1), \ldots, y(s_n))^T$ at spatial locations $s_1, \ldots, s_n$. Once knot locations ω_j, $j = 1, \ldots, m$, and kernel choice $k(s)$ are specified, the remaining model formulation is trivial. Assuming $y(s_i)$ is equal to the spatial process $z(s_i)$ plus an independent measurement error results in the likelihood

$$L(y|x, \lambda_y) \propto \lambda_y^{\frac{n}{2}} \exp\left\{ -\tfrac{1}{2}\lambda_y(y - Kx)^T(y - Kx)\right\}$$

where $K_{ij} = k(\omega_j - s_i)$. For priors we have

$$\pi(x|\lambda_x) \propto \lambda_x^{\frac{m}{2}} \exp\left\{-\tfrac{1}{2}\lambda_x x^T x\right\}$$

$$\pi(\lambda_x) \propto \lambda_x^{a_x - 1} \exp\{-b_x \lambda_x\}$$

$$\pi(\lambda_y) \propto \lambda_y^{a_y - 1} \exp\{-b_y \lambda_y\}.$$

This results in the posterior distribution

$$\pi(x, \lambda_x, \lambda_y|y) \propto \lambda_y^{a_y + \frac{n}{2} - 1} \exp\left\{-\lambda_y[b_y + .5(y - Kx)^T(y - Kx)]\right\}$$

$$\times \lambda_x^{a_x + \frac{m}{2} - 1} \exp\left\{-\lambda_x[b_x + .5x^T x]\right\}.$$

The full conditionals then have the basic forms:

$$\pi(x|\cdots) \propto \exp\left\{-\tfrac{1}{2}[\lambda_y x^T K^T K x - 2\lambda_y x^T K^T y + \lambda_x x^T x]\right\}$$

$$\pi(\lambda_x|\cdots) \propto \lambda_x^{a_x+\frac{m}{2}-1} \exp\left\{-\lambda_x[b_x + .5x^T x]\right\}$$

$$\pi(\lambda_y|\cdots) \propto \lambda_y^{a_y+\frac{n}{2}-1} \exp\left\{-\lambda_y[b_y + .5(y - Kx)^T(y - Kx)]\right\}.$$

These densities are recognized as standard forms so that a Gibbs sampler can be easily implemented by cycling through the parameters and making the draws below.

$$x|\cdots \sim N((\lambda_y K^T K + \lambda_x I_m)^{-1}\lambda_y K^T y, (\lambda_y K^T K + \lambda_x I_m)^{-1})$$

$$\lambda_x|\cdots \sim \Gamma(a_x + \tfrac{m}{2}, b_x + .5x^T x)$$

$$\lambda_y|\cdots \sim \Gamma(a_y + \tfrac{n}{2}, b_y + .5(y - Kx)^T(y - Kx))$$

As an example, we use $n = 18$ data points from spatial locations evenly spaced between 0 and 10 as shown in Figure 6.28. The prior model for $z(s)$ is constructed using $m = 20$ knot locations evenly spaced between -2 and 12. Finally, the kernel $k(s)$ is a normal density with an sd of 2. The resulting posterior estimate of $z(s)$ is shown in Figure 6.28, along with the marginal posterior distributions for λ_y and λ_x. The subsequent figure — Figure 6.29 — shows the same posterior summaries using the same data and an identical formulation, except that the kernel sd is 1 — half as wide as the one used for Figure 6.28. The true value for λ_y is 15, which is consistent with the formulation with the wider kernel. Note the narrow kernel used in the second formulation leads to a very wiggly estimate of $z(s)$ and a very large error precision λ_y. Hence the estimated spatial process nearly interpolates the data.

Clearly, the kernel width can have a dramatic effect on the analysis. This width parameter could be formally incorporated into the analysis by assigning it a prior and including it in the resulting MCMC. This will typically lead to a substantial amount of additional computation. Another alternative is to use a cross-validation approach to settle on the kernel width. A couple of alternatives to these well-established approaches for selecting the kernel width are discussed later in this section. The first is to modify the prior distribution on the latent process x; the second is to use more than one kernel in the basis construction.

6.4.2 Using a MRF model for x

If we take the formulation described in Section 6.4.1 and replace the iid prior for x with a GMRF prior

$$\pi(x|\lambda_x) \propto \lambda_x^{\frac{m}{2}} \exp\left\{-\tfrac{1}{2}\lambda_x x^T W x\right\},$$

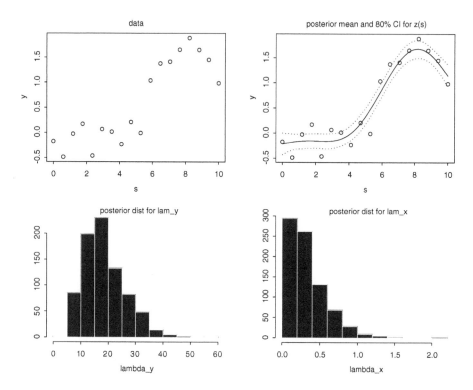

Figure 6.28 Data and posterior estimates. Top left: data y for estimating the underlying spatial process $z(s)$. Top right: posterior mean and pointwise 80% credible intervals for $z(s)$. Bottom: histograms of the posterior realizations for λ_y (left) and λ_x (right). Here the kernel was normal with an sd of 2.

we get the posterior

$$\pi(x, \lambda_x, \lambda_y | y) \propto \lambda_y^{a_y + \frac{n}{2} - 1} \exp\left\{-\lambda_y[b_y + .5(y - Kx)^T(y - Kx)]\right\}$$
$$\times \lambda_x^{a_x + \frac{m}{2} - 1} \exp\left\{-\lambda_x[b_x + .5x^T W x]\right\}.$$

The resulting Gibbs sampler cycles through the steps

$$x | \cdots \sim N((\lambda_y K^T K + \lambda_x W)^{-1} \lambda_y K^T y, (\lambda_y K^T K + \lambda_x W)^{-1})$$
$$\lambda_x | \cdots \sim \Gamma(a_x + \tfrac{m}{2}, b_x + .5 x^T x)$$
$$\lambda_y | \cdots \sim \Gamma(a_y + \tfrac{n}{2}, b_y + .5(y - Kx)^T(y - Kx)),$$

which are no more difficult than those from the original independent x formulation.

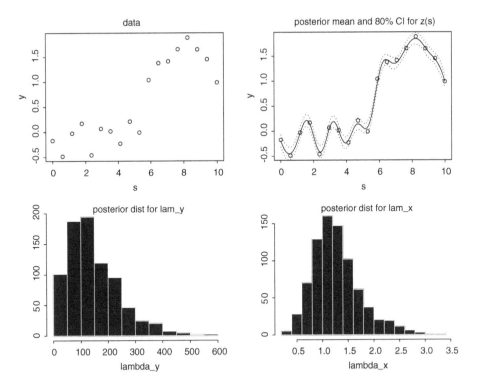

Figure 6.29 Data and posterior estimates. Top left: data y for estimating the underlying spatial process $z(s)$. Top right: posterior mean and pointwise 80% credible intervals for $z(s)$. Bottom: histograms of the posterior realizations for λ_y (left) and λ_x (right). Here the kernel was normal with an sd of 1 — half of that used in the previous figure.

The advantage of allowing dependence in the latent process is that the induced prior distribution for $z(s)$ is now

$$\pi(z) \propto \lambda_x^{\frac{m}{2}} \exp\{-\tfrac{1}{2}\lambda_x z^T KWK^T z\}.$$

This is an intrinsic process and λ_x controls the regularity in $z(s)$ in a way similar to kernel width. To see this, consider the posterior analyses shown in Figure 6.30.

Here the data are $n = 12$ observations from spatial locations evenly spaced between 0 and 10 as shown by the circle plotting symbols in Figure 6.30. The data were generated by adding white noise to a spatial process $z(s)$ that was constructed using $m = 20$ knot locations evenly spaced between -2 and 12, and a normal kernel $k(s)$ with an sd of 2.5.

The top plot is the result from the formulation using the iid prior for x along with a narrow kernel (sd $= .6$). The middle plot is the result from the

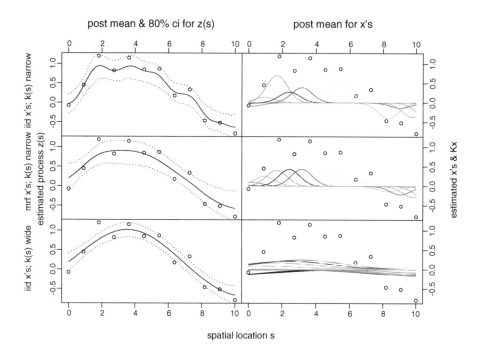

Figure 6.30 Posterior estimates of the underlying spatial process $z(s)$ under three different formulations. Top: iid prior for x with a narrow kernel (sd = .6); Middle: locally linear GMRF prior for x with a narrow kernel (sd = .6) Bottom: iid prior for x with a wide kernel (sd = 2.5). The data were generated by adding noise to a spatial process generated using the wide kernel. The GMRF formulation uses the dependence in x to overcome the overly narrow specification of the kernel width. The right hand column of plots shows the posterior mean estimate of $k(s - \omega_j)x_j$ under each of the formulations.

formulation using the locally linear GMRF prior for x along with the same narrow kernel. For the sake of comparison, the bottom plot gives the result from the formulation using the iid prior for x along with the correct, wide kernel. Clearly, the formulation corresponding to the top plot overfits the data because the kernel is too narrow (by more than a factor of 4). However, the GMRF formulation overcomes this misspecification of the kernel width by controlling the dependence between components of x. The smoothness in the posterior reconstruction of $z(s)$ is quite similar to that of the formulation in the bottom plot that uses the correct model for the true underlying process $z(s)$ from which the data were generated.

Note that if the smoothing kernel $k(s)$ is specified to be wider than the data warrant, there is no way the MRF prior on x can overcome this misspecification. For more details on such models, see Lee et al.(2005).

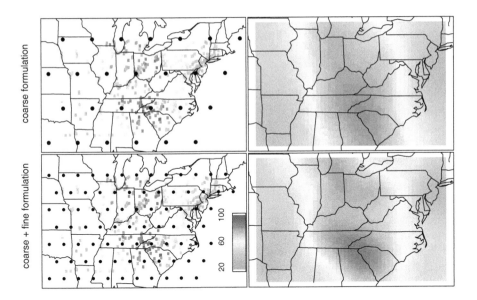

Figure 6.31 (**SEE COLOR INSERT FOLLOWING PAGE 142**) A multiresolution ozone model. The ozone measurements from a single summer day in the eastern U.S. (shown in both of the left hand plots) are modeled as the sum of two process convolution models, one coarse and one fine. The knot locations for the coarse model are shown in the top left plot; the knot locations for the fine process are shown in the bottom left plot. The resulting posterior mean for $z_{\text{coarse}}(s)$ is given in the top right plot; that of the multiresolution model $z(s) = z_{\text{coarse}}(s) + z_{\text{fine}}(s)$ is shown in the bottom right plot. The kernels for the two processes are normal, each of whose sd corresponds to their respective knot spacing.

6.4.3 A multiresolution example

With these tools, it is easy to specify a spatial model that is the sum of multiple processes, each with different spatial dependence properties. To be concrete, consider the ozone measurements taken on a summer day over the eastern U.S. shown in Figure 6.31. We can use a spatial process $z(s)$ to predict ozone measurements at sites where no monitors exist. Our multiresolution model decomposes the field into a coarse resolution component that varies slowly as a function of spatial distance, and a fine resolution component that changes more quickly with spatial distance:

$$z(s) = z_{\text{coarse}}(s) + z_{\text{fine}}(s).$$

Separate convolution priors are used for coarse and fine spatial model components. The coarse process uses $m_c = 27$ locations $\omega_1^c, \ldots, \omega_{m_c}^c$ on a hexagonal grid as shown in Figure 6.31. The latent process x_c is given an independent normal prior with precision λ_c:

$$x_c = (x_{c1}, \ldots, x_{cm_c})^T \sim N(0, \tfrac{1}{\lambda_c} I_{m_c}).$$

The coarse kernel $k_c(s)$ is normal with an sd that is equal to the grid spacing shown in the top left plot of Figure 6.31.

The fine process uses $m_f = 87$ locations $\omega_1^f, \ldots, \omega_{m_f}^f$ on a hexagonal grid as shown in Figure 6.31. The latent process x_f is given an independent normal prior with precision λ_f:

$$x_f = (x_{f1}, \ldots, x_{fm_f})^T \sim N(0, \tfrac{1}{\lambda_f} I_{m_f}).$$

The fine kernel $k_f(s)$ is normal with a sd that is equal to the grid spacing shown in the bottom left plot of Figure 6.31.

Now the data, which consist of $n = 510$ observations, can be modeled as a sum of the two spatial processes plus white noise:

$$y = K_c x_c + K_f x_f + \epsilon$$
$$= Kx + \epsilon$$

where

$$K = \begin{pmatrix} K_c K_f \end{pmatrix} \text{ and } x = \begin{pmatrix} x_c \\ x_f \end{pmatrix}.$$

If we define

$$W_x = \begin{pmatrix} \lambda_c I_{m_c} & 0 \\ 0 & \lambda_f I_{m_f} \end{pmatrix}$$

then this formulation is almost identical to previous specifications. Only now there are two precision parameters governing x. The resulting Gibbs sampler implementation is then

$$x| \cdots \sim N((\lambda_y K^T K + W_x)^{-1} \lambda_y K^T y, (\lambda_y K^T K + W_x)^{-1})$$
$$\lambda_c| \cdots \sim \Gamma(a_x + \tfrac{m_c}{2}, b_x + .5 x_c^T x_c)$$
$$\lambda_f| \cdots \sim \Gamma(a_x + \tfrac{m_f}{2}, b_x + .5 x_f^T x_f)$$
$$\lambda_y| \cdots \sim \Gamma(a_y + \tfrac{n}{2}, b_y + .5(y - Kx)^T (y - Kx)).$$

The posterior means for the coarse resolution process $z_{\text{coarse}}(s)$ and the multiresolution process $z(s)$ are shown in Figure 6.31. This model, which effectively uses two different kernel widths, can capture more complicated dependence structure in a spatial process $z(s)$. It also can result in a model with a single kernel width if the data support this by moving the posterior distribution for one of λ_c or λ_f to a very large value. This approach has links to wavelets (see Vidakovic, 1999) and could be modified to have more resolutions if the application calls for it.

data
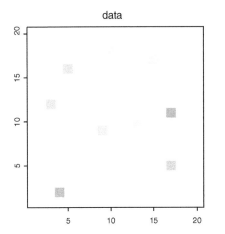

true field
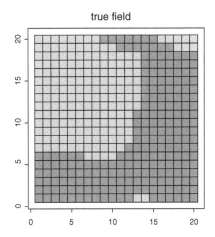

Figure 6.32 Observations from a binary field. The left plot shows $n = 9$ noisy measurements from the true binary field $z^*(s)$ shown on the right. The light denotes $z^*(s) = 1$; dark denotes $z^*(s) = 0$. The data points $y(s_i)$ are created by adding a $N(0,1)$ error to the true field $z^*(s_i)$.

6.4.4 A binary spatial model

Here we show how these convolution-based models are quite useful for modeling non-Gaussian spatial fields. Here we focus on a binary spatial model, although these ideas could be extended to other non-Gaussian fields as well. This binary model borrows from the ideas of De Oliviara (2000).

Consider an example shown in Figure 6.32 where there exists a binary field $z^*(s)$ over $\mathcal{S} = [0,20] \times [0,20]$. We obtain $n = 9$ noisy measurements $y = (y_1, \ldots, y_n)^T$ of this field at spatial locations $s_1, \ldots, s_n$ shown in the figure. The sampling model is

$$y_i = z^*(s_i) + \epsilon_i, \ i = 1, \ldots, n, \quad \text{where } \epsilon \sim N(0, I_n).$$

The binary field $z^*(s)$ is modeled as a Gaussian field $z(s)$, constructed via process convolution, and then assigning $z^*(s) = I[z(s) > 0]$ as shown in Figure 6.33. The knot locations are a 5×5 lattice shown in Figure 6.33 so that $m = 25$; the smoothing kernel is Gaussian and is also shown in the figure. Take z^* to denote the value of the binary process at the data locations $z^* = (z^*(s_1), \ldots, z^*(s_n))^T$ and recall that z^* is a deterministic function of the m-vector x.

The resulting posterior distribution for this formulation is simply the product of the likelihood

$$L(y|z^*) \propto \exp\{-\tfrac{1}{2}(y - z^*)^T (y - z^*)\}$$

and the independent normal prior for x

$$\pi(x) \propto \exp\{-\tfrac{1}{2}x^T x\}$$

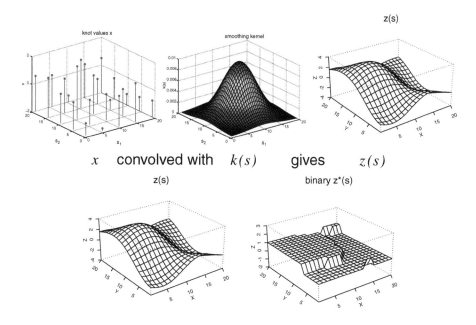

Figure 6.33 Constructing a binary field $z^*(s)$ by clipping a Gaussian field $z(s)$. The Gaussian process $z(s)$ is constructed by smoothing out a lattice of iid normal realizations x. This resulting field is then "clipped" so that $z^*(s) = I[z(s) > 0]$.

giving

$$\pi(x|y) \propto \exp\{-\tfrac{1}{2}(y - z^*)^T(y - z^*) - \tfrac{1}{2}x^T x\}.$$

The full conditional distributions resulting from this formulation are

$$\pi(x_j|x_{-j}, y) \propto \exp\{-\tfrac{1}{2}(y - z^*)^T(y - z^*) - \tfrac{1}{2}x_j^2\}, \quad j = 1, \ldots, m.$$

Because z^* is a nonlinear function of x, this density does not have a recognizable form. Hence a Gibbs sampler is not an easy option here. A simple alternative is to use Metropolis updating (Metropolis et al., 1953) to generate a MCMC sample from this posterior.

A Metropolis implementation for sampling from $\pi(x|y)$ can be carried out by first initializing x at 0 and then cycling through full conditionals updating each x_j in turn, according to the Metropolis rules:

- Generate proposal $x'_j \sim U[x_j - r, x_j + r]$.
- Compute acceptance probability

$$\alpha = \min\left\{1, \frac{\pi(x'_j|x_{-j}, y)}{\pi(x_j|x_{-j}, y)}\right\}.$$

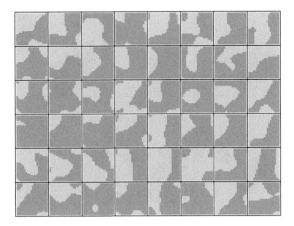

Figure 6.34 Posterior realizations of the binary process $z^*(s)$.

- Update x_j to new value:

$$x_j^{\text{new}} = \begin{cases} x_j' & \text{with probability } \alpha \\ x_j & \text{with probability } 1 - \alpha. \end{cases}$$

Here the sampler ran for $T = 1000$ scans, giving realizations $x^1, \ldots, x^T$ from the posterior. The output from the first 100 scans was discarded for burn-in. Note, we tuned the proposal width r so that the proposal x_j' is accepted about half the time.

A subset of posterior realizations for $z^*(s)$ are shown in Figure 6.34. There is clearly a wide range of configurations for $z^*(s)$ that are consistent with the data. The resulting posterior mean estimate for $z^*(s)$ is shown in Figure 6.35. The process convolution formulation for $z^*(s)$ works very well with the

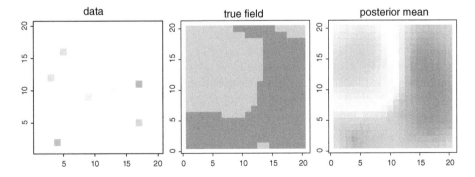

Figure 6.35 Data (left), true binary field $z^*(s)$ (middle), and posterior mean estimate of $z^*(s)$ (right). The posterior mean estimate is simply the average of the MCMC draws of $z^*(s)$.

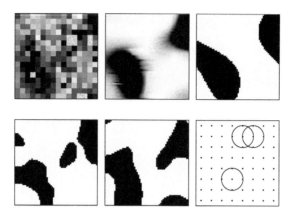

Figure 6.36 Data, posterior mean, posterior realizations, and knot locations for the archeology application of Besag et al. (1991). Here, the binary process is a "clipped" Gaussian field constructed from a process convolution model. The bottom right frame shows the knot locations and a circle that corresponds to 1 sd of the bivariate smoothing kernel $k(s)$.

Metropolis MCMC implementation. One reason for this is that the individual x_j's control just a local piece of the entire process $z^*(s)$. Hence updates to a x_j are effectively updates to a local region of $z^*(s)$.

6.4.4.1 Archeology application

This same model can be applied to the archeology application of Besag et al. (1991). In this application, as well as many other archeological investigations, enhanced soil phosphate content, due to decomposition of organic mater, is often found at known locations of archeological activity. In such areas, measurements of phosphate concentration can aid in determining specific sites of archeological activity.

The application centers on a square plot divided into a 16×16 grid of sites, each measuring 10 m $\times$ 10 meters; the layout and data are shown in Figure 6.36. As in the earlier analysis, we assume the existence of an underlying process, $z^*(s)$, with $z^*(s_i) = 1$ or 0, according to whether there is or is not previous activity at site s_i. The goal here is not only to produce a classification, but also to explore the posterior distribution of $z^*(s)|y$ and to determine posterior probability of previous activity at each site.

The sampling model is adapted from the original application and has the form:

$$y(s_i)|z^*(s_i) \sim N(\mu(z^*(s_i)), 1) \quad \text{where } \mu(0) = 1, \ \mu(1) = 2.$$

Recall $z(s) = \sum_{j=1}^{m} x_j k(s - \omega_j)$ and $z^*(s) = I[z(s) > 0]$. An $m = 8 \times 8$ lattice of knot locations $\omega_1, \ldots, \omega_m$ is used as shown in Figure 6.36. The sd of $k(s)$ is equal to the minimum knot spacings. A brief posterior summary for $z^*(s)$

Figure 6.37 Left: Nick Metropolis, a computing pioneer and inventor of the Monte Carlo method. Right: the MANIAC I computer.

is given in Figure 6.36, which shows posterior realizations of $z^*(s)$ along with its posterior mean.

6.4.4.1.1 Historical note on Metropolis

Nick Metropolis (pictured in Figure 6.37) was involved with the original Manhattan Project at Los Alamos during World War II. He, along with John von Neumann, were computing pioneers at Los Alamos. He is the inventor of the Monte Carlo method as well as MCMC, which was first carried out in 1953 on the MANIAC I computer (also pictured in Figure 6.37). This machine was the first computer that could carry out instructions based on previous computations. The MANIAC I also ran the first computer game — chess.

6.5 Convolution-based space-time models

The convolution-based models can easily be extended in two basic ways to model space-time processes. The first simply defines the latent process $x(s,t)$ to reside on knot values taken over a lattice in space and time, and then defines the kernel $k(s,t)$ to smooth over this augmented space. Mathematically, this approach is the same as that of the previous section, only an additional dimension is added. The second approach allows the latent spatial process to evolve over time, so that at any given time, it can be convolved with the spatial kernel $k(s)$ to give a spatial field $z(s,t)$ at time t. Both approaches will be demonstrated with an application.

6.5.1 A space-time convolution model

The most straightforward extension of the spatial convolution specification to space-time models extends the spatial domain to a space-time domain $\mathcal{S} \times \mathcal{T}$. The latent knot process $x(s,t)$ is defined over m points $\{(\omega_1, \tau_1), \ldots, (\omega_m, \tau_m)\}$

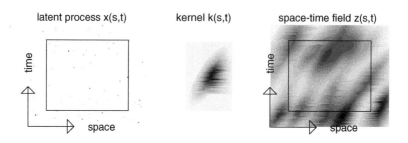

Figure 6.38 A convolution-based space-time model. Here, a space time-model $z(s,t)$ (right) is constructed by defining a latent process $x(s,t)$ over space and time (left) and convolving it with the kernel $k(s,t)$ (middle).

within $\mathcal{S} \times \mathcal{T}$. In addition, the smoothing kernel $k(s,t)$ is also defined over the space-time domain. The space-time process $z(s,t)$ is now defined as

$$z(s,t) = \int_{\mathcal{S} \times \mathcal{T}} k((\omega,\tau) - (s,t))dx(\omega,\tau)$$

$$= \sum_{j=1}^{m} k((\omega_j,\tau_j) - (s,t))x(\omega_j,\tau_j)$$

$$= \sum_{j=1}^{m} k_{st}(\omega_j,\tau_j)x_j$$

where $x_j = x(\omega_j, \tau_j)$, $j = 1, \ldots, m$. This constructive model is shown in Figure 6.38.

We now use this model to build an interpolative space time model $z(s,t)$ for ocean temperatures on a manifold of constant potential density. This application is taken from Higdon (1998). The data consist of $n = 3987$ temperature measurements taken between 1908 and 1988 shown in Figure 6.39. We take $y = (y_1, \ldots, y_n)^T$ to denote the recorded temperatures at irregularly sampled locations $(s_1, t_t), \ldots, (s_n, t_n)$. We center the data so that $\bar{y} = 0$. We assume a simple model for which the data are equal to a smooth process $z(s,t)$ plus white noise.

The smooth space-time process $z(s,t)$ is constructed by smoothing a latent space-time process x with a spatially varying kernel $k_s(s,t)$. The latent process resides on a regular grid over space and time. The spatial locations are shown in the right-hand plot of Figure 6.39. The kernel $k_{st}(\omega, \tau)$ at any given spatial location has the product form $k_s(\omega)k_t(\tau)$, where $k_s(\cdot)$ varies with spatial location and $k_t(\cdot)$ is a fixed, 1-d normal kernel with an sd of 7 years. This product form for $k_{st}(\omega, \tau)$ implies a separable covariance rule that is the product of a purely temporal covariance rule and a purely spatial covariance

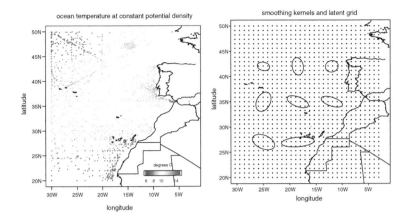

Figure 6.39 (SEE COLOR INSERT) Left: measured temperatures between 1908 and 1988 along a manifold of constant potential density, which corresponds to depths well below 1000 m. The data consist of 3987 measurements. Right: spatial locations of the latent space-time process $x(s, t)$. The time spacings are every 7 years. Also shown are ellipses that correspond to 1 sd of the spatially varying Gaussian smoothing kernels $k_s(\cdot)$.

rule. The spatially varying kernel $k_s(\cdot)$ is bivariate normal. The 1 sd ellipse is shown for eight spatial locations in Figure 6.39. Hence, the space-time process $z(s, t)$ is given by

$$z(s, t) = \sum_{j=1}^{m} k_{st}(\omega_j, \tau_j) x_j.$$

The details of estimating and specifying the kernel are given in Higdon (1998). The resulting formulation is summarized below.

Likelihood:

$$L(y|x, \lambda_y) \propto \lambda_y^{\frac{n}{2}} \exp\left\{-\tfrac{1}{2}\lambda_y (y - Kx)^T (y - Kx)\right\}$$

where $K_{ij} = k_{s_i t_i}(\omega_j, \tau_j)$.

Priors:

$$\pi(x|\lambda_x) \propto \lambda_x^{\frac{m}{2}} \exp\left\{-\tfrac{1}{2}\lambda_x x^T x\right\}$$

$$\pi(\lambda_x) \propto \lambda_x^{a_x - 1} \exp\{-b_x \lambda_x\}$$

$$\pi(\lambda_y) \propto \lambda_y^{a_y - 1} \exp\{-b_y \lambda_y\}$$

Posterior:

$$\pi(x, \lambda_x, \lambda_y | y) \propto \lambda_y^{a_y + \frac{n}{2} - 1} \exp\left\{-\lambda_y[b_y + .5(y - Kx)^T(y - Kx)]\right\}$$
$$\times \lambda_x^{a_x + \frac{m}{2} - 1} \exp\left\{-\lambda_x[b_x + .5x^T x]\right\}$$

This posterior is exactly in the same form as the analogous posterior from the purely spatial convolution formulations. The Gibbs sampler implementation is then

$$x | \cdots \sim N((\lambda_y K^T K + \lambda_x I_m)^{-1} \lambda_y K^T y, (\lambda_y K^T K + \lambda_x I_m)^{-1})$$

$$x_j | \cdots \sim N\left(\frac{\lambda_y r_j^T k_j}{\lambda_y k_j^T k_j + \lambda_x}, \frac{1}{\lambda_y k_j^T k_j + \lambda_x}\right)$$

$$\lambda_x | \cdots \sim \Gamma(a_x + \tfrac{m}{2}, b_x + .5x^T x)$$

$$\lambda_y | \cdots \sim \Gamma(a_y + \tfrac{n}{2}, b_y + .5(y - Kx)^T(y - Kx))$$

where k_j is the j-th column of K and $r_j = y - \sum_{j' \neq j} k_{j'} x_{j'}$. Note that given the large number of knots used here, the latent process will most likely need to be updated using the single site updating corresponding to the second line above. The resulting posterior mean for $z(s, t)$ is shown for selected times in Figure 6.40. For an alternative formulation of a space-time model for this data, see Stroud et al. (1999).

6.5.2 A spatial convolution of a temporally evolving latent process

Rather than rely on the kernel $k_{st}(\omega, \tau)$ to induce the temporal dependence in $z(s, t)$, an appealing alternative is to put the temporal dependence in the latent process $x(s, t)$. The space-time field $z(s, t)$ can then be obtained by a purely spatial convolution at time t. This is depicted in Figure 6.41. The advantage here is that the convolution is purely spatial, which can lead to substantial computational savings. Such an approach is typically practical only when observations come in at regular times. Here, the space-time process $z(s, t)$ is defined on the domain $S \times T$ where $T = 1, 2, \ldots, m_t$. The discrete latent process $x(\omega, t)$ is defined at space-time knot locations $\{\omega_1, \ldots, \omega_{m_s}\} \times \{1, \ldots, m_t\}$, where $m_t \cdot m_s = m$. Finally, a spatial smoothing kernel $k(s)$ is defined to smooth out the latent process separately for each time. More generally, a family of smoothing kernels $k_{st}(\omega)$ can be defined that varies with each space-time point (s, t). The space-time process is then constructed as defined below:

$$z(s, t) = \sum_{j=1}^{m_s} k(\omega_j - s) x(\omega_j, t).$$

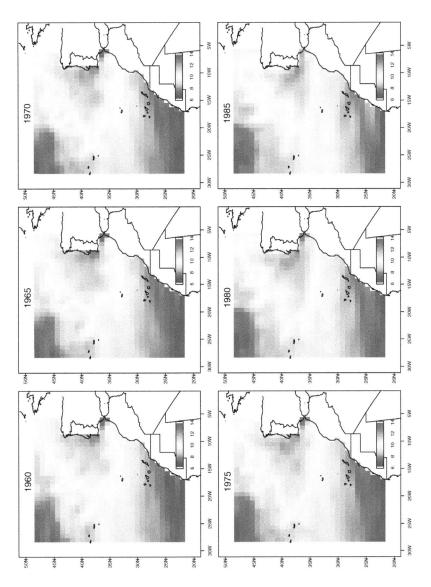

Figure 6.40 (**SEE COLOR INSERT**) Posterior mean estimate of the temperature surface $z(s, t)$ at six different times.

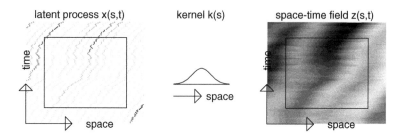

Figure 6.41 A convolution-based space-time model. Here, a space time-model $z(s,t)$ (right) is constructed by defining a latent process $x(s,t)$ that is evolving over time (left) and convolving it with a purely spatial kernel $k(s)$ (middle).

If a kernel that varies with space and time is used, the above construction generalizes to

$$z(s,t) = \sum_{j=1}^{m_s} k_{st}(\omega_j) x_{jt},$$

where x_{jt} is shorthand for $x(\omega_j, t)$. In the upcoming ozone example, a single, circular, bivariate Gaussian kernel is used for $k(s)$.

The spatial dependence is incorporated into the prior distribution for the latent process $x(\omega, t)$. Arguably, the simplest extension to the independence model is to define a locally linear GMRF prior for each $x_j = (x_{j1}, \ldots, x_{jm_t})^T$. Thus,

$$\pi(x_j | \lambda_x) \propto \lambda^{\frac{m_t}{2}} \exp\left\{-\tfrac{1}{2}\lambda_x x_j^T W x_j\right\},$$

where

$$W_{ij} = \begin{cases} -1 & \text{if } |i - j| = 1 \\ 1 & \text{if } i = j = 1 \text{ or } i = j = m_t \\ 2 & \text{if } 1 < i = j < m_t \\ 0 & \text{otherwise.} \end{cases}$$

So, for the entire latent process

$$x = (x_{11}, x_{21} \ldots, x_{m_s 1}, x_{12}, \ldots, x_{m_s 2}, \ldots, x_{1m_t}, \ldots, x_{m_s m_t})^T,$$

the prior density becomes

$$\pi(x | \lambda_x) \propto \lambda^{\frac{m_t m_s}{2}} \exp\left\{-\tfrac{1}{2}\lambda_x x^T (W \otimes I_{m_s}) x\right\}.$$

This formulation is applied to a sequence of 60 days' worth of ozone measurements — five of which are shown in Figure 6.42. Here the data $y = (y_1, \ldots, y_n)^T$ consist of $n_s = 510$ readings taken daily at spatial locations

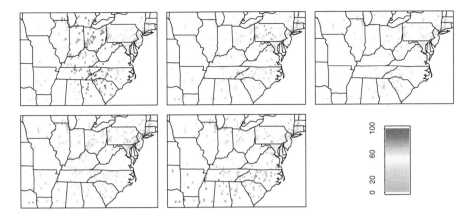

Figure 6.42 (**SEE COLOR INSERT**) Ozone concentrations taken for five consecutive summer days over the eastern United States. This is a subset of the 60 days of measurements used in the example.

$s_1, \ldots, s_{n_s}$, for $m_t = 60$ consecutive days, so that $n = 510 \cdot 60$. For a single day t, the likelihood for the n_s-vector of observations $y_t = (y_{1t}, \ldots, y_{nt})^T$ is

$$L(y_t | x_t, \lambda_y) \propto \lambda_y^{\frac{n}{2}} \exp\left\{-\tfrac{1}{2}\lambda_y(y_t - K^t x_t)^T (y - K^t x_t)\right\}$$

where $K_{ij}^t = k(\omega_j - s_i)$. For the entire set of n observations $y = (y_1^T, \ldots, y_{n_t}^T)^T$, the likelihood is

$$L(y | x, \lambda_y) \propto \lambda_y^{\frac{n}{2}} \exp\left\{-\tfrac{1}{2}\lambda_y(y - Kx)^T (y - Kx)\right\}$$

where $K = \text{diag}(K^1, \ldots, K^{n_t})$.

The prior for x is determined by the $m_s = 27$ knot locations and spatial smoothing kernel used in the coarse model formulation shown in the top left plot of Figure 6.31. For the duration of $m_t = 60$ days, the total number of knots is $m = m_s \cdot m_t = 27 \cdot 60$. So the full prior specification can be written

$$\pi(x | \lambda_x) \propto \lambda_x^{\frac{m}{2}} \exp\left\{-\tfrac{1}{2}\lambda_x x^T W x\right\}$$

$$\pi(\lambda_x) \propto \lambda_x^{a_x - 1} \exp\{-b_x \lambda_x\}$$

$$\pi(\lambda_y) \propto \lambda_y^{a_y - 1} \exp\{-b_y \lambda_y\}.$$

where $a_x = a_y = 1$ and $b_x = b_y = .001$. This results in the posterior density

$$\pi(x, \lambda_x, \lambda_y | y) \propto \lambda_y^{a_y + \frac{n}{2} - 1} \exp\left\{-\lambda_y[b_y + .5(y - Kx)^T (y - Kx)]\right\}$$

$$\times \lambda_x^{a_x + \frac{m}{2}} \exp\left\{-\lambda_x[b_x + .5x^T W x]\right\}.$$

The full conditionals densities are

$$\pi(x|\cdots) \propto \exp\{-\tfrac{1}{2}[\lambda_y x^T K^T K x - 2\lambda_y x^T K^T y + \lambda_x x^T W x]\}$$

$$\pi(\lambda_x|\cdots) \propto \lambda_x^{a_x+\frac{m}{2}-1} \exp\left\{-\lambda_x[b_x + .5x^T W x]\right\}$$

$$\pi(\lambda_y|\cdots) \propto \lambda_y^{a_y+\frac{n}{2}-1} \exp\left\{-\lambda_y[b_y + .5(y - Kx)^T(y - Kx)]\right\},$$

all of which have standard forms so that the Gibbs sampler implementation cycles through the draws

$$x|\cdots \sim N((\lambda_y K^T K + \lambda_x W)^{-1}\lambda_y K^T y, (\lambda_y K^T K + \lambda_x W)^{-1})$$

or

$$x_{jt}|\cdots \sim N\left(\frac{\lambda_y r_{tj}^T k_{tj} + n_j \bar{x}_{\partial j}}{\lambda_y k_{tj}^T k_{tj} + n_j \lambda_x}, \frac{1}{\lambda_y k_{tj}^T k_{tj} + n_j \lambda_x}\right)$$

$$\lambda_x|\cdots \sim \Gamma(a_x + \tfrac{m}{2}, b_x + .5x^T x)$$

$$\lambda_y|\cdots \sim \Gamma(a_y + \tfrac{n}{2}, b_y + .5(y - Kx)^T(y - Kx))$$

where k_{tj} is j-th column of K^t, $r_{tj} = y_t - \sum_{j'\neq j} k_{tj'} x_{tj'}$, n_j = number of neighbors of x_{jt}, and $\bar{x}_{\partial jt}$ is the mean of neighbors of x_{jt}. The full conditionals are for both the multivariate and single site updates of x. For large problems, the multivariate update of x is not feasible, while the univariate update of each x_{jt} will be simple regardless of how large m is. The posterior mean estimate for the space-time field $z(s,t)$, $t = 1, \ldots, 5$, is shown in Figure 6.43. Note that this particular formulation is amenable to the dynamic linear model framework of West and Harrison (1997), which can also ease the computational burden of exploring the posterior. Here the observation and evolution equations are

$$y_t = K^t x_t + \epsilon_t, \quad x_t = x_{t-1} + \varepsilon_t$$

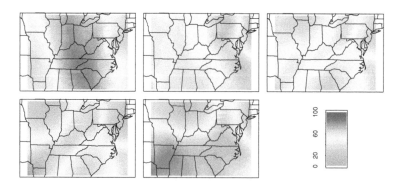

Figure 6.43 (SEE COLOR INSERT) Posterior mean field for ozone concentration for the 5 days shown in Figure 6.42.

where ϵ_t and ε_t are spatially and temporally uncorrelated. Updating x given the precisions can be carried out efficiently via forward filtering backward simulation.

6.6 Combining simulations and experimental observations

It is common that computer codes based on mathematical descriptions of the physical process over space and time exist to simulate the system of interest. Such simulations can account for rich space-time dependence without the need to cook up a stochastic description of the covariance in such a process. In this section we consider a very simple space-time system from Lorenz (1996). Given data generated from this model, we estimate the field at unobserved space-time locations using three different approaches: a Gaussian process model estimated from simulation output, a Bayesian inverse modeling approach, and an MRF model motivated by the partial differential equation (PDE) system itself.

Estimated fields derived from simulations of actual physical processes typically contain uncertainties due to:

- Observation noise,
- Uncertain initial/boundary conditions,
- Uncertain coefficients/parameters in the mathematical description, and
- Inadequate mathematical representation of the physical system being modeled.

Hence, given observations of this process, there will necessarily be uncertainty regarding what this process is at unobserved space-time locations.

6.6.1 The L96 model

We consider a deterministic, nonlinear system from Lorenz (1996) that is a continuous time model defined over discrete, periodic spatial locations $s \in \{1, 2, \ldots, 40\}$:

$$\frac{\partial}{\partial t} z(s, t) = [z(s + 1, t) - z(s - 2, t)] z(s - 1, t) - z(s, t) + F.$$

Given $z(s, t = 0)$, the system can be propagated forward via Euler integration:

$$z(s, t + dt) = z(s, t) + dt\{[z(s + 1, t) - z(s - 2, t)] z(s - 1, t) - z(s, t) + F\}$$

— of course, more sophisticated integration schemes will give more efficient simulations. To ensure stability of the simulation, we use time increments of $dt = .001$. The forcing term F is set to 8. Figure 6.44 shows realizations from this model given an initial field $z(s, t = 0)$ at $n_t = 51$ regularly spaced time increments of $\Delta t = .02$, starting at $t = 0$ and ending at $t = 1$.

We generate a "true" realization $z(s, t)$ by starting with an initial field $z(s, 0)$ at time $t = 0$ and propagating it until $t = 1$ is reached. We restrict

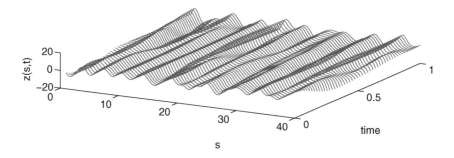

Figure 6.44 Output from the L96 model given an initial field $z(s,0)$ at regularly spaced time increments $t = 0, .02, \ldots, 1$.

ourselves to the space-time set $\mathcal{S} \times \mathcal{T} = \{1, 2, \ldots, 40\} \times \{0, .02, \ldots, 1\}$ so that the true value $z(s,t)$ can be thought of as an $n_s \times n_t = 40 \times 51$ image shown in Figure 6.45. From this ground truth, we obtain $n = 79$ noisy measurements

$$y_{d_i} = z(s_{d_i}, t_{d_i}) + \epsilon_{d_i}, \quad i = 1, \ldots, n$$

where the measurement error $\epsilon \sim N(0, \frac{1}{4} I_n)$. The measurement locations (s_{d_i}, t_{d_i}) are also shown in Figure 6.45. We assume that we have a good estimate of the error precision. Hence we either treat λ_y as known ($\lambda_y = 4$), or we use an informative $\Gamma(4, 1)$ prior for λ_y which has mean 4.

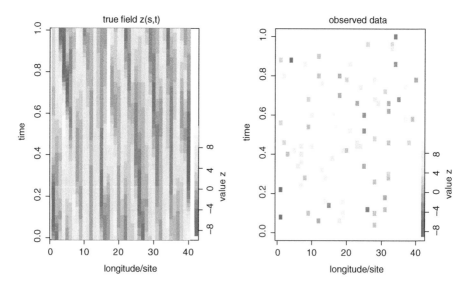

Figure 6.45 The true field $z(s,t)$ generated from the L96 model (left) and the noisy data (right) from which we will estimate $z(s,t)$. The observation errors are independent $N(0, \frac{1}{4})$ random deviates.

If we take z to be the $n_s n_t$-vector containing the entire field, we can define T to be an $n \times n_s n_t$ incidence matrix that assigns observation sites d_i to lattice locations j. Thus the likelihood can be written

$$L(y|z) \propto \lambda_y^{\frac{n}{2}} \exp\{-\tfrac{1}{2}\lambda_y(y - Tz)^T(y - Tz)\}.$$

6.6.2 A Gaussian process model

We can use simulations from the L96 model to obtain estimates of the mean and covariance of $z(s,t)$ over the $n_s \times n_t$ lattice. This is easily accomplished by taking a long simulation and "cutting" it into pieces of size $n_s \times n_t$. Each of these pieces can be thought of as draws $z_1, z_2, \ldots, z_M$ from a prior model for $z(s,t)$.

We can then use these realizations to build a GP prior for $z(s,t)$ over an $n_s \times n_t$ lattice over space and time

$$z \sim N(\mu_z, \Sigma_z).$$

The mean is set to the sample mean of the draws

$$\mu_z = \frac{1}{M}\sum_{j=1}^{M} z_j.$$

Here, μ_z is sufficiently close to 0 that we set it to 0. Then the $n_s n_t \times n_s n_t$ covariance matrix is set to

$$\Sigma_z = \frac{1}{M}\sum_{j=1}^{M}(z_j - \mu_z)(z_j - \mu_z)^T.$$

Hence we have exactly the setup we had before in the soft conditioning case of Section 6.2.

$$L(y|z) \propto \lambda_y^{\frac{n}{2}} \exp\{-\tfrac{1}{2}\lambda_y(y - Tz)^T(y - Tz)\}$$
$$\pi(z) \propto |\Sigma_z|^{-\frac{1}{2}} \exp\{-\tfrac{1}{2}z^T\Sigma_z^{-1}z\}$$

Therefore we know the posterior distribution for z has the form

$$z|y \sim N(\Lambda^{-1}\lambda_y T^T y, \Lambda^{-1}),$$

where $\Lambda = (\lambda_y T^T T + \Sigma_z^{-1})$.

The posterior mean field is shown in Figure 6.46. Also shown is the residual field (true value of z - posterior mean). The root mean square (RMS) error of the residuals is 3.03. Figure 6.47 shows the fitted values against the true field. Also shown in this figure are the pointwise 80% credible intervals for the initial field $z_0 = (z(1,0), z(2,0), \ldots, z(n_s,0))^T$. This initial field is of interest because it is the uncertain initial condition that determines the rest of the space-time field. The next approach focuses on estimating this initial condition z_0.

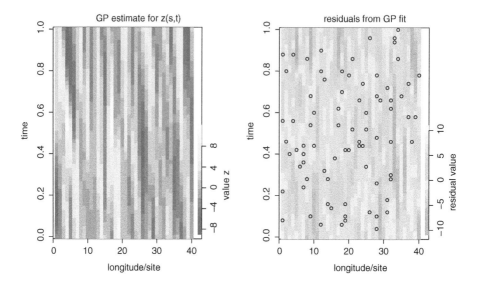

Figure 6.46 (**SEE COLOR INSERT**) Left: posterior mean for $z(s,t)$ estimated under the Gaussian process formulation. Right: the residual field obtained by subtracting the posterior mean estimate from the true field shown in Figure 6.45. The circles denote locations where measurements were obtained.

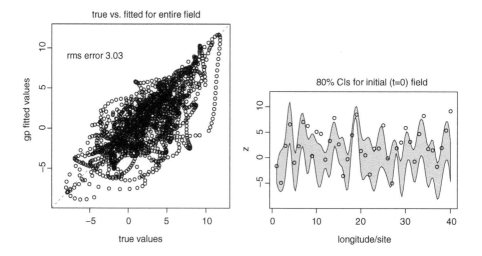

Figure 6.47 Left: true $z(s,t)$ plotted against the posterior mean from the Gaussian process formulation. Right: pointwise 80% credible intervals for the initial condition $z_0 = (z(1,0), z(2,0), \ldots, z(n_s,0))^T$. The circles denote the true values for z_0.

6.6.3 Bayesian inverse formulation

The previous GP formulation uses the simulator only to estimate the mean and covariance of the GP model. Presumably, the simulation model can be much more informative about what fields are plausible given the observed data. An approach that makes much stronger use of the simulator is the Bayesian inverse formulation. This comes at a price — a huge number of simulation runs will be required.

The initial state for the space-time field $z(s,t)$ is given by the n_s-vector $z_0 = (z(1,0), \ldots, z(40,0))^T = z(s, t = 0)$. Given z_0, the remaining field z is determined by the L96 model

$$z = \eta(z_0),$$

where the simulator output z is defined over the $n_s \times n_t = 40 \times 51$ lattice. Hence the likelihood — which requires an L96 simulation to evaluate — is

$$L(y|z_0, \lambda_y) \propto \lambda_y^{\frac{n}{2}} \exp\left\{-\tfrac{1}{2}\lambda_y(y - T\eta(z_0))^T(y - T\eta(z_0))\right\}.$$

A locally linear GMRF prior is specified for z_0:

$$\pi(z_0|\lambda_0) \propto \lambda_0^{\frac{n_s}{2}} \exp\left\{-\frac{1}{2}\lambda_0 \sum_{i=1}^{n_s}(z_{0i} - z_{0i+1})^2\right\}.$$

Note the sum goes to n_s here because the spatial sites are assumed to be periodic. In the sum $n_s + 1$ taken to be 1, making sites 1 and n_s neighbors. The prior specification is completed by specifying priors for the precision parameters:

$$\pi(\lambda_y) \propto \lambda_y^{a_y - 1} \exp\left\{-b_y\lambda_y\right\}, \quad a_y = 4; \ b_y = 1,$$

$$\pi(\lambda_0) \propto \lambda_0^{a_0 - 1} \exp\left\{-b_0\lambda_0\right\}, \quad a_0 = 1; \ b_0 = .001.$$

This results in the posterior distribution:

$$\pi(z_0, \lambda_y, \lambda_0|y) \propto L(y|\eta(z_0), \lambda_y) \times \pi(z_0|\lambda_0) \times \pi(\lambda_y) \times \pi(\lambda_0)$$

$$\propto \lambda_y^{\frac{n}{2}} \exp\left\{-\frac{1}{2}\lambda_y(y - T\eta(z_0))^T(y - T\eta(z_0))\right\}$$

$$\times \lambda_0^{\frac{n_s}{2}} \exp\left\{-\frac{1}{2}\lambda_0 \sum_{i=1}^{n_s - 1}(z_{0i} - z_{0i+1})^2\right\}$$

$$\times \lambda_y^{a_y - 1} \exp\left\{-b_y\lambda_y\right\} \times \lambda_0^{a_0 - 1} \exp\left\{-b_0\lambda_0\right\}.$$

The resulting full conditionals are then

$$\pi(z_{0i}|\cdots) \propto \exp\left\{-\frac{1}{2}\lambda_y(y - T\eta(z_0))^T(y - T\eta(z_0))\right\}$$

$$\times \exp\left\{-\frac{1}{2}\lambda_0 \sum_{j\in\partial i}(z_{0i} - z_{0j})^2\right\}$$

$$\lambda_y|\cdots \sim \Gamma(a_y + n/2, b_y + .5(y - T\eta(z_0))^T(y - T\eta(z_0)))$$

$$\lambda_0|\cdots \sim \Gamma\left(a_0 + n_s/2, b_0 + .5\sum_{i=1}^{n_s}(z_{0i} - z_{0i+1})^2\right).$$

Note that evaluating $\pi(z_{0i}|\cdots)$ involves running the L96 simulator. Also this full conditional will require a Metropolis update in the MCMC, while the precisions can be updated via Gibbs.

As with the previous GP formulation, we consider some posterior output. The posterior mean field is shown in Figure 6.48. Also shown is the residual field (true value of z-posterior mean). The RMS error here is 0.69, which is much smaller than what was obtained under the GP formulation. This is due to stronger linkage to the simulator model. This strong linkage is problematic if the simulator does not adequately model reality, in which case the GP formulation may be advantageous, even without computing

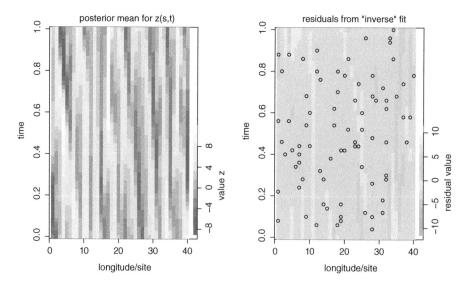

Figure 6.48 (**SEE COLOR INSERT**) Left: posterior mean for $z(s,t)$ estimated under the inverse formulation. Right: the residual field obtained by subtracting the posterior mean estimate from the true field shown in Figure 6.45. The circles denote locations where measurements were obtained.

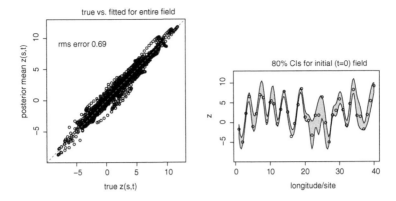

Figure 6.49 Left: true $z(s,t)$ plotted against the posterior mean from the inverse formulation. Right: pointwise 80% credible intervals for the initial condition $z_0 = (z(1,0), z(2,0), \ldots, z(n_s, 0))^T$. The circles denote the true values for z_0.

considerations. Figure 6.49 shows the fitted values against the true field. Also shown in this figure are the pointwise 80% credible intervals for the initial field $z_0 = (z(1,0), z(2,0), \ldots, z(n_s, 0))^T$. Again, these intervals are narrower as compared to those of the GP formulation and they are much closer to the true z_0.

6.6.4 PDE-based MRF formulation

The inverse specification is ideal if the simulator adequately matches the actual physical system being observed and if it runs quickly enough so that MCMC is possible. When this is not the case, one is forced to replace the simulator with some sort of fast model. This necessarily will introduce additional uncertainty regarding our estimate of the true underlying field z. Here we replace the deterministic L96 simulator with MRF model whose full conditionals are constructed to match the steps in the L96 simulator model. Hence the PDE equation is replaced with a cruder, stochastic PDE (SPDE). This SPDE-based MRF model is easy to work with computationally and readily incorporates information from the observed data. The price paid here is that it is an approximation to the actual L96 system.

For the PDE-based MRF model for the $n_s \times n_t$ field z, it is convenient to define z_t to be the n_s-vector $z_t = (z(1,t), \ldots, z(n_s, t))^T$, for $t = 0, \Delta t, \ldots, 1$. We can now model z as a Markov process:

$$\pi(z|\lambda_0, \lambda_z) = \pi(z_0|\lambda_0) \times \prod_{t=1}^{n_t - 1} \pi(z_{t+\Delta t}|z_t).$$

The locally linear GMRF model previously used for z_0 is also used here.

$$\pi(z_0|\lambda_0) \propto \lambda_0^{\frac{n_s}{2}} \exp\left\{-\frac{1}{2}\lambda_0 \sum_{s=1}^{n_s}(z_{s0} - z_{s+1\,0})^2\right\}.$$

For $z_{t+\Delta t}|z_t$, the new field $z_{t+\Delta t}$ is equal to the old field z_t plus a deterministic increment $\mu_{t+\Delta t}$, plus some white noise

$$z_{t+\Delta t}|z_t = \mu_{t+\Delta t} + \delta,$$

where $\mu_{t+\Delta t}$ is a n_s-vector obtained from the L96 step and $\delta \sim N(0, \frac{1}{\lambda_z}I_{n_s})$. Specifically, $\mu_{t+\Delta t} = (\mu_{1\,t+\Delta t}, \ldots, \mu_{40\,t+\Delta t})^T$, where

$$\mu_{s\,t+\Delta t} = z_{st} + \Delta t\{[z_{s+1\,t} - z_{s-2\,t}]z_{s-1\,t} - z_{s\,t} + F\}.$$

Thus, $\pi(z_{t+\Delta t}|z_t)$ is given by

$$\pi(z_{t+\Delta t}|z_t, \lambda_z) \propto \lambda_z^{\frac{n_s}{2}} \exp\left\{-\frac{1}{2}\lambda_z(z_{t+\Delta t} - \mu_{t+\Delta t})^T(z_{t+\Delta t} - \mu_{t+\Delta t})\right\}.$$

The full conditional density for z_{st} can be determined by restricting attention to terms in $\pi(z|\lambda_0, \lambda_z)$ that include the term z_{st}. Hence,

$$\pi(z_{st}|z_{\partial st}, \lambda_0, \lambda_z) \propto$$

$$\exp\left\{-\frac{1}{2}\lambda_0 \sum_{j\in\partial^0 s}(z_{s0} - z_{j0})^2\right\}^{I[t=0]}$$

$$\times \exp\left\{-\frac{1}{2}\lambda_z(z_{st} - \mu_{st})^2\right\}^{I[t>0]}$$

$$\times \exp\left\{-\frac{1}{2}\lambda_y \sum_{j\in\partial^* s}(z_{j\,t+\Delta t} - \mu_{j\,t+\Delta t})^2\right\}^{I[t<1]}$$

where $\partial^0 s$ denotes the neighborhood of s for the locally linear prior for z_0, and $\partial^* s$ denotes the neighborhood of s for the MRF scheme to incorporate the PDE structure. Note that evaluating $\pi(z_{st}|z_{\partial st}, \lambda_0, \lambda_z)$ is a simple, local calculation — but nonlinear. For $0 < t < 1$, the dependence structure for z_{st} is shown in Figure 6.50.

Now, the PDE-based MRF formulation can be summarized as follows. Likelihood/sampling model:

$$L(y|z) \propto \lambda_y^{\frac{n}{2}} \exp\{-\frac{1}{2}\lambda_y(y - Tz)^T(y - Tz)\}.$$

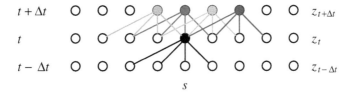

Figure 6.50 Dependence graph for z_{st}, a single component of the PDE-based MRF process z. z_{st}, shown by the central, dark circle, is influenced by $z_{t-\Delta t}$ at the four sites connected to z_{st}. Likewise, the four lightly shaded sites at time $t + \Delta t$ are affected by z_{st}.

SPDE prior for z:

$$\pi(z|\lambda_0, \lambda_z) \propto \lambda_0^{\frac{p}{2}} \exp\left\{-\frac{1}{2}\lambda_0 \sum_{s=1}^{p}(z_{s0} - z_{s+1\,0})^2\right\}$$
$$\times \prod_{t>0} \lambda_z^{\frac{p}{2}} \exp\left\{-\frac{1}{2}\lambda_z(z_t - \mu_t)^T(z_t - \mu_t)\right\}.$$

Priors for precision parameters:

$$\pi(\lambda_y) \propto \lambda_y^{a_y-1} \exp\left\{-b_y\lambda_y\right\}, \quad a_y = 4; \quad b_y = 1$$
$$\pi(\lambda_0) \propto \lambda_0^{a_0-1} \exp\left\{-b_0\lambda_0\right\}, \quad a_0 = 1; \quad b_0 = .001$$
$$\pi(\lambda_z) \propto \lambda_z^{a_z-1} \exp\left\{-b_z\lambda_z\right\}, \quad a_z = 1; \quad b_z = .001$$

The full conditionals from this formulation are then

$$\pi(z_{st}|\cdots) \propto \exp\left\{-\frac{1}{2}\lambda_y(y_{st} - z_{st})^2\right\}^{I[\text{obs at }(s,t)]}$$
$$\times \exp\left\{-\frac{1}{2}\lambda_0 \sum_{j\in\partial^0 s}(z_{s0} - z_{j0})^2\right\}^{I[t=0]}$$
$$\times \exp\left\{-\frac{1}{2}\lambda_z(z_{st} - \mu_{st})^2\right\}^{I[t>0]}$$
$$\times \exp\left\{-\frac{1}{2}\lambda_y \sum_{j\in\partial^* s}(z_{j\,t+\Delta t} - \mu_{j\,t+\Delta t})^2\right\}^{I[t<1]}$$

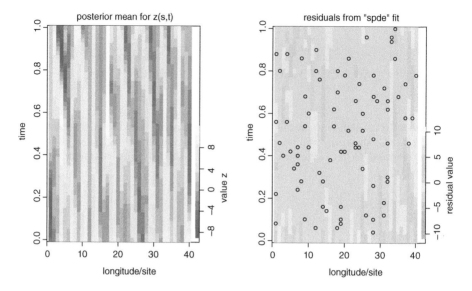

Figure 6.51 (**SEE COLOR INSERT**) Left: posterior mean for $z(s,t)$ estimated under the PDE-based MRF formulation. Right: the residual field obtained by subtracting the posterior mean estimate from the true field shown in Figure 6.45. The circles denote locations where measurements were obtained.

$$\lambda_y | \cdots \sim \Gamma(a_y + n/2, b_y + .5(y - Tz)^T(y - Tz))$$

$$\lambda_0 | \cdots \sim \Gamma\left(a_0 + p/2, b_0 + .5 \sum_{i=1}^{p}(z_{0i} - z_{0i+1})^2\right)$$

$$\lambda_z | \cdots \sim \Gamma\left(a_z + 50p/2, b_z + .5 \sum_{t>0}(z_t - \mu_t)^T(z_t - \mu_t)\right).$$

The MCMC implementation uses Metropolis updates for z_{st} and Gibbs updates for the precision parameters.

Again, posterior output is shown in Figures 6.51 and 6.52. Generally, the posterior reconstruction is much better than that of the GP formulation and only slightly worse than that of the inverse formulation. This performance is quite remarkable given that no evaluations of the L96 simulator were carried out here.

This PDE-based MRF formulation originated in geophysical applications (Wikle et al., 2001; Berliner, 2003). The MRF constructed here is essentially a very coarse version of the L96 PDE system. The prior MRF by itself would not be stable if it were propagated forward because the time step is far too large. However, the observed data are enough to ensure a stable reconstruction. This PDE-based MRF prior is effectively a crude simulator for which MCMC can be carried out. For other approaches that utilize crude simulation models, see Kennedy and O'Hagan (2000), Craig et al. (2001), or Higdon et al. (2003).

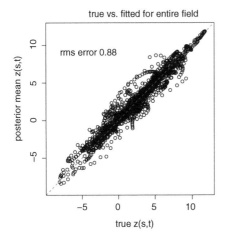
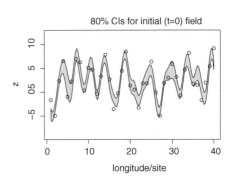

Figure 6.52 Left: true $z(s,t)$ plotted against the posterior mean from the PDE-based MRF formulation. Right: pointwise 80% credible intervals for the initial condition $z_0 = (z(1,0), z(2,0), \ldots, z(n_s, 0))^T$. The circles denote the true values for z_0.

A Kalman filtering approach for the L96 system is given in Bengtsson et al. (2003).

6.7 Discussion

This chapter introduced some basic approaches for constructing space-time models. There was a deliberate attempt here to keep the models simple so that basic modeling ideas would be a bit more clear.

One theme in the chapter has been the flexibility of Gaussian formulations. They are well-developed and understood processes that offer a number of computational advantages. Even for modeling non-Gaussian processes, Gaussian systems can still serve as useful components for the eventual model.

Another facet of space-time modeling is the need to consider the details of the application at hand when developing a model formulation. Features that may require special consideration include large data sets, special dependence structure, availability of a simulation model, speed of the simulation model, and the amount of accuracy required.

References

Anderson, T.W. (1984). *An Introduction to Multivariate Statistical Analysis*, John Wiley & Sons, New York.

Banerjee, S., Carlin, B.P. and Gelfand, A.E. (2003). *Hierarchical Modeling and Analysis for Spatial Data*, Chapman & Hall/CRC, Boca Raton.

Bengtsson, T., Snyder, C., and Nychka, D. (2003). Toward a nonlinear ensemble filter for high-dimensional systems, *Journal of Geophysical Research*, 108: 1–10.

Berliner, L.M. (2003). Physical-statistical modeling in geophysics, *Journal of Geophysical Research*, 108: D–24.

Besag, J. (1986). On the statistical analysis of dirty pictures, *Journal of the Royal Statistical Society (Series B)*, 48: 259–302.

Besag, J. and Kooperberg, C.L. (1995). On conditional and intrinsic autoregressions, *Biometrika*, 82: 733–746.

Besag, J., Green, P.J., Higdon, D.M., and Mengersen, K. (1995). Bayesian computation and stochastic systems (with discussion), *Statistical Science*, 10: 3–66.

Besag, J., York, J., and Mollié, A. (1991). Bayesian image restoration, with two applications in spatial statistics (with discussion), *Annals of the Institute of Statistical Mathematics*, 43: 1–59.

Chilés, J.-P. and Delfiner, P. (1999). *Geostatistics: Modeling Spatial Uncertainty*, Wiley, New York.

Cleveland, W.S. (1979). Robust locally weighted regression and smoothing scatterplots, *Journal of the American Statistical Association*, 74: 829–836.

Craig, P.S., Goldstein, M., Rougier, J.C., and Seheult, A.H. (2001). Bayesian forecasting using large computer models, *Journal of the American Statistical Association*, 96: 717–729.

Cressie, N.A.C. (1991). *Statistics for Spatial Data*, Wiley-Interscience.

De Oliveira, V. (2000). Bayesian prediction of clipped Gaussian random fields, *Computational Statistics & Data Analysis*, 34(3): 299–314.

Dongarra, J.J., Bunch, J.R., Moler, C.B., and Stewart, G.W. (1978). *LAPACK Users Guide*, SIAM Publications, Philadelphia.

Gelman, A. (2005). Prior distributions for variance parameters in hierarchical models, *Bayesian Analysis*, 1:515–533.

Hastie, T., Tibshirani, R., and Friedman, J.H. (2001). *The Elements of Statistical Learning: Data Mining, Inference, and Prediction*, Springer, New York.

Higdon, D. (1998). A process-convolution approach to modeling temperatures in the north Atlantic Ocean, *Journal of Environmental and Ecological Statistics*, 5: 173–190.

Higdon, D.M., Lee, H., and Holloman, C. (2003). Markov chain Monte Carlo-based approaches for inference in computationally intensive inverse problems, *in* J.M. Bernardo, M.J. Bayarri, J.O. Berger, A.P. Dawid, D. Heckerman, A.F.M. Smith, and M. West (Eds.), *Bayesian Statistics 7. Proceedings of the Seventh Valencia International Meeting*, Oxford University Press, pp. 181–197.

Hodges, J.S., Carlin, B.P., and Fan, Q. (2003). On the precision of the conditionally autoregressive prior in spatial models, *Biometrics*, 59:317–322.

Kennedy, M. and O'Hagan, A. (2000). Predicting the output from a complex computer code when fast approximations are available, *Biometrika*, 87: 1–13.

Knorr-Held, L. (2003). Some remarks on Gaussian random field models for disease mapping, in *Highly Structured Stochastic Systems,* P. Green, N. Hjort, and S. Richardson (Eds.), Oxford University Press, pp. 260–264.

Lee, H.H.K., Higdon, D., Calder, K., and Holloman, C. (2005). Efficient models for correlated data via convolutions of intrinsic processes, *Statistical Modelling*, 5: 53–74.

Lorenz, E.N. (1996). Predictability, a problem partially solved, *Proceedings of the Seminar on Predictability*, Vol. 1, European Center for Medium-Range Weather Forecasts, Reading, Berkshire, pp. 1–18.

Metropolis, N., Rosenbluth, A., Rosenbluth, M., Teller, A., and Teller, E. (1953). Equations of state calculations by fast computing machines, *Journal of Chemical Physics*, 21: 1087–1091.

Press, W.H., Flannery, B.P., Teukolsky, S.A., and Vetterling, W.T. (2002). *Numerical Recipes in C++, The Art of Scientific Computing, second edition*, Cambridge University Press, New York.

Rue, H. and Held, L. (2005). *Gaussian Markov Random Fields: Theory and Applications*, Vol. 104 of *Monographs on Statistics and Applied Probability*, Chapman & Hall, London.

Rue, H. and Tjelmeland, H. (2002). Fitting Gaussian Markov random fields to Gaussian random fields, *Scandinavian Journal of Statistics*, 29: 31–49.

Ryti, R.T. (1993). Superfund soil cleanup: developing the Piazza Road remedial design, *Journal of the Air and Waste Management Association*, 24: 381–391.

Stein, M. (1999). *Interpolation of Spatial Data: Some Theory for Kriging*, Springer-Verlag, New York.

Stroud, J., Müller, P., and Sanso, B. (2001). Dynamic models for spatio-temporal data, *Journal of the Royal Statistical Society*, Series B., 63:673–689.

Vidakovic, B. (1999). *Statistical Modeling by Wavelets*, John Wiley & Sons, New York.

Wackernagel, H. (1995). *Multivariate Geostatistics. An Introduction with Applications*, Springer-Verlag.

West, M. and Harrison, J. (1997). *Bayesian Forecasting and Dynamic Models (second edition)*, Springer-Verlag, New York.

Wikle, C.K., Milliff, R.F., Nychka, D., and Berliner, L.M. (2001). Spatio-temporal hierarchical Bayesian modeling: tropical ocean surface winds, *Journal of the American Statistical Association*, 96: 382–397.

Index

T - #0393 - 071024 - C312 - 234/156/14 - PB - 9780367390112 - Gloss Lamination